Reviews of Physiology Biochemistry and Pharmacology 128

Springer-Verlag Berlin Heidelberg GmbH

Reviews of

128 Physiology Biochemistry and Pharmacology

Editors

M.P. Blaustein, Baltimore H. Grunicke, Innsbruck

E. Habermann, Gießen D. Pette, Konstanz

G. Schultz, Berlin M. Schweiger, Berlin

With 17 Figures and 3 Tables

Springer

ISSN 0303-4240

ISBN 978-3-662-30997-1 ISBN 978-3-540-68502-9 (eBook)
DOI 10.1007/978-3-540-68502-9

Library of Congress-Catalog-Card Number 74-3674

Originally published by Springer-Verlag Berlin Heidelberg New York in 1996
Softcover reprint of the hardcover 1st edition 1996

Production: PRO EDIT GmbH, D-69126 Heidelberg
SPIN: 10128274 27/3136-5 4 3 2 1 0 – Printed on acid-free paper

Contents

Indexed in Current Contents

Genes Mediating Glucocorticoid Effects and Mechanisms of Their Regulation

S. Geley, M. Fiegl, B.L. Hartmann, and R. Kofler

Institute for General and Experimental Pathology, University of Innsbruck Medical School, Fritz-Pregl-Str. 3, 6020 Innsbruck, Austria

Contents

1 Introduction

Glucocorticoids (GC) are steroid hormones produced in the adrenal zona fasciculata under the stimulatory influence of anterior pituitary-derived adrenocorticotropic hormone (ACTH). GC influence a variety of important physiological phenomena including carbohydrate, protein and lipid metabolism, water and electrolyte balances, inflammatory and immune reactions, and regulation of endocrine networks. GC are also essential to the adaptation process of an organism to stressful situations by maintaining a permanent energy supply for brain functions, which is mainly based on the induction of gluconeogenesis and peripheral mobilization of amino acids. These powerful activities demand that GC production and expression of their receptor are tightly regulated. Therapeutically, GC are used widely in the management of such diverse disorders as cancer, chronic and acute inflammation, autoimmune and endocrine diseases, AIDS, transplant rejections, various shock forms and other pathological conditions (reviewed in Labhart 1986).

As discussed in more detail throughout this review, GC mediate their effects by influencing gene expression at essentially all known levels of regulation (reviewed in Yamamoto 1985; Beato 1989). Due to their lipophilicity, GC diffuse through the cell membrane and bind to their specific receptor (GR), a ligand-activated nuclear transcription factor of the "Zn-finger" type (Evans 1988; Hollenberg et al. 1985). The GR has been molecularily cloned (Encio and Detera-Wadleigh 1991), its functional domains assessed by mutational analyses (Giguere et al. 1986; Webster et al. 1988; Hollenberg and Evans 1988), and the three dimensional structure of its DNA-binding domain complexed with a consensus GC response element (GRE) determined by X-ray crystallography (Luisi et al. 1991). The receptor presumably resides within the cytoplasm complexed with several proteins, including heat shock proteins (Schlatter et al. 1992; Hutchison et al. 1994; Srinivasan et al. 1994; Yang and DeFranco 1994; Bohen and Yamamoto 1993) and an immunophilin (Pratt et al. 1993; Peattie et al. 1992; Tai et al. 1992; Pratt 1993). Upon ligand-binding, the GR undergoes "activation," a poorly understood process involving conformational changes and alterations in the phosphorylation state that ultimately allows the receptor to translocate into the nucleus and interact with specific *cis*-regulatory DNA sequences of responsive genes (reviewed in Berg 1989; Ortí et al. 1992). These *cis*-acting elements (GREs) tether the activated GR to a promoter, which results in increased transcription by interaction with other transcription factors (Bastian and Nordeen 1991; Chandler et al. 1983; Strähle et al. 1988; Tsai et al. 1988; Yoshinaga et al. 1992; Muchardt and Yaniv 1993) and coactivators (Muchardt and Yaniv 1993; Singh et al. 1995). GC-mediated downregulation of gene transcription (Drouin et al. 1993; Akerblom 1988) is a less well understood phenomenon. One strategy of GR mediated transrepression is occlusion of DNA-binding sites for transactivating factors by binding of the GR to *cis*-regulatory elements that deviate to various degrees from a consensus GRE. Such elements resembling GREs but conferring GC-dependent repression of transcription have been termed "negative" GREs (Strömstedt et al. 1991; Ray et al. 1990). Another mechanism involves *cis*-acting elements termed "composite" GREs that, in addition to a typical GRE, contain adjacent binding sites for other transcription factors, for example, AP-1. Depending on the second transcription factor and/or (as in the case of AP-1) its composition, GC may either enhance or repress transcription (Diamond et al. 1990). While these mechanisms involve GR binding to *cis* elements in the DNA, interference of the activated GR with other transcription factors due to protein-protein interactions represents a potential third mechanism of GC-mediated gene repression that does not require GR binding to DNA. This form of regulation has been reported for several transcription factors, including AP-1 (Jonat et al. 1990; Yang-Yen et al. 1990; Schüle et al. 1990), NFκB (Scheinman et al. 1995; Ray and Prefontaine 1994; Caldenhoven et al. 1995), OTF-1 (Kutoh et al. 1992), and Spi-1/Pu.1 (Guathier et al. 1993). Apart from the above mentioned mechanisms, GC can influence gene transcription indirectly by regulating the expression of other regulatory transcription factors, thereby influencing a host of genes responsive to those factors (reviewed in Schuchard et al. 1993).

GC effects are, however, not restricted to transcriptional regulation. On the posttranscriptional level, GC may enhance (Diamond and Goodman 1985; Petersen et al. 1989) or reduce (Lee et al. 1988; Peppel et al. 1991) mRNA stability. This form of regulation apparently depends on conserved sequences in the 3' untranslated region that might be recognized by GC-regulated specific nucleases or protective RNA-binding proteins. Translational control of gene expression by GC has also been reported (Verdi and Campagnoni 1990; Raz et al. 1989) and may result from binding of regulatory proteins to specific recognition sequences in mRNA species. All the above-mentioned mechanisms require the presence of functional GR. Effects that are not mediated by GR may occur at high GC concentrations and probably result from unspecific perturbation of cell membranes by the lipophilic steroid. Such effects cannot be inhibited by GR antagonists, such as RU486 (Baulieu 1991), and may have little if any significance for the in vivo situation.

While individual gene regulation by GC has been studied in great detail, relatively little is known about the molecular mechanisms by which GC control complex biological phenomena or exert therapeutic effects. Corresponding analyses are difficult because GC simultaneously regulate a host of genes that may all contribute in some way or another to a particular effect. As a further obstacle, some of these genes, and/or their regulation by GC, may still be unknown. In the last years, investigators from many different research areas have identified a large number of proteins whose expression is under apparent GC control. The present study reviews some of these publications with particular emphasis on instances where GC regulation has been substantiated by molecular evidence. In the course of our investigation, several thousand primary references were collected and analyzed. Obviously, space limitations allow citation of only selected key references and/or recent publications that provide additional references for more detailed informations. We apologize to our colleagues that even relevant references had to be omitted.

In the following, GC-regulated genes are discussed as they have been implicated in particular biological phenomena. Obviously, this assignment is artificial, and several genes might participate in more than one phenomenon. In addition to this attempt to explain complex biological GC effects by the coordinate regulation of individual genes, we review some fundamental mechanisms underlying GC regulation of single genes on the basis of representative model systems. Table 1 provides a quick alphabetic summary of all apparently GC-regulated genes discussed in this review, as well as some others that have not been dealt with in the text.

2
Effects on Endocrine Networks

GC influence the expression of a host of hormones and growth factors as well as cytokines, neurotransmitters and other biologically active humoral mediators and some of their receptors. GC can thus be regarded as key players in the regulation of endocrine and other intercellular signaling networks. This regulatory activity can either be direct or indirect

Table 1. Summary of GC-regulated genes and proteins interacting with GR

Designation of genes and proteins	Dealt with in Sect.
α_1-Acid glycoprotein	5
α_1-Inhibitor III	5
α_1-Microglobulin (Pierzchalski et al. 1992)	
α_2-Macroglobulin (Standke et al. 1994)	
$\alpha_{2\mu}$-Globulin	5
α-Fetoprotein (Rabek et al. 1994)	
Acetylcholine receptor	6
ADP ribosylation factor (Duman et al. 1990)	
Adenovirus gene expression	8
Adipsin (Spiegelman et al. 1989)	
Adrenergic receptors α_{1B}, β_1, β_2, β_3	6
Alcohol dehydrogenase	3
Aldehyde dehydrogenase (mitochondrial)	3
Alkaline phosphatase	4
5-Aminolevulinate synthase (Srivastava et al. 1992)	
Aminopeptidase A (Stefanovic et al. 1991)	
Amylase	3
Angiotensinogen	4
Angiotensin receptor	4
Apolipoproteins	3
Arginase (Gotoh et al. 1994)	
Argininosuccinate lyase	3
Aspartate aminotransferase	3
Atrial natriuretic peptide	4
β-Casein	7
β-Galactoside α2,6-sialyltransferase	3
β-Galactoside-binding protein (Goldstone and Lavin 1991)	
β-Endorphin	6
Bcl-2	7
Bombesin receptor (Rosewicz et al. 1994)	
Bone sialoprotein	4
Brain-derived neurotrophic factor (Smith et al. 1995)	
Branched-chain keto acid dehydrogenase (Chicco et al. 1994)	
C/EBP	7
CA125 tumor marker (Nakai et al. 1991)	
cAMP response element-binding protein	7
Calbindin-D28k	7
Calcitonin and calcitonin-related peptide	4

Table 1. Continue.

Designation of genes and proteins	Dealt with in Sect.
Calcitonin receptor	4
Calmodulin	7
Cannabinoid receptor (Mailleux and Vanderhaeghen 1993)	
Casein kinase II (Suzuki et al. 1992)	
Cathepsin-B and -D (Oursler et al. 1993)	
CEF-4	5
Cholinephosphate cytidyltransferase	3
Chorionic gonadotropin	2
Chromogranin A	6
Collagenase and collagenase type IV	5
Collagene types I, III, IV, V	4
Colony factor-1 receptor (C-*fms*) (Sapi et al. 1995)	
Complement C4b-binding protein	5
Complement factor B, C3, H	5
Corticotropin-releasing hormone	2
Corticosteroid-binding protein (Berdusco et al. 1994)	
C-reactive protein (Ku and Mortensen 1993)	
Cyclo-oxygenase	5
Cysteine-rich intestinal protein (Levenson et al. 1994)	
Cytochrome *c* oxidase (Van Itallie 1990)	
Cytochrome P450 enzymes	2
DNA methyltransferase (Rouleau et al. 1995)	
EGR-1	7
Elastin	4
Endothelial-leukocyte adhesion molecule-1	5
Endothelin (Schrey et al. 1992)	
Endothelin receptor (Stanimirovic et al. 1994)	
Enkephalin	6
Epidermal growth factor	7
EGF receptor	7
Epoxide hydrolase (Bell et al. 1990)	
Epstein-Barr virus gene expression	8
Erythropoietin (Lee-Huang et al. 1993)	
Fatty acid synthase	3
Fatty acid binding protein	3
Fc-ε receptor type II (CD23)	5
Fc-γ receptor type I (CD64)	5
Fibrinogen	5

Table 1. Continue.

Designation of genes and proteins	Dealt with in Sect.
Fibronectin	4
Follicle-stimulating hormone	2
Formyl peptide receptor (Perez et al. 1992)	
Fos	7
γ-Glutamyl transferase III	
γ-Glutamylcysteine synthetase (Lu et al. 1992)	
GAP-43 (Perrone-Bizzozero et al. 1993)	
GATA-1	7
Glial fibrillary acidic protein	6
Glucocorticoid receptor (GR)	2
Glucocortin	3
Glucokinase (Fernandez-Mejia and Davidson 1992)	
Glucose-6-phosphatase	3
Glucose-6-phosphate dehydrogenase	3
GLUT 1, 4 glucose transporters	3
Glutamine synthase	3
Glutathione *S*-transferase	7
Glycerol-3-phosphate dehydrogenase	3
Gonadotropin-releasing hormone	2
G protein subunits of Gαs, Gαi, Gβ	7
Granulocyte-macrophage colony-stimulator	5
Growth hormone	2
Guanylin (Hill et al. 1995)	
Haptoglobin	5
Heat shock protein 90 (McGuire et al. 1992)	
Heme oxygenase 1, 2	5
Heparin binding growth factor	7
Hepatitis virus gene expression	8
Hepatocyte growth factor	7
Hepatocyte growth factor receptor (c-*met*)	7
HER-2/neu growth factor receptor	7
Herpes virus gene expression	8
Histidine decarboxylase (Kawai et al. 1992)	
Hormone sensitive lipase	3
3-Hydroxy-3-methylglutaryl CoA reductase	3
15-Hydroxy-prostaglandin dehydrogenase	5
Immunoglobulin A	5
Immunoglobulin E	5

Table 1. Continue.

Designation of genes and proteins	Dealt with in Sect.
Insulin	3
Insulin receptor	3
Insulin receptor substrate-1	3
Insulin-like growth factor I	7
Insulin-like growth factor binding protein	7
Integrin α_1 and β_1 (Z. Zhang et al. 1993)	
Intercellular adhesion molecule-1	5
Interferon-β, -γ	5
Interferon consensus sequence-binding protein	7
Interleukin 1, 2, 4, 5, 6, 8	5
Interleukin receptor for interleukin-1, 2	5
JE (cytokine)	5
Jun	7
K^+ channel (kv1) (Takimoto et al. 1993)	
Kallikrein	5
KC (cytokine)	5
Kininogen T and K	5
Lipocortin I	5
Lipoprotein lipase	3
Luteinizing hormone	2
Lysozyme (Panarelli et al. 1994)	
Macrophage-colony-stimulating factor	5
Major histocompatibility complex I, II	5
Maltase-glycosamylase	3
Manganese superoxide dismutase (Suzuki et al. 1993)	
Mannose receptor (Shepherd et al. 1994)	
Matrilysin	5
Melanin-concentrating hormone (Presse et al. 1992)	
Metallothionein I, IIA, III (Plisov et al. 1995; Karin et al. 1984; Palmiter et al. 1992)	
Methionine tRNA	3
Monoacylglycerol acyltransferase	3
Monocyte chemoattractant protein-1 (Villiger et al. 1992)	
Myc	7
Monoacylglycerol acyltransferase	3
Multidrug resistence gene (Zhao et al. 1993)	
Myelin basic protein	6
Myelin proteolipid protein	6

Table 1. Continue.

Designation of genes and proteins	Dealt with in Sect.
Na^+ channel (amiloride-sensitive)	3
Na^+/H^+ exchanger (NHE-3) (Yun et al. 1993)	
Na^+/K^+ ATPase (Grillo et al. 1994)	
Nerve growth factor	6
Nerve growth factor receptor	6
Neuropeptide Y	6
Neurotrophin-3 (Smith et al. 1995)	
Neutral endopeptidase	5
Nexin-1 (Guttridge et al. 1993)	
Nuclear factor κB	7
Nitric oxide synthase	5
Nur77 (orphan nuclear receptor)	2
Ornithine decarboxylase	7
Osteocalcin	4
OTF-1	7
Papilloma virus genes E6, E7	7/8
Parathyroid hormone	4
Parathyroid hormone-related peptide	4
Peptidyl-glycine α-amidating mono-oxygenase (Grino et al. 1990)	
Peroxisomal proliferator-activated receptor (Steineger et al. 1994)	
Phenylalanin hydroxylase	3
Phenylethanolamin *N*-methyltransferase	6
Phosphatidylinositol 3-kinase	7
Phosphoenolpyruvate carboxykinase	3
6-Phosphofructo-2-kinase/fructose-2,6-bisphosphatase	3
Phospholipase A_2, D	3, 5
Pit-1	7
Platelet-derived growth factor A-chain (Nakano et al. 1993)	
Plasminogen activator	5
Plasminogen activator inhibitor type I	5
Polyoma virus gene expression	8
Prolactin	2
Proliferin	2
Pro-opiomelonocorticotropin	2
Protein kinase C	7
Prostaglandin endoperoxide synthases	5
Ras	7
Retinoid X receptor	2

Table 1. Continue.

Designation of genes and proteins	Dealt with in Sect.
Retroviral gene expression (HIV, HTLV-III)	8
Ribosomal RNA	3
Ribosome protein rpL32	3
SCG10 (neural-specific gene)	6
sgk (serum and GC-regulated kinase)	7
snk (serum induced kinase)	7
Serine dehydratase	3
Serine pyruvate/alanine glyoxylate aminotransferase (Uchida et al. 1994)	
Somatostatin (Liu et al. 1994)	
Stromelysin	5
Substance P receptor (Ihara and Nakanishi 1990)	
Sucrase-isomaltase	3
Sulfated glycoprotein-2 (TRMP-2)	7
Surfactant associated proteins A, B	3
Surfactant proteolipids C, D	3
Tenascin	4
Transin (Fillmore et al. 1992)	
Thyroid hormone receptor (c-*erbA* proto-oncogene)	2
Thyroid-stimulating hormone	2
Thyrotropin-releasing hormone	2
Tissue inhibitor of metalloproteinases	5
Transforming growth factor-α, -β	5
Transglutaminase	7
Trehalase	3
Tryptophane oxygenase	3
Tumor necrosis factor-α	5
Tyrosine aminotransferase	3
Tyrosine hydroxylase	3
Urate oxidase (Raab et al. 1991)	
Vasopressin	4
Vesiculo stomatitits virus gene expression	8
Vitamin D receptor	2
Vitellogenin (Slater et al. 1991)	
Whey acidic protein	7

by influencing a cell's responsiveness to other stimuli, a phenomenon known as the "permissive" action of GC. In the following, we review some of the better defined GC effects on the expression of typical hormones and their receptors, namely those regulating the hypothalamic-pituitary-adrenal/gonadal/thyroid axis, the steroids, and the glycoprotein and protein hormones of the pituitary and placenta. Regulation of additional genes involved in endocrine circuits can be found in Sect. 4. Cytokines, neurotransmitters and some growth factors are discussed along with their receptors in Sects. 5, 6, and 7, respectively.

2.1
Regulation of the Hypothalamic-Pituitary-Adrenal/Gonadal/Thyroid Axis and Steroid Hormone Biosynthesis

GC control their own synthesis, and perhaps that of other steroids and thyroid hormones, via regulation of the expression of hormones acting on the hypothalamic-pituitary-adrenal/gonadal/thyroid axis. On the hypothalamic level, transactivation of the human corticotropin-releasing hormone (CRH) gene promoter by cAMP response element-binding (CREB) proteins is repressed by GC (Van et al. 1990). Interestingly, in human placenta cytotrophoblasts, GC have an opposite effect, i.e., they stimulate CRH mRNA levels (Robinson et al. 1988), a discrepancy that may be explained by alternative usage of the two promoter-like structures in the CRH gene (Vamvakopoulos and Chrousos 1993). Although not directly related to adrenal steroid biosynthesis, two other hypothalamic releasing factors, gonadotropin- (Chandran et al. 1994) and thyrotropin- (Yang and Tashjian 1993) releasing hormones, are also under apparent GC control. On the pituitary gland level, GC downregulate pro-opiomelanocorticotropin (POMC) gene transcription and thus ACTH production. This phenomenon has been well studied, and the POMC promoter has become a model for negative regulation of transcription by GC (Drouin et al. 1987, 1989; Riegel et al. 1991). In vitro and in vivo analyses using transgenic animals revealed that the fragment from nucleotide −480 to −34 of the rat POMC promoter is sufficient for tissue-specific and GC-induced downregulation of POMC expression (Tremblay et al. 1988), and evidence was obtained indicating that only one of the six potential GR-binding sites identified in the rat POMC promoter mediated repression of transcription (Drouin et al. 1989). This GR binding site, centered at position −63, differs significantly from a consensus GRE half site

(TGTTCT), and has been termed a negative GRE (nGRE, CGTCCA). This nGRE element cannot mediate transcriptional activation by GC when it is inserted upstream of an enhancerless promoter-driven reporter plasmid, suggesting that GR interactions with nGREs differ from interactions with GREs and entail transcriptional repression rather than induction (Drouin et al. 1987). Further analyses of transcriptional repression of the rat POMC promoter indicated that a GR homotrimer could bind the nGRE, thus interrupting the interactions between positively cooperating transcription factors acting on the POMC promoter (Drouin et al. 1993). Finally, POMC gene repression by GC in pituitary corticotropes is abrogated by corticotropin-releasing factor (CRF) via induction of c-*fos* (Autelitano and Sheppard 1993). The latter may lead to increased intracellular AP-1 levels that, in many systems, interfere with GR activity (see Sect. 1 and several other examples in this review).

In addition to their effects on the hypothalamus and the pituitary gland, GC influence steroid biosynthesis directly by regulating the expression of numerous steroidogenic enzymes that belong primarily to the large cytochrome P450 family. In the following, we use the nomenclature proposed by Nebert et al. (Nebert et al. 1991) and include the gene assignment in parentheses to avoid confusion. The steroid biosynthesis enzymes P450scc (CYP11A1; Hales et al. 1990), 3β-hydroxysteroid dehydrogenase (Trudel et al. 1991; Agular et al. 1992) and P450$17\alpha$ (CYP17) are regulated by GC, the latter via GREs in the 5' flanking region of its gene (Picado-Leonard and Miller 1987). Several other cytochrome P450 enzymes, some of which play a role in steroid synthesis or degradation in the liver, appear to be GC regulated as well. These include rat P450c (CYP1A1), P450d (CYP1A2), P450b (CYP2B1) and P450e (CYP2B2; Mathis et al. 1989; Silver et al. 1990; Jaiswal et al. 1990; Rao et al. 1990), the murine P450IIIA1 (CYP3A1; Burger et al. 1992; Yanagimoto et al. 1992; Eliasson et al. 1994) or the closely related P450RL33/cDEX (Komori and Oda 1994), P450IIC12 (CYP2C12; Wright and Morgan 1991), human P450IIC8 (with multiple GREs in the 5' region of the CYP2C8 gene; Ged and Beaune 1991) and aromatase (CYP19; Zhao et al. 1995).

2.2
Steroid Receptor Gene Expression

GC not only participate in the regulation of steroid biosynthesis and degradation but also modulate the sensitivity of target cells by regulating steroid receptor levels. This is particularily true for the GR itself (reviewed in Schmidt and Meyer 1994). Most tissues downregulate their GR content after GC treatment (Kalinyak et al. 1989; Burnstein and Cidlowski 1992; Patchev et al. 1994), while in human T-cell leukemic and myeloma lines, upregulation of GR levels by GC has been reported (Denton et al. 1993; Gomi et al. 1990). The GR autoinduction observed in human leukemic cells could be a requirement for GC-induced apoptosis in such cells (see Sect. 7), similar to *Xenopus* tadpole tail cells where autoinduction of nuclear receptors has been found to occur during metamorphosis (reviewed in Tata et al. 1993). In addition to cell type-dependent regulation of GR expression by GC, GR is regulated developmentally (Kalinyak et al. 1989) and in a cell division cycle-dependent manner (Hsu et al. 1992; Hu et al. 1994). The molecular basis for regulation of GR gene expression is not well characterized and several, possibly species-specific, mechanisms have been discussed. In the mouse, GR expression is controlled by at least three promoters, one of which seems to be preferentially used in T-lymphocytes (Strähle et al. 1992). Rat GR mRNA also shows high variability at the 5' end, suggesting expression from different transcriptional start sites and/or promoters (Gearing et al. 1993). The human GR gene promoter has been cloned and does not contain typical GREs, nGREs or even TATA and CAAT boxes (Zong et al. 1990). Analyses of promoter regions sensitive to downregulation by GC led to the identification of the recognition sequence motif (localized to positions –671 to –710) for the specific transcription factor GRF-1. GRF-1 has subsequently been cloned and shown to repress the human GR gene promoter (LeClerc et al. 1991). In addition to the complex transcriptional regulation, GR gene expression is subject to posttranscriptional and posttranslational control mechanisms as well: GC caused GR downregulation in cells expressing transfected GR cDNA from a GC insensitive promoter, suggesting that, at least in this case, GC-dependent GR downregulation occurs by a posttranscriptional mechanism possibly involving GRE-like sequence elements present in the GR cDNA itself (Burnstein et al. 1990; Bellingham et al. 1992; Alksnis et al. 1991).

Posttranslationally, GR activity might be regulated by phosphorylation (Mason and Housley 1993; Bodwell et al. 1995) and accessory molecules [e.g., heat shock proteins (Patchev et al. 1994) and immunophilins] as already mentioned in the Introduction. More recently, another potentially regulatory GR binding protein has been identified: Calreticulin, a major calcium storage protein located in the endoplasmatic reticulum and also found in the nucleus, seems to be able to interact with the DNA binding domain of the GR and, if overexpressed, inhibits GRE-mediated transcriptional activation from GC-sensitive promoters (Burns et al. 1994).

Other members of the steroid receptor superfamily, such as the vitamin D (Lee et al. 1991; Gensure et al. 1993), thyroid hormone (c-*erbA* proto-oncogene; Maroder et al. 1989), retinoid X (Wan et al. 1994), and nur77 orphan nuclear receptors (Davis and Lau 1994), might also be GC-regulated. In the case of the thyroid hormone receptor α gene (C-*erbA1*), a GC-responsive promoter was identified (Laudet et al. 1993), and this regulation may participate in GC-induced differentiation of neuronal PC12 cells (Munoz et al. 1993b).

2.3 Pituitary and Placental Glycoprotein Hormones

Other GC-affected endocrine systems include the pituitary and placental glycoprotein hormones (GPH), i.e., luteinizing, follicle-stimulating and thyroid-stimulating hormone, and chorionic gonadotropin, all of which are heterodimers of a common α-chain and a hormone specific β-chain. The α-GPH promoter, analyzed in JEG3 chorioncarcinoma cells, represents another model for GC-mediated transcriptional repression. A 168-bp fragment (α-168) of the α-GPH 5'-flanking region was found to contain the necessary sequences for cell-specific expression and GC-mediated repression. α-168 contains two CAMP-responsive elements (CREs) and one region upstream of the CREs known to bind a tissue-specific factor that cooperates with CREB proteins. DNA footprinting using purified rat liver GR to protect the α-GPH promoter revealed three GR binding sites. The strongest of these footprints overlaps the most proximal CRE, which might explain negative interference by GR occupancy of enhancing promoter elements (Akerblom 1988). Detailed analysis of GR structural requirements for this repressional activity pointed to the importance of an intact DNA- and ligand-binding domain, whereas the N-terminal transactivating domain seemed to be dispensable in mediating α-GPH repression (Oro et al. 1988). Interestingly however, one of several DNA-binding domain GR mutants (G442) tested, retained more than 50% of the wild-type GR transrepressional activity on the α-168 promoter, although this mutant G442 is incapable of transactivating an MMTV-CAT construct. Thus, the structural GR DNA-binding domain requirements differ between transactivation and transrepression, as is also reflected in the deviations of the α-GPH promoter GR-binding sites from the consensus GRE (Akerblom 1988; Oro et al. 1988). These studies led to the hypothesis that binding of the GR to an imperfect GRE might result in disruption of cooperating transcription factors (e.g., CREB interacting with tissue-specific transcription factors) by steric hindrance. Alternatively, the intact GR DNA-binding domain may directly mediate transrepression without binding to specific DNA elements of the α-GPH promoter. The latter model is further supported by observations in JEG3 cells of potential direct interactions between CREB (or unidentified transcription factors) and GR. Regulation of the α-GPH promoter is tissue-specific since it is activated in fibroblast cell lines, pointing to additional tissue-specific factors modulating GC activity (Chatterjee et al. 1991 and references therein). In addition to regulating the promoter of the common α-chain, GC might influence individual GPH levels via controlling expression of the hormone-specific β-chain, as documented for human chorionic gonadotropin (Ringler et al. 1989).

2.4 Pituitary and Placental Protein Hormones

Similar to the GPH α-chain gene, GC and other steroids repress the transcription of the prolactin gene, although the mechanism of repression is not entirely clear. Hormonal regulation requires *cis*-acting elements in the promoter region of the gene (Somasekhar and Gorski 1988; Day and Maurer 1989) and can be conferred upon a heterologous promoter by a 34-bp nGRE element (Sakai et al. 1988; Cairns et al. 1993). This element acts per se as a transcriptional enhancer, and the hormone-receptor complex supposedly competes with, or inactivates, the factor(s) acting upon this element. Nevertheless, the DNA-binding domain of the GR may not be required for this GC effect (Adler et al. 1988). In extrapituitary sites, such as the uterus and T-lymphocytes, prolactin gene expression is directed by an alternative promoter that contains several half sites for GR binding and is located approximately 6 kb upstream of the pituitary specific start site (Gellersen et al. 1994).

GC are further powerful modulators of growth hormone secretion (Wehrenberg et al. 1990) and induce growth hormone receptor mRNA (Heinrichs et al. 1994). The human growth hormone gene contains a GRE in its first intron that mediates transcriptional induction (Slater et al. 1985), although GC regulation seems to act primarily at a posttranscriptional level (Diamond and Goodman 1985; Strobl et al. 1989). A third member of the pituitary and placental protein hormone family, placental proliferin, is transcriptionally regulated by GC (Mordacq and Linzer 1989). The GC-sensitive element in the mouse proliferin gene is built up of a GRE closely linked to an AP-1 site and has been termed a composite GRE. This composite GRE mediates either GC-dependent induction or repression, which depends on the presence and composition of AP-1 (Miner and Yamamoto 1991).

3 Effects on Carbohydrate, Protein, and Lipid Metabolism

The energy supply of an organism adapting to changing environmental conditions is maintained, in part, by the tightly regulated action of GC on carbohydrate, protein, and lipid metabolism. These effects, central to their physiological role, contribute to a great extent to the unwanted side effects of GC treatment.

3.1 Carbohydrate Metabolism

GC stimulate gluconeogenesis, i.e., the formation of carbohydrates from amino acids, by stimulation of gluconeogenetic liver enzymes, peripheral mobilization and degradation of protein, diminishing peripheral glucose utilisation, and promoting storage of glucose as glycogen in the liver (Labhart 1986). This activity may, in some patients treated with high GC doses, lead to "steroid diabetes," a mild form of diabetes mellitus, that might be explained on the molecular level by regulation of various enzymes participating in carbohydrate metabolism and perhaps also by interfering with insulin action. Expression of insulin (Fernandez-Mejia and Davidson 1992) as well as transcription (Saad et al. 1994) and alternative splicing (Norgren et al. 1994) of the insulin receptor gene appear GC-regulated, with the human insulin receptor gene containing *cis*-acting GREs and other response elements that mediate transcriptional induction (Lee and Tsai 1994; Lee et al. 1992). GC further regulate the expression of the downstream target insulin receptor substrate-1 (IRS-1) (Saad et al. 1994; Turnbow et al. 1994; Giorgino et al. 1993), and apparently participate in the control of phosphorylation of this protein and the receptor itself (Giorgino et al. 1993). Recent analyses in Fao hepatoma cells demonstrated that the cause of GC-induced insulin resistance probably lies downstream of the insulin receptor, IRS-1 or IRS-1-activated PI-3 kinase (Saad et al. 1995).

Gluconeogenesis is regulated by induction of rate-limiting enzymes and GC increase the transcriptional rate (Imai et al. 1990; Mitchell et al. 1994) of the phosphoenolpyruvate carboxykinase (PEPCK) gene, the key enzyme for gluconeogenesis, and enhance the stability of its mRNA (Petersen et al. 1989). The repressive insulin effect on transcription is, however, dominant over induction by GC (Forest et al. 1990). PEPCK gene regulation is primarily exerted at the transcriptional level, and the GC effect can be tissue- and cell-specific with opposite effects possibly depending on the presence of other transcription factors such as CREB, the retinoic acid receptor, the orphan nuclear receptors COUP-TF and HNF-4, or HNF-3 (Xing and Quinn 1993; Lucas et al. 1991; Hall et al. 1995; O'Brien et al. 1995). Analysis of mice transgenic for a PEPCK promoter controlled reporter gene revealed that sequences within 2 kb of the promoter are sufficient for tissue-specific regulation and expression (Eisenberger et al. 1992; Short et al. 1992). The GC regulatory unit was positioned around –400 and encompassed two GREs (Short et al. 1992). Furthermore, in vitro analysis revealed direct binding of the GR to CREB that binds to the E/CRE site (Imai et al. 1993). This protein-protein interaction suggests interference of transcriptional activators during hormonal induction and might explain the synergistic action of glucagon (acting through elevation of cAMP) and GC (Hall et al. 1995). This model was, however, questioned by in vivo genomic footprint analyses of H4IIE hepatoma cells that failed to demonstrate alterations of protein occupancy of the PEPCK promoter following dexamethasone treatment (Faber et al. 1993).

Carbohydrate metabolism may further be affected by GC effects on the expression of a number of additional metabolic enzymes: In the rat, hepatic 6-phosphofructo-2-kinase/fructose-2,6-bisphosphatase expression is increased by GC, at least in part, on the transcriptional level via one or more GREs in the 5'-flanking region and the first intron of its gene (Marker et al. 1989; Lange et al. 1992). Similar to PEPCK regulation, this gene induction is inhibited by insulin (Lemaigre et al. 1994). As reviewed by Edenberg et al. (Edenberg and Brown 1992), alcohol dehydrogenases, a complex system of zinc metalloenzymes, are also induced by GC. The human gene for the class I alcohol dehydrogenase β subunit contains 3 tandemly repeated GREs in its 5'-flanking region (Winter et al. 1990) and, surprisingly, GRE-like sequences in its cDNA (Dong et al. 1988). The promoter of the mouse mitochondrial aldehyde dehydrogenase was also found to contain a GRE (Chang and Yoshida 1994). Further GC-regulated enzymes participating in carbohydrate metabolism include liver β-galactoside α2,6-sialyltransferase that is transcriptionally regulated but lacks a GRE (Wang et al. 1990), the human pancreatic amylase gene whose transcriptional enhancement requires induction of regulatory protein(s) (Logsdon et al. 1987), two transcriptionally induced mouse liver amylase genes (Samuelson et al. 1988; Slater et al. 1993), glycerol-3-phosphate dehydrogenase (Bhandari et al. 1991; Kumar et al. 1989), glucose-6-phosphate dehydrogenase (Kletzien et al. 1994), glucose-6-phosphatase (Lange et al. 1994), as well as maltase- glycosamylase, sucrase-isomaltase, and trehalase (Galand 1988). Aside from regulating various involved enzymes, GC might affect carbohydrate metabolism by inhibiting peripheral glucose uptake, e.g., in adipocytes and thymocytes. This effect has been attributed to induction of several proteins, including a 17-kDa protein termed glucocortin (Colbert and Young 1986), and to the reduction of glucose transporter mRNA in cultured adipocytes (Garvey et al. 1989). More recently, GC induction of glucose transporters in primary cultures of brown adipocytes (Shimizu et al. 1994) and in L6 muscle cells (Sivitz and Pasley 1995) has also been reported.

3.2
Protein Metabolism and Translation

The catabolic activity of GC has long been recognized, in part as side effects of chronic GC treatment such as osteoporosis, muscular waste, skin atrophy, and growth retardation in children. GC lead to increased peripheral protein degradation (negative nitrogen balance; Brillon et al. 1995), which is not fully compensated for by simultaneously increased protein synthesis (Labhart 1986) in the liver. The molecular mechanisms underlying these GC-regulated events have been addressed by studying GC regulation of key enzymes in amino acid degradation. Thus, tyrosine aminotransferase gene activity in the liver is stimulated by GC and glucagon, but repressed by insulin via two major enhancer regions. These contain GREs, CREs and additional binding sites for hepatocyte nuclear factors, and mediate developmental and tissue-specific hormone induction (Tsai et al. 1989; Reik et al. 1991; Rigaud et al. 1991; Ganss et al. 1994; Szapary et al. 1993; Espinás et al. 1994; Reik et al. 1994). Aspartate aminotransferase (AAT) mRNA is increased by GC as well, although only the cytosolic, not the mitochondrial form, appears to be regulated (Pavé-Preux et al. 1988). Recent analysis of the cytosolic AAT promoter revealed the presence of a novel GRE (GRE A), which differs from the consensus GRE (GGTACA *NNN* TGTTCT) in that the two half sites are separated by 8 bp (GRE A: GGTACA *GAAAGACC* TGTTCT). Detailed analyses showed that a GR tetramer is likely to bind to this novel element to transactivate cytosolic AAT (Garlatti et al. 1994). Transcription of the rat tryptophane oxygenase gene is induced by GC via an upstream GRE-containing regulatory region (Danesch et al. 1987) that has been used to study synergistic interactions between various transcription factors and the GR in eukaryotic gene regulation (Schüle et al. 1988a, b). Mouse and human phenylalanine hydroxylase genes also contain GREs in their 5 flanking regions (McDowall and Fischer 1995). GC-regulated expression of γ-glutamyl transferase (Barouki et al. 1982), serine dehydratase (Noda et al. 1988; Su and Pitot 1992), tyrosine hydroxylase (Fossom et al. 1992), and argininosuccinate lyase (Husson et al. 1990) has also been reported.

To enhance clearance of NH_3 molecules generated in the course of protein degradation, GC increase glutamine synthase transcription, as exemplified in muscle (Feng et al. 1990) and leukemic cells (Harmon and Thompson 1982). The mouse and chicken genes for this important enzyme in nitrogen metabolism and neurotransmitter generation were found to contain GREs in their regulatory regions (Bhandari et al. 1988; Zhang and Young 1991). GC induction of transcription, at least in chicken embryo retina, requires cell contact (Vardimon et al. 1988; Reisfeld and Vardimon 1994) and seems to be strongly influenced by the balance between the levels of c-*jun* and GR: The ability of GC to induce glutamine synthase expression correlates inversely with levels of c-*jun* (Berko-Flint et al. 1994), and elevated GR expression overcomes GC nonresponsiveness in these cells (Zhang and Young 1993), which might reflect the previously noted antagonistic effect between AP-1 and GR. Two other explanations for this example of developmentally regulated gene expression by GC have recently been suggested, i.e., developmental stage-specific spatial expression of 2 different GR isotypes with molecular weights of 90 kDa and 95 kDa, respectively (Gorovits et al. 1994), and differential expression of a retinal C/EBP-like protein that might influence inducibility by GC (Ben-Or and Okret 1993).

GC further influence, in a general manner, the first step in protein metabolism, i.e., the translation of mRNA into polypeptide chains, although the molecular mechanism of this GC effect is poorly understood. Ribosomal RNA gene transcription is stimulated by GC in rat liver (Webb et al. 1989; Flusser et al. 1989; Suzuki et al. 1991), myoblasts (Glibetic et al. 1993), and hypothalamus (Jones et al. 1990), yet is suppressed in mouse lymphosarcoma cells (Mahajan and Thompson 1990a; Huang and Hershey 1989). Mammalian ribosomal RNA gene promoters contain a potential GRE (Misseyanni et al. 1991), but the effect of GC on ribosomal RNA gene transcription may be primarily mediated by regulating the amount or activity of TFIC, a transcription factor tightly associated with RNA polymerase I (Mahajan and Thompson 1991). The expression of another ribosome component, the mouse ribosome protein rpL32, might also be influenced by GC, since its gene contains a GRE-like sequence in its promoter (Yoganathan and Sells 1992). Apart from regulating the expression of components in the general translation machinery, GC appear to participate in tRNA production, as exemplified by the report of a GRE in, and enhanced transcription of, the methionine tRNA gene (Li and Kmiec 1990).

3.3 Lipid and Lipoprotein Metabolism

Most of the GC effects on lipid metabolism (increased output of free fatty acids, fat deposition) have been regarded as consequences of its action on carbohydrate metabolism (Labhart 1986). Chronic GC treatment causes fat redistribution ("buffy hump", "moon face"), which is due to the different sensitivity of tissues to the permissive lipolytic activity of GC and the lipogenic activity of reactively induced insulin. Several reports, however, document a direct regulatory effect on genes affecting lipid biosynthesis, metabolism, and transport. The rate-limiting enzyme in cholesterol synthesis, 3-hydroxy-3-methylglutaryl (HMG) CoA reductase, is suppressed by GC in liver (Simonet and Ness 1989) and adrenal glands (Lehoux et al. 1989; Black et al. 1988), although the major transcriptional repressor of the corresponding gene is cholesterol itself (Goldstein and Brown 1990). The suppressive GC effect, at least in rat liver, is due to reduction of mRNA half-life, probably mediated by conserved sequences in the 3' untranslated region (Simonet and Ness 1989). We have observed GC reduction of HMG-CoA reductase mRNA levels in a mouse macrophage-like cell line (Helmberg et al. 1990). The possible significance of this phenomenon in immune regulation will be discussed further below (Sect. 5). Fatty acid metabolism might be affected by GC in several ways. First, GC stimulate fatty acid biosynthesis by transcriptionally activating fatty acid synthase, the central enzyme in this metabolic pathway (Xu et al. 1993). Second, GC have a stimulatory effect on fatty acid-binding protein expression (Cook et al. 1988), which might support the former GC effect. Third, monoacylglycerol acyltransferase, central to hepatic triglyceride biosynthesis, is decreased by a combination of insulin and GC (Coleman 1993). Finally, several enzymes that generate free fatty acids from complex lipids appear to be GC-regulated. These include hormone sensitive lipase (Slavin et al. 1994), lipoprotein lipase and several phospholipases, such as phospholipase A_2 (see Sect. 5) and D (Kobayashi et al. 1988). The stimulatory effect of GC on the extracellular lipoprotein lipase is probably mediated by 2 GREs in the upstream region of its gene (Deeb and Peng 1989), and transcriptional activation has been observed in various tissues, including human macrophages (Domin et al. 1991) and rat and human adipocytes (Ong et al. 1992). In addition to its effects on lipid biosynthesis, GC seem to influence lipid transport by regulating several apolipoproteins, e.g., in rat liver (Staels et al. 1991) and human breast cancer cells (Simard et al. 1992).

Of particular clinical significance is the accelerating effect of prenatally administered GC on lung maturation in premature infants, which is due to its stimulation of the surface active agent surfactant (Mendelson and Boggaram 1991). Surfactant consists largely of phospholipids (mainly phosphatidylcholine), the surfactant specific glycoproteins (SP)-A, SP-B, and the hydrophobic proteolipids SP-C and SP-D. GC regulate the major surfactant- associated protein SP-A in a biphasic fashion, i.e., at nanomolar concentrations, SP-A mRNA is increased in fetal lung, while at higher concentrations and after prolonged exposure, a decrease is observed (Liley et al. 1988; Odom et al. 1988). This was explained on the molecular level by an increase in SP-A gene transcription at low GC concentrations that is overcome by a reduction in mRNA stability at higher GC concentrations (Boggaram et al. 1989, 1991). Transcriptional induction of the SP-A gene by GC might further be influenced by cAMP, which also stimulates SP-A gene activity (Alcorn et al. 1993). An enhancing effect of GC on pulmonary SP-B and SP-C gene transcription has also been reported (O'Reilly et al. 1991). The biosynthesis of the major lipid component of surfactant, phosphatidylcholine, is also stimulated by dexamethasone via activation of choline phosphate cytidyltransferase. This activation seems to be caused by fatty acids that are produced by fatty acid synthase transcriptionally activated by GC (reviewed in Rooney et al. 1994). The critical role of GC in lung maturation has been corroborated in mice made deficient either for CRF or the GR. The homozygous deficient offsprings of these gene-targeted mice die within hours after birth due to severe atelectasis possibly caused by reduced surfactant protein expression as well as a reduced expression of the amiloride-sensitive Na^+ channel, that has been shown to be regulated by GC (Champigny et al. 1994).

4 Effects on Extracellular Matrix, Bone Metabolism, Electrolytes, and Water Homeostasis

GC influence the expression of various extracellular matrix proteins and have potent effects on bone metabolism. Hence, hypercortisolemic individuals often show signs of dermal tissue weakening ("striae distensae"), impaired wound healing and loss of bone ("steroidogenic osteoporosis"). Furthermore, GC have the potential to affect blood electrolyte balance and water homeostasis by two essentially different pathways, mediated either by the GR itself or by the mineralocorticoid receptor (MR). Cortisol and other steroids with GC and mineralocorticoid activity can activate the MR under certain circumstances and, thus, may regulate genes normally controlled by aldosterone. Activation of the MR by cortisol is normally prevented by the coexpression of MR and 11β-hydroxysteroid dehydrogenase (11β-HSD; Funder 1993). 11β-HSD converts cortisol to cortisone, thereby inactivating cortisol in mineralocorticoid target tissues. The weak aldosteronelike effect (sodium retention and potassium excretion) of therapeutic GC doses (Labhart 1986) may, nevertheless, be explained, at least partly, by activating MR. Some direct GC effects on extracellular matrix proteins, bone metabolism, electrolytes, and water homeostasis are discussed below.

4.1 Extracellular Matrix Proteins

The major extracellular matrix proteins, collagen types I, III, IV, V and interstitial collagen, are suppressed by GC (Russell et al. 1989; Ohyama et al. 1990; Perez et al. 1992). This effect, which is associated with reduced collagen mRNA levels, might explain some of the GC effects on wound healing, and is not seen in human fibroblast from keloids and fetal dermis (Russell et al. 1989). The promoter of the procollagen α_2 type I gene has been mapped for GC-sensitive regions, and potential GR binding sites could be identified. Three potential GREs have been identified in the human elastin promoter, whose functionality was tested in transgenic mice (Kähäri 1994). Treatment with GC showed hormone-dependent enhancement of reporter gene expression at sites of elastin expression proving the functionality of the GREs and probably explaining the GC-dependent induction of elastin during rat fetal lung development. Analyses of human fibroblast in vitro, however, showed a negative GC effect on elastin expression (Russell et al. 1995). GC also influence the expression of fibronectin (Odenthal et al. 1992; Guller et al. 1994), tenascin (Ekblom et al. 1993), and the composition of glycosaminoglycans. The expression of extracellular matrix-degrading enzymes, involved in tissue remodeling, inflammation, and cell migration is influenced by GC as well and is discussed below.

4.2
Bone Metabolism

Of particular clinical importance is the catabolic activity of GC on bone metabolism. GC stimulate calcium and phosphate excretion and, in addition to induction of amino acid mobilization, can cause osteoporosis, the most common and severe unwanted side effect of long-term GC therapy, by delaying bone remodeling (reviewed in Labhart 1986; Orwoll and Klein 1995). These effects can, in part, be explained at the molecular level by the influence of GC on parathyroid hormone, the single most important upregulator of serum calcium concentration and inducer of bone resorption, and on its counterregulator calcitonin, a major inhibitor of bone resorption. Thus, GC increase preproparathyroid hormone mRNA steady state levels (Peraldi et al. 1990), but decrease transcription of the related gene for the parathyroid hormone-related peptide (Ikeda et al. 1989; Lu et al. 1989). This gene contains three promoters and, although no GRE was identified up to 3.5 kb 5' of the transcription start site, all three promoters are GC-repressed (Glatz et al. 1994). GC further regulate expression of calcitonin and calcitonin gene-related peptide, which both derive from the same gene by alternative splicing, by apparently complex mechanisms: First, dependent on the cell type, transcription may either be repressed or stimulated by GC, and both activities are mediated via the same 18-bp region containing possible binding sites for AP-1, helix-loop-helix transcription factors, and a GR half site (Tverberg and Russo 1992). Second, GC regulate the splicing mechanism, thereby affecting the steady state levels of the alternatively spliced transcripts of this gene (Russo et al. 1992; Cote et al. 1992). Finally, evidence for induction of calcitonin receptor by GC has also been presented (Kurokawa et al. 1991).

Apart from affecting humoral regulators of bone resorption and neosynthesis, GC influence several other proteins participating in bone metabolism. For example, basal and vitamin D-induced expression of osteocalcin, the major noncollagenous bone protein, is inhibited by GC on the transcriptional level, and both human and rat genes contain a vitamin D-responsive element, and a GRE overlapping the TATA box (Morrison et al. 1989; Strömstedt et al. 1991; Heinrichs et al. 1993). The repressive GC effect on this, and possibly other GC-repressed genes, is elegantly explained by competition for DNA binding sites between the GR and the TATA box binding transcription factor, TFIID (Strömstedt et al. 1991). In the rat osteocalcin promoter additional GR binding sites have been identified and analyzed (Aslam et al. 1995). The human bone sialoprotein gene, another structural protein of the bone matrix, contains a GRE in its promoter (Kim et al. 1994), and transcription from the gene for alkaline phosphatase, an indicator of osteoblastic activity, is regulated by GC in different rat tissues, including bone, and in human osteoblasts (Zernik et al. 1991; Subramaniam et al. 1992).

Finally, as also discussed in Sect. 7, GC have a marked effect on osteoblast differentiation. This and other postulated effects of GC on bone cells may, in part, relate to the rapid GC regulation of early response genes, including the AP-1 components Jun and Fos in osteoblasts (Subramaniam et al. 1992; Shalhoub et al. 1992). GC effects on c-*fos* expression may be particularly important due to the crucial role of c-*fos* in the differentiation and proliferation of such cells, as revealed in mice made transgenic or deficient for this proto-oncogene (Rüther et al. 1987; Grigoriadis et al. 1994).

4.3 Electrolytes and Water Homeostasis

The renin-angiotensin-aldosterone system, central to the regulation of sodium and potassium blood levels and, thus, blood volume and pressure, are influenced by GC at multiple check points. In addition to the previously noted effect on adrenal steroid and mineralocorticoid biosynthesis (see Sect. 2), GC have a permissive role on interleukin (IL)-1 stimulated angiotensinogen gene transcription (Ron et al. 1990) and directly induce expression of this GRE-containing gene in rat hepatocytes (Ohtani et al. 1992) and diencephalon (Deschepper and Flaxman 1990). Interestingly, GC induce a novel transcript with a long 5'-untranslated region from a second promoter (Ben-Ari et al. 1989). In addition to the angiotensinogen gene, GC also induce expression of at least one of the angiotensin receptor isotypes (Uno et al. 1994). While these regulatory GC effects might lead to sodium and water retention, the opposite might be expected from their stimulatory action on atrial natriuretic peptide transcription in man and rats (Shields et al. 1988; Gardner et al. 1988; Argentin et al. 1991) and apparent upregulation of one of its receptor subtypes (Lanier-Smith and Currie 1991). These events, in concert with an inhibitory effect of GC on vasopressin expression observed under certain circumstances (Verbeeck et al. 1991; Hu et al. 1993), might contribute to the enhancement of the glomerular filtration rate often observed during GC therapy (Labhart 1986). Whether the aldosterone-like or aldosterone-antagonistic actions of GC predominate in a particular instance might finally depend, among other factors, on the type of GC used, their dosage, and the duration of treatment.

5
Effects on Inflammation and the Immune System

Perhaps the most frequent therapeutic application of GC exploits their anti-inflammatory and immunosuppressive activities, which subdue symptoms of acute and chronic inflammation. A large number of molecular mechanisms for these GC actions have been proposed, some of which are reviewed below. Although anti-inflammatory and immunosuppressive effects may complement each other and have a common molecular basis, some mechanisms account primarily for anti-inflammatory effects while others predominantly explain immunosuppression. In line with this concept, we first summarize the effect of GC on the production of potent mediators of inflammation, their metabolism in the inflamed tissue, and other issues related to inflammation. We then discuss the regulation of cytokines and their receptors as well as other membrane proteins, including cell adhesion molecules and major histocompatibility complex proteins that may all play a role in inflammation, immune responses, or both. Finally, an attempt is made to relate GC regulation of individual genes to the complex phenomenon of immunosuppression.

5.1 Arachidonic Acid Pathway

The anti-inflammatory GC effect is attributed to a significant degree to the suppressive activity on the arachidonic acid pathway at multiple levels, thereby preventing generation of potent inflammatory mediators, such as prostaglandins and leukotriens (Sebaldt et al. 1990; Pasmanik et al. 1991). As one possible mechanism, GC induce expression of lipocortin I in a number of tissues from guinea pigs, rats, and humans. Lipocortin I is a member of a multigene family that, aside from numerous recently identified activities, has been recognized as mediator of anti-inflammatory GC effects (for reviews see Goulding and Guyre 1992; Parente and Solito 1994; Flower and Rothwell 1994). It inhibits, possibly via substrate binding and sequestration (Davidson et al. 1987), phospholipase A_2, the enzyme starting the entire cascade by liberating arachidonic acid from membrane phospholipids. Although recombinant lipocortin I mimics some of the anti-inflammatory actions of GC (Cirino et al. 1987), its in vivo role as mediator of this GC effect is debatable (Whitehouse 1989; Parente et al. 1990).

Apart from regulating lipocortin I, GC directly control the expression of several enzymes in the arachidonic acid cascade. In peritoneal macrophages from lipopolysaccharide (LPS)-treated mice (Masferrer et al. 1990) and in different human cell cultures, cyclo-oxygenase expression is repressed by GC via reduction of the respective mRNA levels (Bailey et al. 1988) and/or by inhibition of translation (Raz et al. 1989; Koehler et al. 1990). Interestingly, lipocortin I has been implicated in the latter phenomenon (Pash and Bailey 1988). Additional evidence for GC-mediated suppression of cyclo-oxygenase and possible isoforms has recently been published (O'Banion et al. 1992b; Masferrer et al. 1992; Kujubu and Herschman 1992; Crofford et al. 1994). GC further suppress the cytokine-induced group II phospholipase A_2 (Nakano et al. 1990; Mühl et al. 1992; Wu et al. 1994; Miyashita et al. 1995), thereby implicating a mechanism for GC inhibition of phospholipase A_2 other than the somewhat controversial lipocortin pathway. Other possibly GC-regulated enzymes of the arachidonic acid cascade include 15-hydroxyprostaglandin dehydrogenase (Xun et al. 1991) and prostaglandin endoperoxide synthases (Sirois et al. 1992; Kawaguchi et al. 1994; Simonson et al. 1991; Zakar et al. 1995).

5.2 Other Mediators of Inflammation

In addition to their effects on the prostaglandin and leukotriene synthesis, GC interfere with expression of several other mediators and enzymes that are directly or indirectly implicated in the inflammation process, including those derived from, or involved in, the kallikrein-kinin system, the complement cascade, and the blood clotting and fibrinolytic systems. While in most instances the described regulatory phenomena agree with the anti-inflammatory activity of GC, this is not always the case. For instance, in the kallikrein-kinin system, GC induce transcription from the rat T-kininogen gene in hepatoma cells and thus provide a potential precursor for bradykinin, one of the most potent mediators of inflammation (Anderson and Lingrel 1989). However, an opposite observation, i.e., suppression of T-kininogen mRNA by GC in adjuvant-treated rats, has also been reported (Howard et al. 1990). The fine structure of the promoter of the rat T1-kininogen gene and of another member of this gene family, the K-kininogen gene, has recently been dissected, and the regions conferring basal and GC (and IL-6) regulated expression defined (Chen and Liao 1993). More in line with its anti-inflammatory activity is the GC-mediated downregulation of kallikrein, the enzyme liberating bradykinin from its precursor, as shown in pancreatic AR42J cells (Rosewicz et al. 1991), and the induction of the bradykinin-degrading enzyme neutral endopeptidase (Borson and Gruenert 1991).

GC effects on the complement cascade might further contribute to their anti-inflammatory effects: The mouse C4b-binding protein gene contains 4 putative GREs that mediate direct transcriptional induction (Moffat et al. 1992). Factor B mRNA levels and complement factor C3 appear down-regulated, whereas factor H mRNA levels are increased in human umbilical vein endothelial and monocytic cells (Lemercier et al. 1992). In human pulmonary epithelial cells, however, induction of C3 gene expression has been observed (Zach et al. 1992).

Fibrinogen and the fibrinolytic system function in a variety of biological processes, such as blood clotting, ovulation, embryo implantation, neoplasia, and neovascularization. In inflammation, this system plays a role through its central enzyme plasmin, which generates inflammatory mediators such as fibrinogen-derived peptides. GC induce transcription of the GRE-containing type 1 plasminogen activator inhibitor gene (Van Zonneveld et al. 1988; Bruzdzinski et al. 1993; Heaton et al. 1992), thereby suppressing this system in humans (Van Zonneveld et al. 1988) and rats (Konkle et al. 1992). On the other hand, plasminogen activator expression is enhanced in several experimental environments (Heaton et al. 1989, 1992; Heaton and Gelehrter 1989; Wang and Leung 1989; Kathju et al. 1994), although contrary findings have been reported (Karlan et al. 1989). GC further induce the three subunits of fibrinogen in *Xenopus* liver parenchymal cells (Bhattacharya and Holland 1991), and the human (Huber et al. 1990) and rat (Baumann et al. 1990) β-fibrinogen gene is directly transcriptionally regulated by GC via upstream sequences.

GC may limit inflammation-associated tissue damage by regulating metalloproteinases and the tissue inhibitor of metalloproteinases (TIMP). Thus, GC suppress stromelysin, collagenase, type IV collagenase and the counterregulatory TIMP (Shapiro et al. 1991; Offringa et al. 1988; Delany and Brinckerhoff 1992; Lyons et al. 1993). In another report, collagenase down regulation was observed as well, while TIMP expression was not affected by GC (Clark 1987). In the case of collagenase, transcriptional GC repression was mediated by mutual inhibitory protein-protein interactions between the GR and AP-1, the major enhancer factor of the collagenase promoter (Jonat et al. 1990; Yang-Yen et al. 1990; Schüle et al. 1990). In fact, this system was among the first in which this essential molecular mechanism of downregulation by GC was described. Expression of matrilysin, a recently discovered extracellular matrix degrading enzyme produced by glandular epithelia, mononuclear phagocytes, and renal mesangial cells, is also negatively regulated by GC (Busiek et al. 1995).

5.3
Acute-Phase Proteins

Numerous stress forms lead to increased levels of certain plasma proteins, collectively termed acute-phase proteins, some of which might participate in inflammatory reactions. GC, in concert with other hepatocyte stimulating factors such as IL-1 and IL-6, participate in the regulation of the acute-phase response in the rat (Kushner 1988) and, to a lesser degree, in man (Baumann 1987). Several acute-phase proteins are either induced or repressed, and the molecular mechanisms of GC regulation have been delineated for some of these. Regulation of the rat α_1-inhibitor III gene, for instance, occurs at the transcriptional level (Abraham et al. 1991), and GRE-like sequences were identified in the positively regulated promoters of the rat and mouse haptoglobin (Marinkovic and Baumann 1990; Pajovic et al. 1994) and α_1-acid glycoprotein (Baumann et al. 1990; Ingrassia et al. 1994; Alam et al. 1993; Ratajczak et al. 1992) genes. Posttranscriptional regulation of the latter gene has also been reported (Vannice et al. 1984). Transcriptional regulation of the $\alpha_{2\mu}$-globulin is indirect and apparently requires induction of a specific, as yet undefined, transcription factor (Addison and Kurtz 1989). The DNA element mediating transcriptional regulation of the $\alpha_{2\mu}$-globulin has been isolated and studied as paradigm for a delayed secondary GC response element (Chan et al. 1991; Hess et al. 1990). In this and other cases of GC-regulated acute-phase proteins (e.g., α_1-acid glycoprotein), induction of CAAT-enhancer binding protein (C/EBP) and cooperativity of C/EBP and GR may play a crucial role (Baumann et al. 1991, 1992; Alam et al. 1993; Williams et al. 1991). Other examples of GC-regulated acute-phase proteins, such as kininogen (α_1-major acute-phase protein) and fibrinogen, have been discussed above or are listed in Table 1.

Another explanation for the anti-inflammatory effects of GC is their inhibition of the nitric oxide synthase induction in vascular endothelial cells, macrophages, hepatocytes, and brain cells (Radomski et al. 1990; Moncada and Palmer 1991; Geller et al. 1993; Jun et al. 1994). Nitric oxide synthase can be induced in macrophages by various stimuli, resulting in enhanced production of nitric oxide, which may play an important role in lowering blood pressure at sites of inflammation. The inducible form has been reported to be inhibited and the constitutive form to be upregulated by GC in rats (Baltrons et al. 1995). The effects of GC on nitric oxide synthase expression in the rat brain depend on neuronal cell specificity. Similar observations have been made for heme oxygenase 1 and 2 (generating carbon monoxide; Weber et al. 1994).

5.4 Adhesion Molecules and Other Cell Surface Proteins

Adhesion molecule expression is essential in directing inflammatory cells to their destined sites of action. Cell adhesion molecules are expressed on endothelial cells in response to various cytokines. Through their negative effects on cytokines, GC suppress adhesion molecule expression indirectly, but direct effects have also been observed. The LPS-induced overexpression of two major adhesion molecules, intercellular adhesion molecule-1 (ICAM-1) and endothelial-leukocyte adhesion molecule-1 (ELAM-1), is suppressed by GC on the protein and mRNA levels (Cronstein et al. 1992). The molecular mechanism of this gene repression has been delineated for the ICAM-1 gene (Van de Stolpe et al. 1994). Its promoter contains 4 tumor phorbol ester (e.g., TPA)-responsive DNA regions with different TPA-responsive enhancer sequences that mediate TPA and tumor necrosis factor α (TNF-α)-dependent transcriptional induction. One of them, an NFκB-like enhancer sequence, is responsible for repression of TNF-α and TPA-induced transactivation, and serves as another model for negative cross-talk between the NFκB transcription factor and the GR. The above mentioned adhesion molecules may also be involved in GC suppression of T-lymphocyte adhesion to endothelial and synovial cells (Eguchi et al. 1992), and thereby act in an immunosuppressive manner.

One way in which GC might influence the immune system more directly concerns the expression of certain membrane proteins that participate in the initiation and/or effector phase of immune responses. Major histocompatibility (MHC) class I and class II antigens, which present processed peptides to the antigen receptors of different types of T-lymphocytes and thereby initiate and perpetuate immune responses (Yewdell and Bennink 1990), seem to be under the regulatory influence of GC. The underlying molecular mechanisms, however, are not well understood and, although GC downregulation of these immunostimulatory proteins seems to be the rule, contrary findings have also been reported. For example, in mouse macrophages (Snyder and Unanue 1982; Warren and Vogel 1985) and B-cells (Dennis and Mond 1986) MHC class II expression is suppressed by GC, while in humans GC either increased (Manyak et al. 1988; Rhodes et al. 1986), decreased (Von Knebel Doeberitz et al. 1990), or had essentially no effect (Watanabe et al. 1990) on MHC class II and class I expression, depending on the studied cell type. GC-dependent repression of MHC II expression was explained by prevention of transactivating factor(s) binding to the X box motif in the Ia β gene (Celada et al. 1993). A second group of GC-regulated immunomodulatory surface proteins are the receptors for the Fc portion of immunoglobulins (Ig). Thus, spontaneous and lymphokine-induced expression of the type II Fc receptor for IgE on human lymphocytes and hematopoietic cell lines (Fischer and König 1990; Naray-Fejes-Toth et al. 1985; Kaufman Paterson et al. 1994) and mouse mast cells (Robin et al. 1985) is suppressed by GC. This, in concert with the subduing effect on various mediators of inflammation (see above), might participate in the beneficial effect of GC on allergic reactions. The interferon-γ induced expression of the type I Fc receptor for IgG is inhibited on human neutrophils and myeloid cell lines, but increased on monocytes (Petroni et al. 1988; Pan et al. 1990). As with MHC genes, the precise mechanisms responsible for the observed phenomena remain to be elucidated.

5.5 Cytokines and Their Receptors

Among the central anti-inflammatory and/or immunosuppressive mechanisms of GC are their effects on a series of cytokines and their receptors that appear suppressed by GC on transcriptional, posttranscriptional and translational levels (for review see Barnes and Adcock 1993). Thus, an important lymphokine and mediator of inflammation, IL-1, is controlled by GC via inhibition of mitogen-induced transcription of the IL-1β gene and decreased stability of its mRNA in monocytes of several species (Knudsen et al. 1987; Lee et al. 1988; Huether et al. 1993; Amano et al. 1993). In contrast, the IL-1 receptor is induced by GC in a protein and mRNA neosynthesis-dependent fashion (Akahoshi et al. 1988; Gottschall et al. 1991), as is the type II receptor, a potential decoy target for IL-1 (Re et al. 1994). In line with the anti-inflammatory activity of GC is their negative effect on expression of the pleiotropic mediator IL-6 in mouse macrophages and human monocytes, endothelial cells, and fibroblasts (Tobler et al. 1992; Amano et al. 1993). DNase I footprinting experiments have shown occlusion by the GR of binding sites for several stimulatory transcription factors in the IL-6 gene promoter region as the molecular mechanism underlying this transcriptional repression (Ray et al. 1990). GC suppression of TNFα in humans (Velasco et al. 1991), mice and guinea pigs (Gadina et al. 1991; Remick et al. 1989) may, at least in part, explain the protective action of GC in endotoxic shock. The suppressive effect of GC is manifested at the levels of both transcription and translation requiring sequences in the promoter and 3' untranslated region, respectively (Han et al. 1991). Interferon-γ overcomes this effect, which might explain the controversial results obtained with GC in treating human septic shock (Luedke and Cerami 1990). This lymphokine itself is, however, subject to GC regulation (Cippitelli et al. 1995), and is suppressed on the mRNA level (Arya et al. 1984) via AU-rich destabilizing sequences in the 3' untranslated region (Peppel et al. 1991). Mouse and human interferon-β mRNA levels are also reduced by GC posttranscriptionally (Gessani et al. 1988), but, on the other hand, induction of interferon-β expression in murine L929 cells has recently been reported and an efficient GRE could be identified in the promoter region (Soury et al. 1995). Granulocyte-macrophage colony-stimulating factor (GM-CSF), a powerful growth and differentiation factor of hematopoietic progenitor cells, is induced by inflammatory stimuli and cytokines. This effect is almost completely suppressed by GC on the RNA and protein levels in human fibroblasts (Tobler et al. 1992) and mouse peritoneal macrophages (Thorens et al. 1987). In contrast, human fibroblast M-CSF production was unaltered by GC (Tobler et al. 1992) and, in human endometrium, GC even induced mRNA and its receptor (Azuma et al. 1990).

Of particular relevance to the immune system, GC inhibit the transcription of the human IL-2 gene in activated T-lymphocytes by interfering with the protein kinase C and calcium-mediated transactivation of this gene (Vacca et al. 1992; Boumpas et al. 1991). The IL-2 enhancer has been studied in great detail, and transcription factor binding sites shown to be essential for basal and inducible transcription have been mapped (reviewed in Umlauf et al. 1993). Typical GREs have not been identified, and GC repression does not depend on direct GR DNA binding. The GR apparently impairs the cooperativity between AP-1 and nuclear factor of activated T-cells (NF-AT; Vacca et al. 1992; Paliogianni et al. 1993b). Hence, as in many other examples throughout this review, the suppressive GR activity is presumably exerted by protein-protein interactions with AP-1 or a similar, yet distinguishable, transcription factor (Northrop et al. 1992). Concerning IL-2 receptor α and β gene expression, the influence of GC is less clear, and controversial results have been published (Lamas et al. 1993; Reed and Nowell 1986; Paliogianni et al. 1993a). Recently, another facet of GC- mediated inhibition of T-cell function has been described. Paliogianni et al. (1993a) showed that, in addition to inhibition of IL-2 expression, GC can also negatively interfere with signals transmitted via the IL-2 receptor, thus inhibiting preactivated T-cells. In particular, they observed decreased phosphorylation of several intracellular proteins, including the cell cycle-regulating retinoblastoma protein. Additional possibly GC-regulated lymphokines and cytokines include IL-4 (Daynes and Araneo 1989; Wu et al. 1991a), IL-5 (Rolfe et al. 1992), the IL-8 family member CEF-4 (Gonneville et al. 1991), IL-8 (Tobler et al. 1992; Wertheim et al. 1993; Mukaida et al. 1994), transforming growth factor α and β (Raja et al. 1991; Centrella et al. 1991; AyanlarBatuman et al. 1991), and the growth factor-induced secreted proteins JE and KC that are related to the IL-2 and IL-8 cytokine families, respectively (Rameh and Armelin 1992; Deng et al. 1994; Poon et al. 1991).

5.6
Effects on the Immune System

The primary overall GC effect on the immune system is immunosuppression. This, and the anti-inflammatory effect, often complement each other, making GC a widely used drug to battle acute and chronic inflammatory disorders with (auto)immune etiology. More specifically, GC alter the ratio of circulating lymphocytes (Lamas et al. 1991; Tuncer et al. 1991) and lymphocyte subsets (Westermann and Pabst 1990), suppress mitogen, lymphokine- or alloantigen-induced in vitro lymphoproliferation (Pukhalsky et al. 1990; Redondo et al. 1988; Sierra-Honigmann and Murphy 1992), inhibit the generation of lymphokine-activated killer cells in vitro (McVicar et al. 1989), and induce apoptosis (programmed cell death) in certain lymphocytes, a subject that is dealt with in more detail in Sect. 7. While the precise molecular explanations for these complex phenomena, collectively termed "immunosuppression," are not known, information regarding some of the underlying mechanisms are emerging.

Thus, GC-suppressed mitogen- or alloantigen-induced lymphoproliferation can be prevented in vitro with a combination of IL-1, IL-6, and interferon-γ (Almawi et al. 1991), pointing to an essential role for GC repression of lymphokine expression in this facet of immunosuppression. The role of lymphokines is further supported by the pronounced effect of GC on IL-2 and its receptor, both of which are known to be essential molecules in lymphocyte proliferation. IL-2 and its receptor and other lymphokines/cytokines possibly participating in GC-mediated effects on the immune system have been discussed above, along with cell surface proteins, in particular MHC and Fc receptor molecules, that may also serve as direct targets for immunomodulatory GC effects. Another molecular mechanism is based on the observation that GC suppress HMG-CoA reductase expression (see Sect. 3). Since other HMG-CoA reductase inhibitors also repress mitogen-stimulated lymphoproliferation (Cuthbert and Lipsky 1990), we have proposed previously (Helmberg et al. 1990) that inhibition of HMG-CoA reductase expression might contribute to GC-mediated immunosuppression.

Finally, while GC dampen most immune responses, they stimulate polyclonal Ig production, e.g., by human peripheral blood lymphocytes (Goodwin and Atluru 1986), increase the synthesis of IgE by IL-4-stimulated human lymphocytes (Wu et al. 1991b) and induce IgA in serum (but not at mucosal surfaces) and secretory components in serum, saliva, and bile of rats (Wira et al. 1990; Wira and Rossoll 1991). The inducing effect of GC on IL-4 (Orson et al. 1986; Daynes and Araneo 1989; Wu et al. 1991a) or, alternatively, the inhibitory GC effect on the production of potent Ig suppressive arachidonic acid metabolites such as leukotriene B4 (Goodwin and Atluru 1986), are possible candidates for the poorly understood molecular mechanisms underlying this phenomenon.

6
Effects on the Central and Peripheral Nervous System

GC influence the expression of a variety of genes expressed exclusively or preferentially in central and peripheral neural tissue. These genes may be involved in some of the complex GC effects on the nervous system observed during chronic GC therapy or in pathophysiological states of GC imbalance. GC can lead to an endocrine psychosyndrome ranging from mild euphoria or depression to all degrees of schizophrenia-like symptoms, electroencephalogram abnormalities and, in animals, to cytoplasmic vacuolization in the regions of the hypothalamic, paraventricular, and supraoptic nuclei (Labhart 1986). In rat and/or primate hippocampus, GC impair the capacity of neurons to survive various insults, a phenomenon termed "GC endangerment" (Sapolsky et al. 1990), somewhat reminiscent of GC-induced apoptosis in lymphocytes (see Sect. 7). This form of neuronal cell death is, however, not associated with the DNA fragmentation (Masters et al. 1989) typically seen in lymphocyte apoptosis. The transmitters of the GC signal, GR and MR, are expressed in various regions of the brain (Funder 1994) and influence neuronal gene expression and excitability (reviewed in Joels and DeKloet 1992). Interestingly, as has been observed in some leukemic cells (Gametchu et al. 1993), atypical receptors for GC can be identified as membrane-associated molecules (Orchinik et al. 1991). As both GR and MR are coexpressed in some regions of the brain, heterodimerization has been proposed to be a means of fine-tuning neuronal gene expression (Trapp et al. 1994). Below we review GC regulation of genes typically expressed in neuronal tissue although, in most instances, their relevance to known GC effects remains unclear. Genes encoding hypothalamic and pituitary releasing factors and hormones are dealt with in Sects. 2 and 4. Other genes primarily expressed elsewhere in the body that may, nevertheless, also participate in neuronal effects, such as IL-1 (Gottschall et al. 1991), α_2- macroglobulin (Higuchi et al.

1994) and others, have been discussed throughout this review. Some issues pertaining to differentiation and proliferation of glial cells are reviewed in Sect. 7.

6.1 Structural Proteins and Neuronal Enzymes

The major myelin protein genes encoding myelin basic protein and proteolipid protein appear to be under posttranscriptional regulation by steroids (Kumar et al. 1989). In the case of the myelin basic protein, translational regulation by GC via an element in the 5'-untranslated region of its mRNA has been reported (Verdi et al. 1989; Verdi and Campagnoni 1990). Another structural brain protein, glial fibrillary acidic protein (O'Callaghan et al. 1989), seems to be GC-regulated also.

6.2 Neurotransmitters and Growth Factors

Several neuronal growth factors, neurotransmitters and/or their receptors appear to be GC-regulated in the central and peripheral nervous system. This holds true for nerve growth factor (NGF; Neveu et al. 1991; Jehan et al. 1993) as well as its receptor (Yakovlev et al. 1990; Foreman et al. 1992), thereby providing an indirect means for GC influence on numerous NGF-regulated genes. These include the neuropeptide Y, which is also directly regulated by GC (Sabol and Higuchi 1990; Sahu et al. 1992; Barnea et al. 1991). Similarily, NGF-induced transcription of the neural-specific gene, SCG10, is inhibited by GC (Stein et al. 1988). Possibly related to their mood-altering effects, GC participate in the regulation of important opoid receptor ligands, such as β-endorphin (Kavelaars et al. 1990) and enkephalin (Naranjo 1986; Lightman and Young 1989; Chao and McEwen 1990). In the case of the enkephalin gene, GRE and GRE-like sequences have been identified in the 5'-upstream region and introns in the hamster (Zhu et al. 1994). Nevertheless, GC have little, if any, effect on basal levels but suppress stress-induced (Harbuz and Lightman 1989; Lightman and Young 1989) and potentiate forskolin-induced (Joshi and Sabol 1991) expression. The molecular mechanism of the latter form of cooperative GC action has been analyzed in more detail. While GC alone has no effect on enkephalin gene transcription, as expected, it strongly increases the magnitude and duration of transcriptional elevation by the cAMP-inducing agent forskolin. Although the element responsible for forskolin induction could be localized within this region, it conferred GC suppression rather than induction upon a reporter construct. The DNA element conferring the cooperative effect could not be identified within an almost 6 kb spanning 5'-flanking region.

GC influence the content and function of chromaffine granules from the adrenal medulla and other sources: The acidic granule glycoprotein chromogranin A is regulated by GC in parathyroid cells (J.-X. Zhang et al. 1993), in corticotropes, where it may act as an autocrine inhibitor of POMC secretion (Wand et al. 1991), and in chromaffine cells itself. The rat chromogranin A promoter has been cloned and a potential novel GRE identified (Rozansky et al. 1994). The rat gene encoding phenylethanolamin *N*-methyltransferase, one of the four enzymes required for catecholamine biosynthesis, is induced by GC and contains a GRE (Ross et al. 1990; Kennedy et al. 1993). Recent analysis of GC effects on the expression of this gene in vivo, however, suggested that GC-regulated transcriptional activation plays a less important role than expected from in vitro studies (Wong et al. 1992).

Expression of major receptors of the sympathetic and parasympathetic peripheral nervous systems, namely the different adrenergic receptor (AR) isotypes and the acetylcholine receptor, is also regulated by GC in various tissues. The hamster β_2-AR gene contains four GREs (Malbon and Hadcock 1988), and elevated transcription may account for increased mRNA steady state levels in various GC-treated tissues (Collins et al. 1988; Stern and Kunos 1988; Takahashi and Iizuka 1991; Hadcock and Malbon 1988). On the other hand, GC downregulate β_1-AR gene expression by transcriptional repression (Kiely et al. 1994), which nicely explains the differential regulation observed in cells expressing β_1-AR and β_2-AR (Fève et al. 1990; Zhong and Minneman 1993). The β_3-AR gene, at least in mouse adipocyte cells, also appears transcriptionally downregulated by GC (Fève et al. 1992). In contrast, GC increase the rate of transcription of the α_1B-AR gene in smooth muscle cells, which might contribute to the increased sensitivity of such cells to catecholamines during GC excess (Sakaue and Hoffman 1991). Finally, induction of acetylcholine receptor on human muscle cells by GC provides a possible molecular mechanism additional to the well-known immunosuppression for the beneficial effect of this drug in autoimmune myasthenia gravis (Kaplan et al. 1990).

7 Effects on Cell Differentiation, Proliferation, and Apoptosis

In a multicellular organism, individual cells have to make several crucial decisions: Whether to enter into the cell cycle and proliferate, to proceed on a path of differentiation from a multipotential undifferentiated state towards a highly specialized tissue component, or, finally, to undergo programmed cell death (often characterized by the morphological hallmarks of apoptosis). As an oversimplification, GC have a largely negative effect on proliferation, support differentiation and, in some systems, induce apoptosis. The role of GC in these processes has largely been obtained from in vitro cell culture systems, which do not represent all of the manifold cell-cell interactions occurring in vivo in intact tissues and organs. The obvious complexity of proliferation, differentiation, and apoptosis, as well as the dependence on imperfect model systems, explain the paucity of detailed molecular data regarding the role of GC in these phenomena. Nevertheless, investigations during the last decade have identified GC regulation of numerous genes that govern these fundamental cellular decisions by participating in signal transduction, cell cycle regulation and transcriptional control. Many of these genes have further been implicated in cancerogenesis, and are known as oncogenes or tumor suppressor genes. In the following, we first provide a brief introduction to some of the models for GC-regulated differentiation, proliferation and apoptosis. In the second part, we summarize GC effects on molecules involved in signal transduction. Obviously, multiple genes participate in all of these models, and one particular gene may function in more than one such process. Thus, as in other sections, the following organization in, and the assignment of different GC-regulated genes to, distinct subheadings is somewhat artificial. Moreover, several genes that might also participate in the subjects of this section, such as some cytokines, hormones and their receptors, have already been discussed.

7.1
Model Systems for GC-Regulated Differentiation, Proliferation and Apoptosis

7.1.1
Differentiation and Development

The potential importance of GC for the development of an organism is supported by numerous findings. One example is provided by mice made deficient for CRF by homozygous inactivation of their CRF genes, rendering them unable to produce GC. Their fetuses, which develop in the complete absence of GC, die shortly after birth due to lung immaturity. This phenomenon may, in part, be explained by the stimulatory effect of GC on surfactant production (see Sect. 3). GC administration to the mother rescues these fetuses and allows them to develop normally to term and continue healthy development without the need for GC substitution (Muglia et al. 1995). The critical role of GC for lung maturation and viability is also seen in GR gene-targeted mice. Mice lacking GR expression die perinatally due to severe atelectasias of the lung, exhibit severe reduction of gluconeogenic enzyme mRNA steady state levels, and show a drastically reduced number of chromaffine cells (Cole et al. 1995). Further indirect support for a vital role of GC in fetal development comes from human inherited GC resistance caused by GR gene defects. Although complete loss of function mutations have been reported (reviewed in Chrousos et al. 1993), they have never been observed in a homozygous state, suggesting that the complete lack of GR might be incompatible with human development. While complete lack of GC function is apparently fatal, increased GC levels during fetal development might cause growth retardation, possibly by GC-induced overexpression of insulin-like growth factor binding proteins (Price et al. 1992). Possible additional effects of GC overexposure are not well established, although this topic is of considerable clinical interest for prenatal therapy in congenital adrenal hyperplasia, a defect in GC biosynthesis (Miller 1994).

While less dramatic than their role in fetal development, GC strongly influence differentiation in many cell types and tissues including osteoblasts, adipocytes, lactating mammary gland, oligodendrocytes, certain hematopoietic cell lines such as HL-60 and U937, the neuronal progenitor cell line PC12, and others. Osteoblast differentiation in vitro is induced by the addition of GC to the growth medium. The effects of GC on osteoblast progenitor cells from rat calvaries depend on the duration of GC treatment, that initially inducing proliferation and, after a prolonged dexamethasone treatment, supports the differentiation of progenitors into cells exhibiting a osteoblastlike gene expression pattern. These cells continue to proliferate, form bone nodules and calcification of extracellular matrix ensues. Short term treatment of progenitor cells (which tend to differentiate spontaneously) leads to more divergent GC effects, depending on the differentiation stage of the cells (Shalhoub et al. 1992).

Differentiation of mouse 3T3-L1 fibroblastic cells by GC, methylisobutylxanthine, and insulin into adipocytes occurs in contact-inhibited dense cell culture. Analysis of altered protein expression by two-dimensional gel electrophoresis revealed changes in over 300 proteins during differentiation, only a minority of which seem to be regulated by GC treatment alone (Wheeler et al. 1994). Identification of the regulated proteins will probably contribute to our understanding of GC-influenced differentiation processes. In differentiated adipocytes, GC have complex transcriptional and posttranslational effects on insulin receptor, IRS-1 and phosphatidylinositol 3-kinase expression and/or activity (Turnbow et al. 1994).

Mammary gland differentiation depends on the action of multiple hormones. The most important role for GC in mammary gland function is their lactogenic activity wherein GC, in concert with other hormones (e.g., insulin, progesterone, prolactin), induce production of milk proteins, such as β-casein (Welte et al. 1993). This induction can also be observed in established cell lines in vitro and may be caused by a GC-induced relief of transcriptional repression (Doppler et al. 1989, 1990; Schmitt-Ney et al. 1991) and GC-treatment alters chromatine structure of the whey acidic milk protein promoter (Li and Rosen 1994). GC-induced differentiation can be blocked by overexpression of *mos*, *ras*, and *src*, which results in enhanced levels of endogenous c-*fos*, that, as an AP-1 component, could counteract GC-induced differentiation and induction of milk proteins (Touray et al. 1991; Qi et al. 1989; Jehn et al. 1992).

The molecular mechanism initiating differentiation in these and other cells remains obscure, but an interesting model has been put forward to explain GC-induced oligodendrocyte differentiation (Barres et al. 1994). Raff and coworkers propose that oligodendrocytes tend to live considerably longer in the absence of lipophilic messengers such as the steroid and thyroid hormones, or retinoic acid. Hydrophilic growth factors are necessary to induce proliferation which oligodendrocytes seem to "count" and limit to eight consecutive divisions in the presence of the lipophilic hormones. The molecular mechanisms underlying these phenomena might be exerted by regulating the intracellular levels of AP-1, which is predicted to decrease during ongoing cell division down to a certain threshold beyond which activated nuclear hormone receptors can effectively antagonize AP-1 to stop proliferation; differentiation would then ensue by default (Barres and Raff 1994). Antagonism between growth factors and lipophilic hormones has also been implicated to influence differentiation of the bipotent sympathoadrenal progenitor cells. Differentiation of these cells is directed either by GC into a chromaffine cell or by NGF into a sympathetic neuron (reviewed in Anderson 1995). This dichotomous capacity of the progenitor cell is preserved in PC12 cells and seems to be strongly influenced by expression of the thyroid hormone receptor α-1 (TRα-1). Presence of TRα-1 and thyroid hormone enhances differentiation, of PC12 cells, which become NGF dependent for survival, while differentiation into a chromaffine cell by GC is blocked (Munoz et al. 1993a).

In some instances, however, GC appear to inhibit cell differentiation. Interestingly, even in these cases, AP-1 antagonism seems to be the underlying molecular mechanism, as exemplified by GC inhibition of differentiation induced by tumor phorbol esters in HL-60, or U937 human myeloid leukemic cells (Hass et al. 1991). In addition GR interactions with the transcription factors EGR-1 and NFκB may also play a role (Hass et al. 1992). Inhibition of differentiation by GC has also been observed in mouse erythroleukemic MEL cells, in which the steroid effect seems to be due to inhibition of the transcription factor GATA-1 (Chang et al. 1993).

7.1.2 Proliferation

In most cases investigated in this respect, GC inhibit proliferation. While this may be beneficial in the treatment of malignancies, it can result in impaired wound healing, reduced lymphocytic responses and other less desirable antiproliferative effects in cases of GC excess. Related to, yet distinct from, the antiproliferative effect of GC is their apoptosis-inducing ability in certain cells of the lymphoid lineage, which is discussed further below. The antiproliferative effect on fibroblasts (reviewed in Durant et al. 1986) and perhaps other cells may, in part, be due to induction of antiproliferative proteins, such as lipocortin in A549 human lung adenocarcinoma cells (Croxtall and Flower 1992, 1994). However, a great number of genes supposedly involved in regulation of proliferation appear downregulated in various GC growth-arrested cells, such as c-*myc* and others (see below). Other downregulated genes or proteins possibly contributing to GC-induced growth arrest include the posttranscriptionally regulated thymidine kinase in mouse L cells (Frost et al. 1993), enzymes involved in polyamine biosynthesis, e.g., ornithine decarboxylase (see below), and glucose transporters (Horner et al. 1987). In some tumor cell lines, the antiproliferative effect of GC could be pinned down to repression of viral oncogene expression, e.g., human papilloma virus genes E6 and E7 in a cervical cancer cell line (Von Knebel Doeberitz et al. 1994). In addition to regulating specific genes responsible for proliferation directly, GC may act via their general suppressive activity on translation (Sect. 3) and transcription. Although rarely, evidence for mitogenic activity of GC has also been reported. Thus, the growth of human AIDS-associated Kaposi sarcoma cells and SC-3 cells is stimulated by GC in vitro (Guo and Antakly 1995), the latter via expression of an autocrine growth factor (Yamanishi et al. 1995).

7.1.3 Apoptosis

As outlined above, GC interference with signal transduction at multiple levels may result in growth arrest, which occurs mainly in the G_1 phase of the cell division cycle (Goya et al. 1993; Sanchez et al. 1993). This arrest may either be reversible (Frost et al. 1994) or result in apoptosis, as in the case of certain cells of the lymphoid lineage (reviewed in Thompson 1994). Thymocyte cell death induced by GC was among the first phenomena in which a morphologically distinct form of death, subsequently termed apoptosis, was discovered (Wyllie et al. 1980). Immature T- and B-lymphocytes are extremely sensitive to GC and undergo rapid apoptosis in vivo and in vitro. This phenomenon might be an important mechanism during thymocyte selection and differentiation (Ashwell et al. 1994), as well as in regulating the activity of peripheral lymphocytes (Gonzalo et al. 1993). Also, GC-induced apoptosis has great clinical relevance in the treatment of various hematological malignancies. Despite numerous publications on molecular events associated with, and possibly contributing to, GC-induced apoptosis (reviewed in Thompson 1994; Hughes and Cidlowski 1994; Zhivotovsky et al. 1994; Nieto and López-Rivas 1992), the precise mechanism how GC induce cell death remains to be a mystery. In particular, it is still unknown whether GC cause cell death by induction of specific "death genes" or repression of "survival genes". Support for the latter concept has been provided by Helmberg et al. (1995) using the rat GR mutant LS7, that differs at two amino acid residues in the C-terminal zinc finger from wild-type GR. This mutant GR has been shown to be severly impaired in transactivating properties, but behaves indistinguishable to wild-type GR in transrepression (Yang-Yen et al. 1990). Stable expression of both the rat GR and the LS7 mutant restored GC sensitivity in human GR negative Jurkat T-cells. Overexpression of the LS7 mutant GR in yeast impaired proliferation (Lefstin et al. 1994) as has been shown previously for the human GR (Wright et al. 1991). Lefstin et al. (1994) demonstrated that the LS7 effect (growth inhibition and transrepression) is mediated by one of the two amino acid substitutions in the second zinc finger. They also provided evidence that this mutant exhibits higher nonspecific DNA binding than the wild-type GR. Searching for complementing mutations in the GR DNA binding domain revealed that mutations that abrogate DNA binding impair the LS7 effect. Mutations in the dimer interface region of the DNA binding domain, however, allowed dissection of transrepression and antiproliferative properties of the LS7 mutant, suggesting that transrepression needs nonspecific DNA binding while antiproliferation needs nonspecific DNA binding and dimerization. Interestingly, Heck et al. (1994) recently also suggested that transcriptional interference of the GR with AP-1 on a synthetic AP-1-regulated promoter might be mediated by monomers rather than dimers. Evidence that transrepression underlies the growth inhibitory properties of GR in human leukemic cells has also been obtained by Nazareth et al. (1991), who showed that expression of a GR fragment, mainly consisting of the DNA-binding domain, is sufficient to induce apoptosis. Transrepression could be a specific event acting by interference of the activated GR with regulatory transcription factors, e.g., AP-1. On the other hand, a high GR expression level could also interfere with the availability of basal transcription factors or coactivator molecules for transcription of a broad range of genes. We are currently investigating the latter possibility, i.e., that general suppression of transcription (see below) and/or translation (Sect. 3) mediated by GC may prevent appropriate cell cycle progression. In cells destined for proliferation by internal (oncogenes) or external (growth factors) stimuli this unspecific repression of transcription may then lead to apoptosis. Finally, while GC induce apoptosis in lymphocytes, they prevent or retard the androgen withdrawal-associated prostatic cell death by influencing the expression of several potential cell death-associated genes (Rennie et al. 1989).

Based on the observation that GC-induced apoptosis in some, but not all, systems is dependent on macromolecule neosynthesis, several attempts to clone death inducing "lysis genes" have been made. Subtractive cDNA and other cloning approaches led to the identification of several induced genes (Baughman et al. 1991, 1992; Goldstone and Lavin 1991), many of as yet unknown function, but until now no gene could be shown to be the essential regulator of GC-induced apoptosis. Nevertheless, these and other genes implicated in GC-induced apoptosis may still play a role in the molecular events associated with this process. The role of calcium in GC-induced apoptosis is still a matter of debate (McConkey et al. 1994). Calmodulin antagonists have been shown to interfere with activation of the GR (Ning and Sánchez 1995), and calmodulin steady state mRNA levels dramatically increase during GC-induced mouse lymphocyte apoptosis (Dowd et al. 1991). Overexpression of calbindin-D28K, a major calcium storage protein of the endoplasmic reticulum, has been shown to block GC-induced apoptosis (Dowd et al. 1992), which supports the role of calcium in rodent apoptosis (McConkey et al. 1994). In addition, *bcl-2* has been implicated in regulating calcium fluxes between subcellular compartments (Baffy et al. 1993; Lam et al. 1994). Calcium might be required to activate a postulated calcium/magnesium-dependent endonuclease (Montague et al. 1994) or other undefined calcium-dependent enzymes. Expression of sulfated glycoprotein-2 (SGP-2, also termed TRPM-2 or clusterin) during GC-induced apoptosis (Bettuzzi et al. 1991; Briehl and Miesfeld 1991) may directly be regulated by GC (Rosemblit and Chen 1994). The role of ornithine decarboxylase, the rate limiting enzyme of polyamine biosynthesis encoded by a GRE-containing gene (Wen et al. 1989), is also controversial (Grassilli et al. 1991). GC further induce tissue transglutaminase (Suto et al. 1993; reviewed in Fesus et al. 1987), an enzyme that has been associated with TGF-β-induced apoptosis, although the significance of this phenomenon in GC modulation of TGF-β-induced apoptosis or in GC-induced apoptosis itself is not clear (Fukuda et al. 1994). Another gene increased during GC-induced lymphocyte apoptosis, glutathione *S*-transferase, has provided further evidence that the apoptotic process might be associated with oxidative stress (Flomerfelt et al. 1993). Interestingly, although at least one glutathione *S*-transferase gene is transcriptionally induced by GC in hamster smooth muscle cells (Fan et al. 1992), the increased expression in the course of GC-induced apoptosis is a GR-independent secondary event (Flomerfelt et al. 1993).

A number of additional genes have been identified as crucial in other systems of apoptosis, including *bcl-2* (the mammalian homolog of the *Caenorhabditis elegans* death inhibitor *ced-9*), p53, IL-1 converting enzyme (ICE; the homolog of *C. elegans* death inducer gene *ced-3*), c-*myc*, poly(ADP)-phosphoribosyltransferase (ADPRT), $p34^{CDC2}$ kinase, AP-1, and others (reviewed in Vaux et al. 1994; Raff 1992; Oltvai and Korsmeyer 1994). Interestingly, with the exception of c-*myc* (see below), it is not certain whether these genes are GC-regulated and what, if any, role they play in GC-induced cell death. Thompson's group (Thulasi et al. 1993) has recently provided evidence that GC downregulation of c-*myc* followed by G_1 arrest could be the essential step in GC-induced apoptosis of human CEM-C7 T-leukemic cells. This conclusion, mainly based on transient transfection technology, has, however, been questioned by others (Wood et al. 1994). The *bcl-2* overexpression prevents GC-induced T- and B-lymphocyte death (Caron-Leslie et al. 1994; Lam et al. 1994; Miyashita and Reed 1992, 1993; Alnemri et al. 1992), and bel-2 expression, e.g., in developing B-cells, correlates with stage-dependent susceptibility to GC (Merino et al. 1994). Only limited evidence is available whether GC action involves regulation of *bcl-2* or *bcl-2* family member expression (Lotem and Sachs 1994). Similarily, p53 expression seems not to be required for GC-induced thymocyte cell death in p53 gene targeted mice (Clarke et al. 1993; Lotem and Sachs 1993). The above mentioned homolog of the *C. elegans* death-inducing gene *ced-3*, ICE, and its related proteases could play a pivotal role as downstream effectors of apoptosis. ADPRT has been shown to be cleaved during GC-induced apoptosis by one of the ICE isozymes, presumably CPP32 (Fernandes-Alnemri et al. 1994, 1995; Alnemri et al. 1995). It will be interesting to learn which of the upstream ICE-like proteases necessary to activate CPP32 are induced by GC. Preliminary data of our group suggest that induction of apoptosis by Apo-1 and GC differ in activation of ADPRT cleaving ICE-like proteases, as Apo-1 induced apoptosis seems to be inhibitable by overexpression of the cowpox virus serpin crmA, whereas GC-induced cell death is not.

Another approach towards a better understanding of GC-induced cell death is based on the analysis of the molecular mechanisms of GC resistance, a serious problem in the therapeutic management of leukemia (reviewed in Kaspers et al. 1994). However, acquired GC resistance in leukemic cell lines analyzed thus far was almost exclusively due to various mutations in the GR gene itself (Ip et al. 1993; Strasser-Wozak et al. 1995; Palmer et al. 1992; Powers et al. 1993; Ashraf and Thompson 1993; Moalli et al. 1993; Krett et al. 1995; Moalli and Rosen 1994). Despite considerable efforts, molecular defects in other genes resulting in acquired GC resistance in the presence of functional GR have not yet been precisely defined (see also Flomerfelt and Miesfeld 1994).

7.2 GC Effects on Molecules Involved in Signal Transduction

7.2.1 Growth Hormones and Their Receptors

GC might affect differentiation, proliferation and apoptosis by controlling the expression of a host of growth-promoting factors and/or their receptors, some of which have been discussed in previous section. Insulin-like growth factor I mRNA levels are reduced by GC in the rat (Luo and Murphy 1989; McCarthy et al. 1990; Lowe et al. 1992), and the 6 isoforms of its binding protein are either up- or downregulated (Orlowski et al. 1991; Okazaki et al. 1994). The human and rat insulin-like growth factor binding protein-1 gene promoters are well conserved and contain a potential GRE (Unterman et al. 1992; Robertson et al. 1994; Suh et al. 1994; Suwanichkul et al. 1994). However, regulation may not only occur at the transcriptional level, but may include cycloheximide-sensitive labile proteins responsible for the rapid mRNA turnover (Ooi et al. 1993). GC further regulate epidermal growth factor in mice (Tuomela et al. 1990) and suppress the mRNA for its receptor in fetal rat lung cells (Oberg and Carpenter 1989), while this message is increased in ovarian cancer cells even though GC exert an antiproliferative effect on those cells (Ferrandina et al. 1992). Other GC-regulated growth factors or receptors include heparin-binding growth factor I, a potent mitogen for human endothelial and epithelial cells (Harris et al. 1989; Nakano et al. 1993), hepatocyte growth factor (Matsumoto et al. 1992; Gohda et al. 1992), its receptor, the c-*met* proto-oncogene (Moghul et al. 1994), and the HER-2/neu growth factor receptor in ovarian cancer cells that is induced not on the mRNA, but on the protein level (Karlan et al. 1994).

7.2.2
Signal Transduction Molecules

In contrast to the large number of GC-regulated growth factors and receptors, relatively few reports have been published on GC regulation of the molecules that transduce the signal from activated membrane receptors to further downstream effector molecules (e.g., transcription factors). One of the central players in mitogenic signal transduction, the product of the *ras* proto-oncogene, may be GC-regulated, since GR binds to and transactivates the mouse Ha-*ras* promoter (Strawhecker et al. 1989). Various subunits of the trimeric G-proteins, equally important signal transducers, apparently respond to GC, such as Gαs, Gαi, and Gβ (Saito et al. 1989; Ros et al. 1989; Kawai and Arinze 1993). Promoter analyses of porcine $G\alpha_{1-2}$ and $G\alpha_{1-3}$ genes suggest that regulation is, at least in part, transcriptonal, with $G\alpha_{1-2}$ being down- and $G\alpha_{1-3}$ being upregulated (Holtzman et al. 1993). Finally, GC regulate signal transducing proteins from the insulin receptor, as mentioned in Sect. 3.

7.2.3
Protein Kinases

Protein phosphorylation is one of the most important mechanisms of posttranslational protein modification, and a host of kinases participate in the transduction of incoming signals. GC have recently been shown to increase protein kinase C activity in some systems, but neither the molecular mechanisms of regulation nor the protein kinase C isotype possibly involved have been defined (Krasil'nikov et al. 1992; Vrtovsnik et al. 1994). Better evidence for GC regulation is available for two other putative novel protein kinases: sgk (for serum and GC-regulated kinase), a putative serine/threonine protein kinase that contains a GRE in its gene promoter and is rapidly induced by GC (Webster et al. 1993), and snk (for serum induced kinase), whose serum induction is apparently repressed by GC (Simmons et al. 1992). Since both were identified as early growth response genes that possibly participate in signaling G_1 to S transition, they may provide a link between GC and cell cycle regulation.

7.2.4
Transcription Factors

GC regulate transcription factor activity by different mechanisms. The first is based on protein-protein interactions of the GR with the respective transcription factor and often, although not always, results in inactivation of both molecules engaged in the interaction. Numerous examples of this mechanism have been discussed throughout this review and are summarized in the Introduction. Another related, but perhaps more general, mechanism of transcriptional repression has been termed "squelching" (Gill and Ptashne 1988), and refers to the ability of a transcription factor, such as the GR, to bind law abundant, limiting general transcription factors or coactivators (Cahill et al. 1994).

In addition to these mechanisms, GC regulate sequence-specific transcription factor expression more directly as well. Expression of c-*myc*, an early growth response proto-oncogene with a presumably central role in proliferation, is repressed in various model systems of GC-induced growth arrest (Yuh and Thompson 1989; Maroder et al. 1990; O'Banion et al. 1992a). As shown in P1798 T-lymphoma cells, the molecular basis of this phenomenon may be regulation of alternative c-*myc* promoter usage by GC (Ma et al. 1992). However, no evidence could be obtained that this phenomenon is due to GC-induced alterations in the *myc* promoter binding of the early response transcription factor E2F (Rhee et al. 1994). Other investigators have suggested that *c-myc* expression is affected by GC primarily at the postranscriptional level (Maroder et al. 1990). The *c-myc* downregulation might explain the GC-induced growth arrest in numerous cells, and may participate in GC-induced apoptosis, as discussed above. While the interaction between GR and AP-1 at the protein level may be more relevant, expression of the AP-1 components c-*jun* and c-*fos* is also regulated by GC. Thus, the increase in c-*jun* and c-*fos* mRNA in human myeloid leukemia cells after induction of differentiation by phorbol esters is repressed by GC at the transcriptional level (Hass et al. 1991). In NIH3T3 cells, this effect has been attributed to GR binding to a serum response element, thereby antagonizing the function of positive transcription factors (Karagianni and Tsawdaroglou 1994). In agreement with this model, a GRE was found 13 bp downstream from the AP-1 site in the rat *fos* promoter (Wang and Howells 1994). Further potential GC-regulated transcription factors are a member of the interferon regulatory factor family, interferon consensus sequence binding protein (Politis et al. 1994), the erythroid differentiation transcription factor GATA-1 (Chang et al. 1993), some isoforms of CREB (Jungmann et al. 1992), Pit-1 (Jong et al. 1994), C/EBP (MacDougald et al. 1994), EGR-1 and NFκB (Hass et al. 1992). GC regulation of multiple members of the steroid receptor gene superfamily is reviewed in Sect. 2.

8 Effect on Virus Gene Expression

The observation, first made in animals and soon extended to humans, that GC regulate gene expression in several endogenous and exogenous viruses, may have far-reaching clinical implications that may not be fully appreciated at present. GC may be used to combat symptoms accidentally occurring in, or causally associated with, certain viral diseases, without necessarily recognizing the potentially harmful effects resulting from the regulation of viral gene expression by GC. Furthermore, although speculative at this point, GC may alter the expression of endogenous retroviral genes, or stimulate otherwise latent viruses, thereby contributing to disease development. The following section summarizes some of the better studied effects of GC on viruses that have been implicated in human disorders, including AIDS and cancer.

8.1 Retroviruses

Perhaps the best studied example of interactions between viruses and GC is their pronounced effect on retroviral gene expression. Thus, the mouse mammary tumor virus long terminal repeat (LTR) belongs to the most intensively investigated promoter/enhancer elements. It contains several GREs and response elements for other transcription factors, and serves as a paradigm for GC-regulated promoters (Chandler et al. 1983; Chalepakis et al. 1988; Beato 1991). Other mouse (Miksicek et al. 1986; Celander and Haseltine 1987) and human (Larsson et al. 1989) endogenous retroviruses contain GREs as well and may, therefore, be GC-regulated. The extent and pathophysiological role, if any, of GC regulation of endogenous retroviral gene expression is still unclear. In mice, we and others have suggested a possible role of GC induction of immunosuppressive retroviral p15E expression (Helmberg et al. 1990; Fiegl et al. 1995). While p15E-like immunosuppressive material is elaborated from human cells as well (Cianciolo et al. 1984; Benomar et al. 1987), it is not known whether GC activate expression of such molecules and, if so, whether this participates in the immunosuppresive activity of GC.

The possible influence of GC on human exogenous retroviruses, including HIV, is of considerable clinical significance, particularly since GC treatment in AIDS patients with *Pneumocystis carinii* pneumonia has attracted a great deal of attention (Palca 1990; NIH-UC Expert Panel Corticosteroids 1990). Such treatment is controversial, since GR-binding sites are present in the HIV LTR (Ghosh 1992), and HIV replication and gene expression are both enhanced via this and a novel intragenic GRE (Soudeyns et al. 1993). Other exogenous human retroviruses, such as HTLV-III (Markham and Gallo 1986) and a novel type D retrovirus (Oda 1988), respond to GC and/or have GRE-like sequences in their LTR.

8.2
Other Pathogenic Human Viruses

GC influence the expression of several other pathogenic viruses as well. Human papilloma viruses (HPV) implicated in the etiology of several tumors, including squamus cell carcinoma of the uterine cervix, one of the most common cancers among women, contain GREs in their control regions (Chan et al. 1989; Mittal et al. 1993). Thus, GC influence the expression of the oncogenic E6 and E7 proteins that bind to, and inactivate, the tumor suppressors p53 and Rb, respectively. However, this regulation is apparently complex and may depend on the chromosomal virus integration site (Von Knebel Doeberitz et al. 1991). In addition, evidence has been reported for upregulation of the E6/E7 promoter (Butz and Hoppe-Seyler 1993) that possibly results in GC dependence of oncogenic transformation (Pater et al. 1988), but also downregulation of E6/E7 transcription (Von Knebel Doeberitz et al. 1994) has been observed. In mice transgenic for an HPV-18 upstream regulatory region-reporter gene fusion construct, GC and phorbol ester interfered with each other in their respective effects, suggesting that, at least in some aspects, HPV regulation may serve as another example of cross-talk between AP-1 and GR (Cid et al. 1993). Site-directed mutagenesis of the GRE in the HPV-18 upstream regulatory region abolished inducibility by GC and enhanced basal activity (Butz and Hoppe-Seyler 1993). Epidemiological studies have indeed suggested a link between steroid exposure and HPV dependent cellular transformation (reviewed in Mittal et al. 1994). Interestingly, GR transactivation activity has been found to increase in the course of cellular transformation in a transgenic mouse model. These mice, transgenic for the bovine papilloma virus type 1 genome, develop fibrosarcomas originating from dermal fibroblasts. While in cells from benign lesions GR transcriptional activity is low, it was found to be markedly increased in fibrosarcoma cells (Vivanco et al. 1995).

As shown in transgenic mice and other systems, Hepatitis B virus contains a stimulatory GRE that mediates GC induction of hepatitis B antigen expression (Tur-Kaspa 1986; Farza et al. 1987; Chou et al. 1992). The etiological agent of mononucleosis, nasopharyngeal carcinoma, and possibly Burkitt's lymphoma, the Epstein-Barr virus, contains a GC-sensitive sequence in its promoter region (Kupfer and Summers 1990; Schuster et al. 1991). Other viruses, such as the vesicular stomatitis virus (Lancz et al. 1990), adenovirus (Mahajan and Thompson 1990b), bovine herpes virus (Rock et al. 1992), and polyoma virus (Mudrak et al. 1994), may also be regulated by GC.

9 Conclusion

Over the last decade, a growing number of GC-regulated genes has been identified that encode proteins critically involved in the structure and function of essentially all organ systems in the body. GC have also been found to influence the expression of a variety of viral proteins that may be responsible for life-threatening diseases, including AIDS and some forms of cancer. The regulation of various combinations of these genes can be assumed to underly the plethora of complex biological effects exerted by GC under physiological conditions, during prolonged GC therapy or in pathological states of dysregulated secretion of the hormone or its receptor. While it is often difficult to correlate complex GC effects with the regulation of individual genes, the improving understanding of GC actions at the molecular level and the increasing number of genes known to be regulated by GC has allowed the suggestion of reasonable models for the molecular basis of GC-regulated phenomena.

Regarding the molecular mechanisms of GC regulation of individual genes, several general aspects have become apparent. As mentioned in the Introduction and outlined in numerous examples throughout this review, GC influence gene expression at essentially all levels. In many instances, gene regulation is indirect, i.e., GC regulate the expression of other, often unknown, proteins that directly or indirectly influence the expression or activity of the particular molecule under scrutinity. In other cases, GC may have a "permissive" effect, meaning that by themselves they have no biological effect, but enable other regulators (hormones, etc.) to exert their activity, e.g., by enhancing expression of the respective receptor or other mechanisms. Apart from these actions, GC influence the expression of a large set of genes in a more direct fashion. Strong and rapid induction of gene expression is obtained by strong *cis* elements (GREs) that have been mapped in many gene promoters. Transcriptional repression, on the other hand, is apparently much more sophisticated. As discussed in numerous examples throughout this review, at least three basic mechanisms can be discerned: Repression mediated by negative GREs, by composite GREs, and by negative interaction of the GR with other transcription factors on the protein level.

The high degree of complexity in the regulation of eukaryotic gene expression suggests that GC regulation, as other forms of gene regulation, might depend on factors such as species, cell type, state of differentiation, presence of inhibitory or costimulatory factors, GC type and concentration, etc. Thus, in addition to sufficient levels of functional GR, these parameters may ultimately determine whether a gene appears induced, repressed, or even unregulated by GC in a particular in vivo or in vitro situation.

References

Abraham LJ, Bradshaw AD, Northemann W, Fey GH (1991) Identification of a glucocorticoid response element contributing to the constitutive expression of the rat liver α1-inhibitor III gene. J Biol Chem 266:18268–18275

Addison WR, Kurtz DT (1989) Identification of nuclear proteins that bind to the glucocorticoid regulatory region of a rat α2μ-globulin gene. J Biol Chem 264:21891–21895

Adler S, Waterman ML, Rosenfeld MG (1988) Steroid receptor-mediated inhibition of rat prolactin gene expression does not require the receptor DNA-binding domain. Cell 52:685–695

Agular B-M, Vinggaard AM, Vind C (1992) Regulation by dexamethasone of the 3β-hydroxysteroid dehydrogenase activity in adult rat Leydig cells. J Steroid Biochem Mol Biol 43:565–571

Akahoshi T, Oppenheim JJ, Matsushima K (1988) Induction of high-affinity interleukin 1 receptor on human peripheral blood lymphocytes by glucocorticoid hormones. J Exp Med 167:924–936
Akerblom IE (1988) Negative regulation by glucocorticoids through interference with cAMP responsive enhancer. Science 241:350
Alam T, An MR, Mifflin RC, Hsieh C-C, Ge X, Papaconstantinou J (1993) Trans-activation of the α1-acid glycoprotein gene acute phase responsive element by multiple isoforms of C/EBP and glucocorticoid receptor. J Biol Chem 268:15681–15688
Alcorn JL, Gao E, Chen Q, Smith ME, Gerard RD, Mendelson CR (1993) Genomic elements involved in transcriptional regulation of the rabbit surfactant protein-A gene. Mol Endocrinol 7:1072–1085
Alksnis M, Barkhem T, Strömstedt P-E, Ahola H, Kutoh E, Gustafsson JA, Poellinger L, Nilsson S (1991) High level expression of functional full length and truncated glucocorticoid receptor in chinese hamster ovary cells. J Biol Chem 266:10078–10085
Almawi WY, Lipman ML, Stevens AC, Zanker B, Hadro ET, Strom TB (1991) Abrogation of glucocorticosteroid-mediated inhibition of T cell proliferation by the synergistic action of IL-1, IL-6, and IFN-gamma. J Immunol 146:3523–3527
Alnemri ES, Fernandes TF, Haldar S, Croce CM, Litwack G (1992) Involvement of BCL-2 in glucocorticoid-induced apoptosis of human pre-B-leukemias. Cancer Res 52:491–495
Alnemri ES, Fernandes-Alnemri T, Litwack G (1995) Cloning and expression of four novel isoforms of human interleukin-1β converting enzyme with different apoptotic activities. J Biol Chem 270:4312–4317
Amano Y, Lee SW, Allison AC (1993) Inhibition by glucocorticoids of the formation of interleukin-1α, interleukin-1β, and interleukin-6: mediation by decreased mRNA stability. Mol Pharmacol 43:176–182
Anderson DJ (1995) Molecular control of cell fate in the neural crest: the sympathoadrenal lineage. Annu Rev Neurosci 16:129–158
Anderson KP, Lingrel JB (1989) Glucocorticoid and estrogen regulation of a rat T-kininogen gene. Nucleic Acids Res 17:2835–2848
Argentin S, Sun YL, Lihrmann I, Schmidt TJ, Drouin J, Nemer M (1991) Distal cis-acting promoter sequences mediate glucocorticoid stimulation of cardiac atrial natriuretic factor gene transcription. J Biol Chem 266:23315–23322
Arya SK, Wong-Staal F, Gallo RC (1984) Dexamethasone-mediated inhibition of human T cell growth factor and gamma-interferon messenger RNA. J Immunol 133:273–276
Ashraf J, Thompson EB (1993) Identification of the activation-labile gene: a single point mutation in the human glucocorticoid receptor presents as two distinct receptor phenotypes. Mol Endocrinol 7:631–642
Ashwell JD, Berger NA, Cidlowski JA, Lane DP, Korsmeyer SJ (1994) Coming to terms with death: apoptosis in cancer and immune development. Immunol Today 15:147–151
Aslam F, Shalhoub V, Van Wijnen AJ, Banerjee C, Bortell R, Shakoori AR, Litwack G, Stein JL, Stein GS, Lian JB (1995) Contributions of distal and proximal promoter

elements to glucocorticoid regulation of osteocalcin gene transcription. Mol Endocrinol 9:679–690

Autelitano DJ, Sheppard KE (1993) Corticotrope responsiveness to glucocorticoids is modulated via rapid CRF-mediated induction of the proto-oncogene c-fos. Mol Cell Endocrinol 94:111–119

AyanlarBatuman O, Ferrero AP, Diaz A, Jimenez SA (1991) Regulation of transforming growth factor-β1 gene expression by glucocorticoids in normal human T lymphocytes. J Clin Invest 88:1574–1580

Azuma C, Saji F, Kimura T, Tokugawa Y, Takemura M, Samejima Y, Tanizawa O (1990) Steroid hormones induce macrophage colony-stimulating factor (MCSF) and MCSF receptor mRNAs in the human endometrium. J Mol Endocrinol 5:103–108

Baffy G, Miyashita T, Williamson JR, Reed JC (1993) Apoptosis induced by withdrawal of interleukin-3 (IL-3) from an IL-3-dependent hematopoietic cell line is associated with repartitioning of intracellular calcium and is blocked by enforced Bcl-2 oncoprotein production. J Biol Chem 268:6511–6519

Bailey JM, Makheja AN, Pash J, Verma M (1988) Corticosteroids suppress cyclooxygenase messenger RNA levels and prostanoid synthesis in cultured vascular cells. Biochem Biophys Res Commun 157:1159–1163

Baltrons MA, Agullo L, Garcia A (1995) Dexamethasone up-regulates a constitutive nitric oxide synthase in cerebellar astrocytes but not in granule cells in culture. J Neurochem 64:447–450

Barnea A, Cho G, Hajibeigi A, Aguila MC, Magni P (1991) Dexamethasone-induced accumulation of neuropeptide-Y by aggregating fetal brain cells in culture: a process dependent on the developmental age of the aggregates. Endocrinology 129:931–938

Barnes PJ, Adcock I (1993) Anti-inflammatory actions of steroids: molecular mechanism. Trends Pharmacol Sci 14:436–441

Barouki R, Chobert MN, Billon MC, Finidori J, Tsapis R, Hanoune J (1982) Glucocorticoid hormones increase the activity of gamma-glutamyltransferase in a highly differentiated hepatoma cell line. Biochim Biophys Acta 721:11–21

Barres BA, Raff MC (1994) Control of oligodendrocyte number in the developing rat optic nerve. Neuron 12:935–942

Barres BA, Lazar MA, Raff MC (1994) A novel role for thyroid hormone, glucocorticoids and retinoic acid in timing oligodendrocyte development. Development 120:1097–1108

Bastian LS, Nordeen SK (1991) Concerted stimulation of transcription by glucocorticoid receptors and basal transcription factors: limited transcriptional synergism suggests mediation by coactivators/adaptors. Mol Endocrinol 5:619–627

Baughman G, Harrigan MT, Campbell NF, Nurrish SJ, Bourgeois S (1991) Genes newly identified as regulated by glucocorticoids in murine thymocytes. Mol Endocrinol 5:637–644

Baughman G, Lesley J, Trotter J, Hyman R, Bourgeois S (1992) Tcl-30, a new T cell-specific gene expressed in immature glucocorticoid-sensitive thymocytes. J Immunol 149:1488–1496

Baulieu EE (1991) On the mechanism of action of RU486. Ann N Y Acad Sci 626:545–560
Baumann H (1987) Interaction among hepatocyte-stimulating factors, interleukin 1, and glucocorticoids for regulation of acute phase plasma proteins in human hepatoma (HepG2) cells. J Immunol 139:4122
Baumann H, Jahreis GP, Morella KK (1990) Interaction of cytokine- and glucocorticoid-response elements of acute-phase plasma protein genes. Importance of glucocorticoid receptor level and cell type for regulation of the elements from rat α1-acid glycoprotein and β-fibrinogen genes. J Biol Chem 265:22275–22281
Baumann H, Jahreis GP, Morella KK, Won K-A, Pruitt SC, Jones VE, Prowse KR (1991) Transcriptional regulation through cytokine and glucocorticoid response elements of rat acute phase plasma protein genes by C/EBP and JunB. J Biol Chem 266:20390–20399
Baumann H, Morella KK, Campos SP, Cao Z, Jahreis GP (1992) Role of CAAT-enhancer binding protein isoforms in the cytokine regulation of acute-phase plasma protein genes. J Biol Chem 267:19744–19751
Beato M (1989) Gene regulation by steroid hormones. Cell 56:335–344
Beato M (1991) Transcriptional control by nuclear receptors. FASEB J 5:2044–2051
Bell PA, Falany CN, McQuiddy P, Kasper CB (1990) Glucocorticoid repression and basal regulation of the epoxide hydrolase promoter. Arch Biochem Biophys 279:363–369
Bellingham DL, Sar M, Cidlowski JA (1992) Ligand-dependent down-regulation of stably transfected human glucocorticoid receptors is associated with the loss of functional glucocorticoid responsiveness. Mol Endocrinol 6:2090–2102
Ben-Ari ET, Lynch KR, Garrison JC (1989) Glucocorticoids induce the accumulation of novel angiotensinogen gene transcripts. J Biol Chem 264:13074–13079
Ben-Or S, Okret S (1993) Involvement of a C/EBP-like protein in the acquisition of responsiveness to glucocorticoid hormones during chick neural retina development. Mol Cell Biol 13:331–340
Benomar A, Wang JM, Taraboletti G, Ghezzi P, Balotta C, Cianciolo GJ, Snyderman R, Dore J-F, Mantovani A (1987) Chemotactic factor and p15E-related chemotaxis inhibitor in human melanoma cell lines with different macrophage content and tumorigenicity in nude mice. J Immunol 138:2372–2379
Berdusco ET, Milne WK, Challis JR (1994) Low-dose cortisol infusion increases plasma corticosteroid-binding globulin (CBG) and the amount of hepatic CBG mRNA in fetal sheep on day 100 of gestation. J Endocrinol 140:425–430
Berg JM (1989) DNA binding specificity of steroid receptors. Cell 57:1065–1068
Berko-Flint Y, Levkowitz G, Vardimon L (1994) Involvement of c-Jun in the control of glucocorticoid receptor transcriptional activity during development of chicken retinal tissue. EMBO J 13:646–654
Bettuzzi S, Troiano L, Davalli P, Tropea F, Ingletti MC, Grassilli E, Monti D, Corti A, Franceschi C (1991) In vivo accumulation of sulfated glycoprotein 2 mRNA in rat thymocytes upon dexamethasone-induced cell death. Biochem Biophys Res Commun 175:810–815

Bhandari B, Beckwith KD, Miller RE (1988) Cloning, nucleotide sequence, and potential regulatory elements of the glutamine synthetase gene from murine 3T3-L1 adipocytes. Proc Natl Acad Sci USA 85:5789–5793
Bhandari B, Saini KS, Miller RE (1991) Glycerol 3-phosphate dehydrogenase gene expression in cultured 3T3-L1 adipocytes: regulation by insulin, dexamethasone and dibutyryl cAMP at the level of mRNA abundance, transcription and mRNA stability. Mol Cell Endocrinol 76:71–77
Bhattacharya A, Holland LJ (1991) Coordinate regulation of fibrinogen subunit messenger RNA levels by glucocorticoids in primary cultures of Xenopus liver parenchymal cells. Mol Endocrinol 5:587–597
Black VH, Brody RI, Martin KO (1988) 3-hydroxy-3-methylglutaryl coenzyme A reductase in outer versus inner cortices of the guinea pig adrenal: effects of adrenocorticotropin and dexamethasone. Endocrinology 122:296–305
Bodwell JE, Hu J-M, Orti E, Munck A (1995) Hormone-induced hyperphosphorylation of specific phosphorylated sites in the mouse glucocorticoid receptor. J Steroid Biochem Mol Biol 52:135–140
Boggaram V, Smith ME, Mendelson CR (1989) Regulation of expression of the gene encoding the major surfactant protein (SP-A) in human fetal lung in vitro. Disparate effects of glucocorticoids on transcription and on mRNA stability. J Biol Chem 264:11421–11427
Boggaram V, Smith ME, Mendelson CR (1991) Posttranscriptional regulation of surfactant protein-A messenger RNA in human fetal lung in vitro by glucocorticoids. Mol Endocrinol 5:414–423
Bohen SP, Yamamoto KR (1993) Isolation of Hsp90 mutants by screening for decreased steroid receptor function. Proc Natl Acad Sci USA 90:11424–11428
Borson DB, Gruenert DC (1991) Glucocorticoids induce neutral endopeptidase in transformed human tracheal epithelial cells. Am J Physiol 260:L83–L89
Boumpas DT, Anastassiou ED, Older SA, Tsokos GC, Nelson DL, Balow JE (1991) Dexamethasone inhibits human interleukin 2 but not interleukin 2 receptor gene expression in vitro at the level of nuclear transcription. J Clin Invest 87:1739–1747
Briehl MM, Miesfeld RL (1991) Isolation and characterization of transcripts induced by androgen withdrawal and apoptotic cell death in the rat ventral prostate. Mol Endocrinol 5:1381–1388
Brillon DJ, Zheng B, Campbell RG, Matthews DE (1995) Effect of cortisol on energy expenditure and amino acid metabolism in humans. Am J Physiol 268:E501–E513
Bruzdzinski CJ, Johnson MR, Goble CA, Winograd SS, Gelehrter TD (1993) Mechanism of glucocorticoid induction of the rat plasminogen activator inhibitor-1 gene in HTC rat hepatoma cells: identification of cis-acting regulatory elements. Mol Endocrinol 7:1169–1177
Burger H-J, Schuetz JD, Schuetz EG, Guzelian PS (1992) Paradoxical transcriptional activation of rat liver cytochrome P-450 3A1 by dexamethasone and the antiglucocorticoid pregnenolone 16α-carbonitrile: analysis by transient transfection into

primary monolayer cultures of adult rat hepatocytes. Proc Natl Acad Sci USA 89:2145–2149

Burns K, Duggan B, Atkinson EA, Famulski KS, Nemer M, Bleackley RC, Michalak M (1994) Modulation of gene expression by calreticulin binding to the glucocorticoid receptor. Nature 367:476–480

Burnstein KL, Jewell CM, Cidlowski JA (1990) Human glucocorticoid receptor cDNA contains sequences sufficient for receptor down-regulation. J Biol Chem 265:7284–7291

Burnstein KL, Cidlowski JA (1992) The down side of glucocorticoid receptor regulation. Mol Cell Endocrinol 83:C1–C8

Busiek DF, Baragi V, Nehring LC, Parks WC, Welgus HG (1995) Matrilysin expression by human mononuclear phagocytes and its regulation by cytokines and hormones. J Immunol 154:6484–6491

Butz K, Hoppe-Seyler F (1993) Transcriptional control of human papillomavirus (HPV) oncogene expression: composition of HPV type 18 upstream regulatory region. J Virol 67:6476–6486

Cahill MA, Ernst WH, Janknecht R, Nordheim A (1994) Regulatory squelching. FEBS Lett 344:105–108

Cairns C, Cairns W, Okret S (1993) Inhibition of gene expression by steroid hormone receptors via a negative glucocorticoid response element: evidence for the involvement of DNA-binding and agonistic effects of the antiglucocorticoid/antiprogestin RU486. DNA Cell Biol 12:695–702

Caldenhoven E, Liden J, Wissink S, Van de Stolpe A, Raaijmakers J, Koenderman L, Okret S, Gustafsson JÅ, Van der Saag PT (1995) Negative cross-talk between RelA and the glucocorticoid receptor: a possible mechanism for the antiinflammatory action of glucocorticoids. Mol Endocrinol 9:401–412

Caron-Leslie L-AM, Evans RB, Cidlowski JA (1994) Bcl-2 inhibits glucocorticoid-induced apoptosis but only partially blocks calcium ionophore or cycloheximide-regulated apoptosis in S49 cells. FASEB J 8:639–645

Celada A, McKercher S, Maki RA (1993) Repression of major histocompatibility complex IA expression by glucocorticoids: the glucocorticoid receptor inhibits the DNA binding of the X box DNA binding protein. J Exp Med 177:691–698

Celander D, Haseltine WA (1987) Glucocorticoid regulation of murine leukemia virus transcription elements is specified by determinants within the viral enhancer region. J Virol 61:269–275

Centrella M, McCarthy TL, Canalis E (1991) Glucocorticoid regulation of transforming growth factor β1 activity and binding in osteoblast-enriched cultures from fetal rat bone. Mol Cell Biol 11:4490–4496

Chalepakis G, Arnemann J, Slater E, Brüller H-J, Gross B, Beato M (1988) Differential gene activation by glucocorticoids and progestins through the hormone regulatory element of mouse mammary tumor virus. Cell 53:371–382

Champigny G, Voilley N, Lingueglia E, Friend V, Barbry P, Lazdunski M (1994) Regulation of expression of the lung amiloride-sensitive Na+ channel by steroid hormones. EMBO J 13:2177–2181

Chan GC-K, Hess P, Meenakshi T, Carlstedt-Duke J, Gustafsson J-A, Payvar F (1991) Delayed secondary glucocorticoid response elements. Unusual nucleotide motifs specify glucocorticoid receptor binding to transcribed regions of α2μ-globulin DNA. J Biol Chem 266:22634–22644

Chan W-K, Klock G, Bernard H-U (1989) Progesterone and glucocorticoid response elements occur in the long control regions of several human papillomaviruses involved in anogenital neoplasia. J Virol 63:3261–3269

Chandler VL, Maler BA, Yamamoto KR (1983) DNA sequences bound specifically by glucocorticoid receptor in vitro render a heterologous promoter hormone responsive in vivo. Cell 33:489–499

Chandran UR, Attardi B, Friedman R, Dong K-W, Roberts JL, DeFranco DB (1994) Glucocorticoid receptor-mediated repression of gonadotropin-releasing hormone promoter activity in GT1 hypothalamic cell lines. Endocrinology 134:1467–1474

Chang C, Yoshida A (1994) Cloning and characterization of the gene encoding mouse mitochondrial aldehyde dehydrogenase. Gene 148:331–336

Chang TJ, Scher BM, Waxman S, Scher W (1993) Inhibition of mouse GATA-1 function by the glucocorticoid receptor: possible mechanism of steroid inhibition of erythroleukemia cell differentiation. Mol Endocrinol 7:528–542

Chao HM, McEwen BS (1990) Glucocorticoid regulation of preproenkephalin messenger ribonucleic acid in the rat striatum. Endocrinology 126:3124–3130

Chatterjee VKK, Madison LD, Mayo S, Jameson JL (1991) Repression of the human glycoprotein hormone α-subunit gene by glucocorticoids: evidence for receptor interactions with limiting transcriptional activators. Mol Endocrinol 5:100–110

Chen HM, Liao WS (1993) Molecular analysis of the differential hepatic expression of rat kininogen family genes. Mol Cell Biol 13:6766–6777

Chicco AG, Adibi SA, Liu W-Q, Morris SM Jr, Paul HS (1994) Regulation of gene expression of branched-chain keto acid dehydrogenase complex in primary cultured hepatocytes by dexamethasone and a cAMP analog. J Biol Chem 269:19427–19434

Chou C-K, Wang L-H, Lin H-M, Chi C-W (1992) Glucocorticoid stimulates hepatitis B viral gene expression in cultured human hepatoma cells. Hepatology 16:13–18

Chrousos GP, Detera-Wadleigh SD, Karl M (1993) Syndromes of glucocorticoid resistance. Ann Intern Med 119:1113–1124

Cianciolo GJ, Phipps D, Snyderman R (1984) Human malignant and mitogen-transformed cells contain retroviral p15E-related antigen. J Exp Med 159:964–969

Cid A, Auewarakul P, Garcia-Carranca A, Ovseiovich R, Gaissert H, Gissmann L (1993) Cell-type-specific activity of the human papillomavirus type 18 upstream regulatory region in transgenic mice and its modulation by tetradecanoyl phorbol acetate and glucocorticoids. J Virol 67:6742–6752

Cippitelli M, Sica A, Viggiano V, Ye J, Ghosh P, Birrer MJ, Young HA (1995) Negative transcriptional regulation of the interferon-gamma promoter by glucocorticoids and dominant negative mutants of c-Jun. J Biol Chem 270:12548–12556

Cirino G, Flower RJ, Browning JL, Sinclair LK, Pepinsky RB (1987) Recombinant human lipocortin 1 inhibits thromboxane release from guinea-pig isolated perfused lung. Nature 328:270–272

Clark SD (1987) Regulation of the expression of tissue inhibitor of metalloproteinases and collagenase by retinoids and glucocorticoids in human fibroblasts. J Clin Invest 80:1280

Clarke AR, Purdie CA, Harrison DJ, Morris RG, Bird CC, Hooper ML, Wyllie AH (1993) Thymocyte apoptosis induced by p53-dependent and independent pathways. Nature 362:849–852

Colbert RA, Young DA (1986) Glucocorticoid-mediated induction of glucocortin: a rapid primary response common to major target tissues. Proc Natl Acad Sci USA 83:72–76

Cole TJ, Blendy JA, Monaghan AP, Krieglstein K, Schmid W, Aguzzi A, Fantuzzi G, Hummler E, Unsicker K, Schütz G (1995) Targeted disruption of the glucocorticoid receptor gene blocks adrenergic chromaffin cell development and severely retards lung maturation. Genes Dev 9:1608–1621

Coleman RA (1993) Hepatic monoacylglycerol acyltransferase activity in HA1 and HA7 hepatoma/hepatocyte hybrid cells: regulation by insulin and dexamethasone and by cell density. Biochim Biophys Acta Lipids Lipid Metab 1165:306–313

Collins S, Caron MG, Lefkowitz RJ (1988) β2-adrenergic receptors in hamster smooth muscle cells are transcriptionally regulated by glucocorticoids. J Biol Chem 263:9067–9070

Cook JS, Lucas JJ, Sibley E, Bolanowski MA, Christy RJ, Kelly TJ, Lane MD (1988) Expression of the differentiation-induced gene for fatty acid-binding protein is activated by glucocorticoid and cAMP. Proc Natl Acad Sci USA 85:2949–2953

Cote GJ, Stolow DT, Peleg S, Berget SM, Gagel RF (1992) Identification of exon sequences and an exon binding protein involved in alternative RNA splicing of calcitonin/CGRP. Nucleic Acids Res 20:2361–2366

Crofford LJ, Wilder RL, Ristimäki AP, Sano H, Remmers EF, Epps HR, Hla T (1994) Cyclooxygenase-1 and -2 expression in rheumatoid synovial tissues. Effects of interleukin-1β, phorbol ester, and corticosteroids. J Clin Invest 93:1095–1101

Cronstein BN, Kimmel SC, Levin RI, Martiniuk F, Weissmann G (1992) A mechanism for the antiinflammatory effects of corticosteroids: the glucocorticoid receptor regulates leukocyte adhesion to endothelial cells and expression of endothelial-leukocyte adhesion molecule 1 and intercellular adhesion molecule 1. Proc Natl Acad Sci USA 89:9991–9995

Croxtall JD, Flower RJ (1992) Lipocortin 1 mediates dexamethasone-induced growth arrest of the A549 lung adenocarcinoma cell line. Proc Natl Acad Sci USA 89:3571–3575

Croxtall JD, Flower RJ (1994) Antisense oligonucleotides to human lipocortin-1 inhibit glucocorticoid-induced inhibition of A549 cell growth and eicosanoid release. Biochem Pharmacol 48:1729–1734

Cuthbert JA, Lipsky PE (1990) Inhibition by 6-fluoromevalonate demonstrates that mevalonate or one of the mevalonate phosphates is necessary for lymphocyte proliferation. J Biol Chem 265:18568–18575

Danesch U, Gloss B, Schmid W, Schütz G, Schüle R, Renkawitz R (1987) Glucocorticoid induction of the rat tryptophan oxygenase gene is mediated by two widely separated glucocorticoid-responsive elements. EMBO J 6:625–630

Davidson FF, Dennis EA, Powell M, Glenney JR (1987) Inhibition of phospholipase A2 by "lipocortins" and calpactins. J Biol Chem 262:1698–1705

Davis IJ, Lau LF (1994) Endocrine and neurogenic regulation of the orphan nuclear receptors Nur77 and Nur-1 in the adrenal glands. Mol Cell Biol 14:3469–3483

Day RN, Maurer RA (1989) The distal enhancer region of the rat prolactin gene contains elements conferring response to multiple hormones. Mol Endocrinol 3:3–9

Daynes RA, Araneo BA (1989) Contrasting effects of glucocorticoids on the capacity of T cells to produce the growth factors interleukin 2 and interleukin 4. Eur J Immunol 19:2319–2325

Deeb SS, Peng R (1989) Structure of the human lipoprotein lipase gene. Biochemistry 28:4131–4135

Delany AM, Brinckerhoff CE (1992) Post-transcriptional regulation of collagenase and stromelysin gene expression by epidermal growth factor and dexamethasone in cultured human fibroblasts. J Cell Biochem 50:400–410

Deng Z-W, Denkinger DJ, Peterson KE, Deuel TF, Kawahara RS (1994) Glucocorticoids negatively regulate the transcription of KC, the mouse homolog of MGSA/GRO. Biochem Biophys Res Commun 203:1809–1814

Dennis GJ, Mond JJ (1986) Corticosteroid-induced suppression of murine B cell immune response antigens. J Immunol 136:1600–1604

Denton RR, Eisen LP, Elsasser MS, Harmon JM (1993) Differential autoregulation of glucocorticoid receptor expression in human T- and B-cell lines. Endocrinology 133:248–256

Deschepper CF, Flaxman M (1990) Glucocorticoid regulation of rat diencephalon angiotensinogen production. Endocrinology 126:963–970

Diamond DJ, Goodman HM (1985) Regulation of growth hormone messenger RNA synthesis by dexamethasone and triiodothyronine. Transcriptional rate and mRNA stability changes in pituitary tumor cells. J Mol Biol 181:41–62

Diamond MI, Miner JN, Yoshinaga SK, Yamamoto KR (1990) Transcription factor interactions: selectors of positive or negative regulation from a single DNA element. Science 249:1266–1272

Domin WS, Chait A, Deeb SS (1991) Transcriptional activation of the lipoprotein lipase gene in macrophages by dexamethasone. Biochemistry 30:2570–2574

Dong Y, Poellinger L, Okret S, Höög JO, von Bahr-Lindström H, Jörnvall H, Gustafsson JA (1988) Regulation of gene expression of class I alcohol dehydrogenase by glucocorticoids. Proc Natl Acad Sci USA 85:767–771

Doppler W, Groner B, Ball RK (1989) Prolactin and glucocorticoid hormones synergistically induce expression of transfected rat β-casein gene promoter constructs in a mammary epithelial cell line. Proc Natl Acad Sci USA 86:104–108

Doppler W, Höck W, Hofer P, Groner B, Ball RK (1990) Prolactin and glucocorticoid hormones control transcription of the β-casein gene by kinetically distinct mechanisms. Mol Endocrinol 4:912–919

Dowd DR, MacDonald PN, Komm BS, Haussler MR, Miesfeld R (1991) Evidence for early induction of calmodulin gene expression in lymphocytes undergoing glucocorticoid-mediated apoptosis. J Biol Chem 266:18423–18426

Dowd DR, MacDonald PN, Komm BS, Haussler MR, Miesfeld RL (1992) Stable expression of the calbindin-D28 K complementary DNA interferes with the apoptotic pathway in lymphocytes. Mol Endocrinol 6:1843–1848

Drouin J, Charron J, Gagner J-P, Jeanotte L, Nemer M, Plante RK, Wrange O (1987) Pro-opiomelanocortin gene: a model for negative regulation of transcription by glucocorticoids. J Cell Biochem 35:293–304

Drouin J, Trifiro MA, Plante RK, Nemer M, Eriksson P, Wrange Ö (1989) Glucocorticoid receptor binding to a specific DNA sequence is required for hormone-dependent repression of pro-opiomelanocortin gene transcription. Mol Cell Biol 9:5305–5314

Drouin J, Sun YL, Chamberland M, Gauthier Y, De Léan A, Nemer M, Schmidt TJ (1993) Novel glucocorticoid receptor complex with DNA element of the hormone-repressed POMC gene. EMBO J 12:145–156

Duman RS, Winston SM, Clark JA, Nestler EJ (1990) Corticosterone regulates the expression of ADP-ribosylation factor messenger RNA and protein in rat cerebral cortex. J Neurochem 55:1813–1816

Durant S, Duval D, Homo-Delarche F (1986) Factors involved in the control of fibroblast proliferation by glucocorticoids. Endocr Rev 7:254–269

Edenberg HJ, Brown CJ (1992) Regulation of human alcohol dehydrogenase genes. Pharmacogenetics 2:185–196

Eguchi K, Kawakami A, Nakashima M, Ida H, Sakito S, Matsuoka N, Terada K, Sakai M, Kawabe Y, Fukuda T, Ishimaru T, Kurouji K, Fujita N, Aoyagi T, Maeda K, Nagataki S (1992) Interferon-alpha and dexamethasone inhibit adhesion of T cells to endothelial cells and synovial cells. Clin Exp Immunol 88:448–454

Eisenberger CL, Nechushtan H, Cohen H, Shani M, Reshef L (1992) Differential regulation of the rat phosphoenolpyruvate carboxykinase gene expression in several tissues of transgenic mice. Mol Cell Biol 12:1396–1403

Ekblom M, Fässler R, Tomasini-Johansson B, Nilsson K, Ekblom P (1993) Downregulation of tenascin expression by glucocorticoids in bone marrow stromal cells and fibroblasts. J Cell Biol 1045:--

Eliasson E, Mkrtchian S, Halpert JR, Ingelman-Sundberg M (1994) Substrate-regulated, cAMP-dependent phosphorylation, denaturation, and degradation of glucocorticoid-inducible rat liver cytochrome P450 3A1. J Biol Chem 269:18378–18383

Encio IJ, Detera-Wadleigh SD (1991) The genomic structure of the human glucocorticoid receptor. J Biol Chem 266:7182–7188

Espinás ML, Roux J, Ghysdael J, Pictet R, Grange T (1994) Participation of Ets transcription factors in the glucocorticoid response of the rat tyrosine aminotransferase gene. Mol Cell Biol 14:4116–4125

Evans RM (1988) The steroid and thyroid hormone receptor superfamily. Science 240:889–895

Faber S, O'Brien RM, Imai E, Granner DK, Chalkley R (1993) Dynamic aspects of DNA/protein interactions in the transcriptional initiation complex and the hormone-responsive domains of the phosphoenolpyruvate carboxykinase promoter in vivo. J Biol Chem 268:24976–24985

Fan W, Trifiletti R, Cooper T, Norris JS (1992) Cloning of aμ-class glutathione S-transferase gene and identification of the glucocorticoid regulatory domains in its 5' flanking sequence. Proc Natl Acad Sci USA 89:6104–6108

Farza H, Salmon AM, Hadchouel M, Moreau JL, Babinet C, Tiollais P, Pourcel C (1987) Hepatitis B surface antigen gene expression is regulated by sex steroids and glucocorticoids in transgenic mice. Proc Natl Acad Sci USA 84:1187–1191

Feng B, Hilt DC, Max SR (1990) Transcriptional regulation of glutamine synthetase gene expression by dexamethasone in L6 muscle cells. J Biol Chem 265:18702–18706

Fernandes-Alnemri T, Litwack G, Alnemri ES (1994) CPP32, a novel human apoptotic protein with homology to Caenorhabditis elegans cell death protein Ced-3 and mammalian interleukin-1β-converting enzyme. J Biol Chem 269:30761–30764

Fernandes-Alnemri T, Litwack G, Alnemri ES (1995) Mch2, a new member of the apoptotic Ced-3/Ice cysteine protease gene family. Cancer Res 55:2737–2742

Fernandez-Mejia C, Davidson MB (1992) Regulation of glucokinase and proinsulin gene expression and insulin secretion in RIN-m5F cells by dexamethasone, retinoic acid, and thyroid hormone. Endocrinology 130:1660–1668

Ferrandina G, Scambia G, Benedetti Panici P, Bonanno G, De Vincenzo R, Rumi C, Bussa S, Genuardi M, Spica Romano V, Mancuso S (1992) Effects of dexamethasone on the growth and epidermal growth factor receptor expression of the OVCA 433 ovarian cancer cells. Mol Cell Endocrinol 83:183–193

Fesus L, Thomazy V, Falus A (1987) Induction and activation of tissue transglutaminase during programmed cell death. FEBS Lett 224:104–108

Fève B, Emorine LJ, Briend-Sutren M-M, Lasnier F, Strosberg AD, Pairault J (1990) Differential regulation of β1- and β2-adrenergic receptor protein and mRNA levels by glucocorticoids during 3T3-F442A adipose differentiation. J Biol Chem 265:16343–16349

Fève B, Baude B, Krief S, Strosberg AD, Pairault J, Emorine LJ (1992) Inhibition by dexamethasone of β3-adrenergic receptor responsiveness in 3T3-F442A adipocytes. Evidence for a transcriptional mechanism. J Biol Chem 267:15909–15915

Fiegl M, Strasser-Wozak E, Geley S, Gsur A, Drach J, Kofler R (1995) Gencocorticoid-mediated immunomodulation: hydrocortisone enhances immunosuppressive endogenous retroviral protein (p15E) expression in mouse immune cells. Clin Exp Immunol 101:259–264

Fillmore HL, Mainardi CL, Hasty KA (1992) Differentiation of PC12 cells with nerve growth factor is associated with induction of transin synthesis and release. J Neurosci Res 31:662–669

Fischer A, König W (1990) Regulation of CD23 expression, soluble CD23 release and immunoglobulin synthesis of peripheral blood lymphocytes by glucocorticoids. Immunology 71:473–479

Flomerfelt FA, Miesfeld RL (1994) Recessive mutations in a common pathway block thymocyte apoptosis induced by multiple signals. J Cell Biol 127:1729–1742

Flomerfelt FA, Briehl MM, Dowd DR, Dieken ES, Miesfeld RL (1993) Elevated glutathione S-transferase gene expression is an early event during steroid-induced lymphocyte apoptosis. J Cell Physiol 154:573–581

Flower RJ, Rothwell NJ (1994) Lipocortin-1: cellular mechanisms and clinical relevance. Trends Pharmacol Sci 15:71–76

Flusser G, Ginzburg V, Meyuhas O (1989) Glucocorticoids induce transcription of ribosomal protein genes in rat liver. Mol Cell Endocrinol 64:213–222

Foreman PJ, Taglialatela G, Jackson GR, Perez-Polo JR (1992) Dexamethasone blocks nerve growth factor induction of nerve growth factor receptor mRNA in PC12 cells. J Neurosci Res 31:52–57

Forest CD, O'Brien RM, Lucas PC, Magnuson MA, Granner DK (1990) Regulation of phosphoenolpyruvate carboxykinase gene expression by insulin. Use of the stable transfection approach to locate an insulin responsive sequence. Mol Endocrinol 4:1302–1310

Fossom LH, Sterling CR, Tank AW (1992) Regulation of tyrosine hydroxylase gene transcription rate and tyrosine hydroxylase mRNA stability by cyclic AMP and glucocorticoid. Mol Pharmacol 42:898–908

Frost GH, Rhee K, Thompson EA Jr (1993) Glucocorticoid regulation of thymidine kinase (TK-1) expression in L929 cells. J Biol Chem 268:6748–6754

Frost GH, Rhee K, Ma T, Thompson EA (1994) Expression of c-myc in glucocorticoid-treated fibroblastic cells. J Steroid Biochem Mol Biol 50:109–119

Fukuda K, Kojiro M, Chiu JF (1994) Differential regulation of tissue transglutaminase in rat hepatoma cell lines McA-RH7777 and McA-RH8994: relation to growth rate and cell death. J Cell Biochem 54:67–77

Funder JW (1993) Mineralocorticoids, glucocorticoids, receptors and response elements. Science 259:1132–1133

Funder JW (1994) Corticosteroid receptors and the central nervous system. J Steroid Biochem Mol Biol 49:381–384

Gadina M, Bertini R, Mengozzi M, Zandalasini M, Mantovani A, Ghezzi P (1991) Protective effect of chlorpromazine on endotoxin toxicity and TNF production in glucocorticoid-sensitive and glucocorticoid-resistant models of endotoxic shock. J Exp Med 173:1305–1310

Galand G (1988) Effect of an antiglucocorticoid (RU-38486) on hydrocortisone induction of maltase-glucosamylase, sucrase-isomaltase and trehalase in brush border membranes of suckling rats. Experientia 44:516–518

Gametchu B, Watson CS, Wu S (1993) Use of receptor antibodies to demonstrate membrane glucocorticoid receptor in cells from human leukemic patients. FASEB J 7:1283–1292

Ganss R, Weih F, Schütz G (1994) The cyclic adenosine 3',5'-monophosphate- and the glucocorticoid-dependent enhancers are targets for insulin repression of tyrosine aminotransferase gene transcription. Mol Endocrinol 8:895–903

Gardner DG, Gertz BJ, Deschepper CF, Kim DY (1988) Gene for the rat atrial natriuretic peptide is regulated by glucocorticoids in vitro. J Clin Invest 82:1275–1281

Garlatti M, Daheshia M, Slater E, Bouguet J, Hanoune J, Beato M, Barouki R (1994) A functional glucocorticoid-responsive unit composed of two overlapping inactive receptor-binding sites: evidence for formation of a receptor tetramer. Mol Cell Biol 14:8007–8017

Garvey WT, Hueckstea dt TP, Lima FB, Birnbaum MJ (1989) Expression of a glucose transporter gene cloned from brain in cellular models of insulin resistance: dexamethasone decreases transporter mRNA in primary cultured adipocytes. Mol Endocrinol 3:1132–1141

Gearing KL, Cairns W, Okret S, Gustafsson J-Å (1993) Heterogeneity in the 5' untranslated region of the rat glucocorticoid receptor mRNA. J Steroid Biochem Mol Biol 46:635–639

Ged C, Beaune P (1991) Isolation of the human cytochrome P-450 IIC8 gene: multiple glucocorticoid responsive elements in the 5' region. Biochim Biophys Acta Gene Struct Expression 1088:433–435

Geller DA, Nussler AK, Di Silvio M, Lowenstein CJ, Shapiro RA, Wang SC, Simmons RL, Billiar TR (1993) Cytokines, endotoxin, and glucocorticoids regulate the expression of inducible nitric oxide synthase in hepatocytes. Proc Natl Acad Sci USA 90:522–526

Gellersen B, Kempf R, Telgmann R, DiMattia GE (1994) Nonpituitary human prolactin gene transcription is independent of Pit-1 and differentially controlled in lymphocytes and in endometrial stroma. Mol Endocrinol 8:356–373

Gensure RC, Fish JM, Walters MR (1993) Dexamethasone downregulates vitamin D receptors in rat kidney, unmasking a high affinity binding site. Biochem Biophys Res Commun 195:1139–1144

Gessani S, McCandless S, Baglioni C (1988) The glucocorticoid dexamethasone inhibits synthesis of interferon by decreasing the level of its mRNA. J Biol Chem 263:7454–7457

Ghosh D (1992) Glucocorticoid receptor-binding site in the human immunodeficiency virus long terminal repeat. J Virol 66:586–590

Giguere V, Hollenberg SM, Rosenfeld MG, Evans RM (1986) Functional domains of the human glucocorticoid receptor. Cell 46:645–652

Gill G, Ptashne M (1988) Negative effect of the transcription activator GAL4. Nature 334:721–724
Giorgino F, Almahfouz A, Goodyear LJ, Smith RJ (1993) Glucocorticoid regulation of insulin receptor and substrate IRS-1 tyrosine phosphorylation in rat skeletal muscle in vivo. J Clin Invest 91:2020–2030
Glatz JA, Heath JK, Southby J, O'Keeffe LM, Kiriyama T, Moseley JM, Martin TJ, Gillespie MT (1994) Dexamethasone regulation of parathyroid hormone-related protein (PTHrP) expression in a squamous cancer cell line. Mol Cell Endocrinol 101:295–306
Glibetic M, Larson DE, Rothblum LI, Sells BH (1993) Dexamethasone stimulates rRNA gene transcription in rat myoblasts. Mol Cell Endocrinol 94:243–252
Gohda E, Kataoka H, Tsubouchi H, Daikilara Y, Yamamoto I (1992) Phorbol ester-induced secretion of human hepatocyte growth factor by human skin fibroblasts and its inhibition by dexamethasone. FEBS Lett 301:107–110
Goldstein JL, Brown MS (1990) Regulation of the mevalonate pathway. Nature 343:425–430
Goldstone SD, Lavin MF (1991) Isolation of a cDNA clone, encoding a human β-galactoside binding protein, overexpressed during glucocorticoid-induced cell death. Biochem Biophys Res Commun 178:746–750
Gomi M, Moriwaki K, Katagiri S, Kurata Y, Thompson EB (1990) Glucocorticoid effects on myeloma cells in culture: correlation of growth inhibition with induction of glucocorticoid receptor messenger RNA. Cancer Res 50:1873–1878
Gonneville L, Martins TJ, Bedard PA (1991) Complex expression pattern of the CEF-4 cytokine in transformed and mitogenically stimulated cells. Oncogene 6:1825–1833
Gonzalo JA, González-García A, Martínez-AC, Kroemer G (1993) Glucocorticoid-mediated control of the activation and clonal deletion of peripheral T cells in vivo. J Exp Med 177:1239–1246
Goodwin JS, Atluru D (1986) Mechanism of action of glucocorticoid-induced immunoglobulin production: role of lipoxygenase metabolites of arachidonic acid. J Immunol 136:3455–3460
Gorovits R, Ben-Dror I, Fox LE, Westphal HM, Vardimon L (1994) Developmental changes in the expression and compartmentalization of the glucocorticoid receptor in embryonic retina. Proc Natl Acad Sci USA 91:4786–4790
Gotoh T, Haraguchi Y, Takiguchi M, Mori M (1994) The delayed glucocorticoid-responsive and hepatoma cell-selective enhancer of the rat arginase gene is located around intron 7. J Biochem (Tokyo) 115:778–788
Gottschall PE, Koves K, Mizuno K, Tatsuno I, Arimura A (1991) Glucocorticoid upregulation of interleukin 1 receptor expression in a glioblastoma cell line. Am J Physiol Endocrinol Metab 261:E362–E368
Goulding NJ, Guyre PM (1992) Regulation of inflammation by lipocortin 1. Immunol Today 13:295–297
Goya L, Maiyar AC, Ge Y, Firestone GL (1993) Glucocorticoids induce a G1/Go cell cycle arrest of Con8 rat mammary tumor cells that is synchronously reversed by steroid withdrawal or addition of transforming growth factor-α. Mol Endocrinol 7:1121–1132

Grassilli E, Bettuzzi S, Monti D, Ingletti MC, Franceschi C, Corti A (1991) Studies on the relationship between cell proliferation and cell death: opposite patterns of SGP-2 and ornithine decarboxylase mRNA accumulation in PHA-stimulated human lymphocytes. Biochem Biophys Res Commun 180:59–63

Grigoriadis AE, Wang Z-Q, Cecchini MG, Hofstetter W, Felix R, Fleisch HA, Wagner EF (1994) c-Fos: a key regulator of osteoclast-macrophage lineage determination and bone remodeling. Science 266:443–448

Grillo C, Piroli G, González SL, Angulo J, McEwen BS, De Nicola AF (1994) Glucocorticoid regulation of mRNA encoding (Na + K)ATPase $\alpha 3$ and $\beta 1$ subunits in rat brain measured by in situ hybridization. Brain Res 657:83–91

Grino M, Guillaume V, Boudouresque F, Conte-Devolx B, Maltese J-Y, Oliver C (1990) Glucocorticoids regulate peptidyl-glycine α-amidating monooxygenase gene expression in the rat hypothalamic paraventricular nucleus. Mol Endocrinol 4:1613–1619

Guathier JM, Bourachot B, Doucas V, Yaniv M, Moreau-Gachelin F (1993) Functional interference between the Spi-1/PU.1 oncoprotein and steroid hormone or vitamin receptor. EMBO J 12:5089–5096

Guller S, Wozniak R, Leibman MI, Lockwood CJ (1994) Negative regulation of placental fibronectin expression by glucocorticoids and cyclin adenosine 3',5'-monophosphate. Ann N Y Acad Sci 734:132–142

Guo WX, Antakly T (1995) AIDS-related Kaposi's sarcoma: evidence for direct stimulatory effect of glucocorticoid on cell proliferation. Am J Pathol 146:727–734

Guttridge DC, Lau AL, Cunningham DD (1993) Protease nexin-1, a thrombin inhibitor, is regulated by interleukin-1 and dexamethasone in normal human fibroblasts. J Biol Chem 268:18966–18974

Hadcock JR, Malbon CC (1988) Regulation of β-adrenergic receptors by "permissive" hormones: glucocorticoids increase steady-state levels of receptor mRNA. Proc Natl Acad Sci USA 85:8415–8419

Hales DB, Sha L, Payne AH (1990) Glucocorticoid and cyclic adenosine 3'5'-monophosphate-mediated induction of cholesterol side-chain cleavage cytochrome P450 (P450SCC) in MA-10 tumor Leydig cells. Increases in mRNA are cycloheximide sensitive. Endocrinology 126:2800–2808

Hall RK, Sladek FM, Granner DK (1995) The orphan receptors COUP-TF and HNF-4 serve as accessory factors required for induction of phosphoenolpyruvate carboxykinase gene transcription by glucocorticoids. Proc Natl Acad Sci USA 92:412–416

Han J, Huez G, Beutler B (1991) Interactive effects of the tumor necrosis factor promoter and 3'-untranslated regions. J Immunol 146:1843–1848

Harbuz MS, Lightman SL (1989) Glucocorticoid inhibition of stress-induced changes in hypothalamic corticotrophin-releasing factor messenger RNA and proenkephalin A messenger RNA. Neuropeptides 14:17–20

Harmon JM, Thompson EB (1982) Glutamine synthetase induction by glucocorticoids in the glucocorticoid-sensitive human leukemic cell line CEM-C7. J Cell Physiol 110:155–160

Harris SE, Smith RG, Zhou H, Mansson P-E, Malark M (1989) Androgens and glucocorticoids modulate heparin-binding growth factor I mRNA accumulation in DDT1 cells as analyzed by in situ hybridization. Mol Endocrinol 3:1839–1844

Hass R, Brach M, Kharbanda S, Giese G, Traub P, Kufe D (1991) Inhibition of phorbol ester-induced monocytic differentiation by dexamethasone is associated with down-regulation of c-fos and c-jun (AP-1). J Cell Physiol 149:125–131

Hass R, Brach M, Gunji H, Kharbanda S, Kufe D (1992) Inhibition of EGR-1 and NF-kappaB gene expression by dexamethasone during phorbol ester-induced human monocytic differentiation. Biochem Pharmacol 44:1569–1576

Heaton JH, Gelehrter TD (1989) Glucocorticoid induction of plasminogen activator and plasminogen activator-inhibitor messenger RNA in rat hepatoma cells. Mol Endocrinol 3:349–355

Heaton JH, Nebes VL, O'Dell LG, Morris SM Jr, Gelehrter TD (1989) Glucocorticoid and cyclic nucleotide regulation of plasminogen activator and plasminogen activator-inhibitor gene expression in primary cultures of rat hepatocytes. Mol Endocrinol 3:185–192

Heaton JH, Kathju S, Gelehrter TD (1992) Transcriptional and posttranscriptional regulation of type 1 plasminogen activator inhibitor and tissue-type plasminogen activator gene expression in HTC rat hepatoma cells by glucocorticoids and cyclic nucleotides. Mol Endocrinol 6:53–60

Heck S, Kullmann M, Gast A, Ponta H, Rahmsdorf HJ, Herrlich P, Cato ACB (1994) A distinct modulating domain in glucocorticoid receptor monomers in the repression of activity of the transcription factor AP-1. EMBO J 13:4087–4095

Heinrichs AA, Bortell R, Rahman S, Stein JL, Alnemri ES, Litwack G, Lian JB, Stein GS (1993) Identification of multiple glucocorticoid receptor binding sites in the rat osteocalcin gene promoter. Biochemistry 32:11436–11444

Heinrichs C, Yanovski JA, Roth AH, Yu YM, Domené HM, Yano K, Cutler GB Jr, Baron J (1994) Dexamethasone increases growth hormone receptor messenger ribonucleic acid levels in liver and growth plate. Endocrinology 135:1113–1118

Helmberg A, Fässler R, Geley S, Jöhrer K, Kroemer G, Böck G, Kofler R (1990) Glucocorticoid-regulated gene expression in the immune system: analysis of glucocorticoid-regulated transcripts from the mouse macrophage-like cell line P388D1. J Immunol 145:4332–4337

Helmberg A, Auphan N, Caelles C, Karin M (1995) Glucocorticoid-induced apoptosis of human leukemic cells is caused by the repressive function of the glucocorticoid receptor. EMBO J 14:452–460

Hess P, Meenakshi T, Chan GC-K, Carlstedt-Duke J, Gustafsson J-Å, Payvar F (1990) Purified glucocorticoid receptors bind selectively in vitro to a cloned DNA fragment that mediates a delayed secondary response to glucocorticoids in vivo. Proc Natl Acad Sci USA 87:2564–2568

Higuchi M, Ito T, Imai T, Iwaki T, Hattori M, Kohsaka S, Niho Y, Sakaki Y (1994) Expression of the alpha 2-macroglobulin-encoding gene in rat brain and cultured astrocytes. Gene 141:155–162

Hill O, Kuhn M, Zucht HD, Cetin Y, Kulaksiz H, Adermann K, Klock G, Rechkemmer G, Forssmann WG, Magert HJ (1995) Analysis of the human guanylin gene and the processing and cellular localization of the peptide. Proc Natl Acad Sci USA 92:2046–2050

Hollenberg SM, Evans RM (1988) Multiple and cooperative Trans-activation domains of the human glucocorticoid receptor. Cell 55:899–906

Hollenberg SM, Weinberger C, Ong ES, Cerelli G, Oro A, Lebo R, Thompson EB, Rosenfeld MG, Evans RM (1985) Primary structure and expression of a functional human glucocorticoid receptor cDNA. Nature 318:635–641

Holtzman EJ, Kinane TB, West K, Soper BW, Karga H, Ausiello DA, Ercolani L (1993) Transcriptional regulation of G-protein alpha i subunit genes in LLC-PK1 renal cells and characterization of the porcine G alpha 1-3 gene promoter. J Biol Chem 268:3964–3975

Horner HC, Munck A, Lienhard GE (1987) Dexamethasone causes translocation of glucose transporters from the plasma membrane to an intracellular site in human fibroblasts. J Biol Chem 262:17696–17702

Howard EF, Thompson YG, Lapp CA, Greenbaum LM (1990) Reduction of T-kininogen messenger RNA levels by dexamethasone in the adjuvant-treated rat. Life Sci 46:411–417

Hsu S, Qi M, DeFranco DB (1992) Cell cycle regulation of glucocorticoid receptor function. EMBO J 11:3457–3468

Hu J-M, Bodwell JE, Munck A (1994) Cell cycle-dependent glucocorticoid receptor phosphorylation and activity. Mol Endocrinol 8:1709–1713

Hu S-B, Tannahill LA, Lightman SL (1993) Regulation of arginine vasopressin mRNA in rat fetal hypothalamic cell culture. Role of protein kinases and glucocorticoids. J Mol Endocrinol 10:51–57

Huang S, Hershey JWB (1989) Translational initiation factor expression and ribosomal protein gene expression are repressed coordinately but by different mechanisms in murine lymphosarcoma cells treated with glucocorticoids. Mol Cell Biol 9:3679–3684

Huber P, Laurent M, Dalmon J (1990) Human β-fibrinogen gene expression. Upstream sequences involved in its tissue specific expression and its dexamethasone and interleukin 6 stimulation. J Biol Chem 265:5695–5701

Huether MJ, Lin G, Smith DM, Murtaugh MP, Molitor TW (1993) Cloning, sequencing and regulation of an mRNA encoding porcine interleukin-1 beta. Gene 129:285–289

Hughes FM Jr, Cidlowski JA (1994) Regulation of apoptosis in S49 cells. J Steroid Biochem Mol Biol 49:303–310

Husson A, Renouf S, Fairand A, Buquet C, Benamar M, Vaillant R (1990) Expression of argininosuccinate lyase mRNA in foetal hepatocytes-Regulation by glucocorticoids and insulin. Eur J Biochem 192:677–681

Hutchison KA, Dittmar KD, Czar MJ, Pratt WB (1994) Proof that hsp70 is required for assembly of the glucocorticoid receptor into a heterocomplex with hsp90. J Biol Chem 269:5043–5049

Ihara H, Nakanishi S (1990) Selective inhibition of expression of the substance P receptor mRNA in pancreatic acinar AR42J cells by glucocorticoids. J Biol Chem 265:22441–22445

Ikeda K, Lu C, Weir EC, Mangin M, Broadus AE (1989) Transcriptional regulation of the parathyroid hormone-related peptide gene by glucocorticoids and vitamin D in a human C-cell line. J Biol Chem 264:15743–15746

Imai E, Stromstedt P-E, Quinn PG, Carlstedt-Duke J, Gustafsson J-Å, Granner DK (1990) Characterization of a complex glucocorticoid response unit in the phosphoenolpyruvate carboxykinase gene. Mol Cell Biol 10:4712–4719

Imai E, Miner JN, Mitchell JA, Yamamoto KR, Granner DK (1993) Glucocorticoid receptor-cAMP response element-binding protein interaction and the response of the phosphoenolpyruvate carboxykinase gene to glucocorticoids. J Biol Chem 268:5353–5356

Ingrassia R, Savoldi GF, Caraffini A, Tironi M, Poiesi C, Williams P, Albertini A, Di Lorenzo D (1994) Characterization of a novel transcription complex required for glucocorticoid regulation of the rat alpha-1-acid glycoprotein gene. DNA Cell Biol 13:615–627

Ip MM, Shea WK, Rowan BG (1993) Mutant glucocorticoid receptors in lymphoma. Ann N Y Acad Sci 684:94–115

Jaiswal AK, Haaparanta T, Luc P-V, Schembri J, Adesnik M (1990) Glucocorticoid regulation of a phenobarbital-inducible cytochrome P-450 gene: the presence of a functional glucocorticoid response element in the 5'-flanking region of the CYP2B2 gene. Nucleic Acids Res 18:4237–4242

Jehan F, Neveu I, Naveilhan P, Brachet P, Wion D (1993) Complex interactions among second messenger pathways, steroid hormones, and protooncogenes of the Fos and Jun families converge in the regulation of the nerve growth factor gene. J Neurochem 60:1843–1853

Jehn B, Costello E, Marti A, Keon N, Deane R, Li F, Friis RR, Burri PH, Martin F, Jaggi R (1992) Overexpression of Mos, Ras, Src, and Fos inhibits mouse mammary epithelial cell differentiation. Mol Cell Biol 12:3890–3902

Joels M, DeKloet RE (1992) Control of neuronal excitability by corticosteroid hormones. Trends Neurosci 15:25–30

Jonat C, Rahmsdorf HJ, Park K-K, Cato ACB, Gebel S, Ponta H, Herrlich P (1990) Antitumor promotion and antiinflammation: down-modulation of AP-1 (Fos/Jun) activity by glucocorticoid hormone. Cell 62:1189–1204

Jones KJ, Harrington CA, Chikaraishi DM, Pfaff DW (1990) Steroid hormone regulation of ribosomal RNA in rat hypothalamus: early detection using in situ hybridization and precursor-product ribosomal DNA probes. J Neurosci 10:1513–1521

Jong MTC, Raaka BM, Samuels HH (1994) A sequence in the rat Pit-1 gene promoter confers synergistic activation by glucocorticoids and protein kinase-C. Mol Endocrinol 8:1320–1327

Joshi J, Sabol SL (1991) Proenkephalin gene expression in C6 rat glioma cells: potentiation of cyclic adenosine 3',5'-monophosphate-dependent transcription by glucocorticoids. Mol Endocrinol 5:1069–1080

Jun C-D, Hoon-Ryu, Um J-Y, Kim T-Y, Kim J-M, Kang S-S, Kim H-M, Chung H-T (1994) Involvement of protein kinase C in the inhibition of nitric oxide production from murine microglial cells by glucocorticoid. Biochem Biophys Res Commun 199:633–638

Jungmann RA, Wang X, Milkowski DM, Short ML (1992) Glucocorticoid induction of CRE-binding protein isoform mRNAs in rat C6 glioma cells. Nucleic Acids Res 20:825–829

Kähäri V-M (1994) Dexamethasone suppresses elastin gene expression in human skin fibroblasts in culture. Biochem Biophys Res Commun 201:1189–1196

Kalinyak JE, Griffin CA, Hamilton RW, Bradshaw JG, Perlman AJ, Hoffman AR (1989) Developmental and hormonal regulation of glucocorticoid receptor messenger RNA in the rat. J Clin Invest 84:1843–1848

Kaplan I, Blakely BT, Pavlath GK, Travis M, Blau HM (1990) Steroids induce acetylcholine receptors on cultured human muscle: implications for myasthenia gravis. Proc Natl Acad Sci USA 87:8100–8104

Karagianni N, Tsawdaroglou N (1994) The c-fos serum response element (SRE) confers negative response to glucocorticoids. Oncogene 9:2327–2334

Karin M, Westphal HM, Beato M (1984) Characterization of DNA sequences through which cadmium and glucocorticoid hormones induce human metallothionein-IIA gene. Nature 308:513

Karlan BY, Rivero JA, Crabtree ME, Littlefield BA (1989) Different mechanisms contribute to simultaneous inhibition of urokinase and tissue-type plasminogen activators by glucocorticoids in human ovarian carcinoma cells. Mol Endocrinol 3:1006–1013

Karlan BY, Jones J, Slamon DJ, Lagasse LD (1994) Glucocorticoids stabilize HER-2/neu messenger RNA in human epithelial ovarian carcinoma cells. Gynecol Oncol 53:70–77

Kaspers GJL, Pieters R, Klumper E, DeWaal FC, Veerman AJP (1994) Glucocorticoid resistance in childhood leukemia. Leukemia Lymphoma 13:187–201

Kathju S, Heaton JH, Bruzdzinski CJ, Gelehrter TD (1994) Synergistic induction of tissue-type plasminogen activator gene expression by glucocorticoids and cyclic nucleotides in rat HTC hepatoma cells. Endocrinology 135:1195–1204

Kaufman Paterson RL, Or R, Domenico JM, Delespesse G, Gelfand EW (1994) Regulation of CD23 expression by IL-4 and corticosteroid in human B lymphocytes: altered response after EBV infection. J Immunol 152:2139–2147

Kavelaars A, Ballieux RE, Heijnen CJ (1990) β-Endorphin secretion by human peripheral blood mononuclear cells: regulation by glucocorticoids. Life Sci 46:1233–1240

Kawaguchi H, Raisz LG, Voznesensky OS, Alander CB, Hakeda Y, Pilbeam CC (1994) Regulation of the two prostaglandin G/H synthases by parathyroid hormone, interleukin-1, cortisol, and prostaglandin E2 in cultured neonatal mouse calvariae. Endocrinology 135:1157–1164

Kawai H, Ohgoh M, Emoto S, Ohmori E, Imanishi N, Yatsunami K, Ichikawa A (1992) Synergistic effects of 12-o-tetradecanoylphorbol-13-acetate and dexamethasone on de novo synthesis of histidine decarboxylase in mouse mastocytoma P-815 cells. Biochim Biophys Acta 1133:172–178

Kawai Y, Arinze IJ (1993) Glucocorticoid regulation of G-protein subunits in neonatal liver. Mol Cell Endocrinol 90:203–209

Kennedy B, Elayan H, Ziegler MG (1993) Glucocorticoid induction of epinephrine synthesizing enzyme in rat skeletal muscle and insulin resistance. J Clin Invest 92:303–307

Kiely J, Hadcock JR, Bahouth SW, Malbon CC (1994) Glucocorticoids down-regulate β1-adrenergic-receptor expression by suppressing transcription of the receptor gene. Biochem J 302:397–403

Kim RH, Shapiro HS, Li JJ, Wrana JL, Sodek J (1994) Characterization of the human bone sialoprotein (BSP) gene and its promoter sequence. Matrix Biol 14:31–40

Kletzien RF, Harris PKW, Foellmi LA (1994) Glucose-6-phosphate dehydrogenase: a "housekeeping" enzyme subject to tissue-specific regulation by hormones, nutrients, and oxidant stress. FASEB J 8:174–181

Knudsen PJ, Dinarello CA, Strom TB (1987) Glucocorticoids inhibit transcriptional and post-transcriptional expression of interleukin 1 in U937 cells. J Immunol 139:4129–4134

Kobayashi M, Bansal VS, Singh I, Kanfer JN (1988) Dexamethasone-induced reduction of phospholipase D activity in the rat: possible role of lipocortin. FEBS Lett 236:380–382

Koehler L, Hass R, DeWitt DL, Resch K, Goppelt-Struebe M (1990) Glucocorticoid-induced reduction of prostanoid synthesis in TPA-differentiated U937 cells is mainly due to a reduced cyclooxygenase activity. Biochem Pharmacol 40:1307–1316

Komori M, Oda Y (1994) A major glucocorticoid-inducible P450 in rat liver is not P450 3A1. J Biochem (Tokyo) 116:114–120

Konkle BA, Schuster SJ, Kelly MD, Harjes K, Hassett DE, Bohrer M, Tavassoli M (1992) Plasminogen activator inhibitor-1 messenger RNA expression is induced in rat hepatocytes in vivo by dexamethasone. Blood 79:2636–2642

Krasil'nikov MA, Bezrukov VM, Shatskaya VA (1992) Glucocorticoid regulation of phospholipid turnover and protein kinase C activity in mouse hepatoma 22 cells. Biochim Biophys Acta Mol Cell Res 1135:91–96

Krett NL, Pillay S, Moalli PA, Greipp PR, Rosen ST (1995) A variant glucocorticoid receptor messenger RNA is expressed in multiple myeloma patients. Cancer Res 55:2727–2729

Krieg AM, Gause WC, Gourley MF, Steinberg AD (1989) A role for endogenous retroviral sequences in the regulation of lymphocyte activation. J Immunol 143:2448–2451

Ku NO, Mortensen RF (1993) Cloning and tissue-specific expression of the gene for mouse C-reactive protein. Biochem J 295:379–386

Kujubu DA, Herschman HR (1992) Dexamethasone inhibits mitogen induction of the TIS10 prostaglandin synthase/cyclooxygenase gene. J Biol Chem 267:7991–7994
Kumar S, Cole R, Chiappelli F, De Vellis J (1989) Differential regulation of oligodendrocyte markers by glucocorticoids: post-transcriptional regulation of both proteolipid protein and myelin basic protein and transcriptional regulation of glycerol phosphate dehydrogenase. Proc Natl Acad Sci USA 86:6807–6811
Kupfer SR, Summers WC (1990) Identification of a glucocorticoid-responsive element in Epstein-Barr virus. J Virol 64:1984–1990
Kurokawa M, Michelangeli VP, Findlay DM (1991) Induction of calcitonin receptor expression by glucocorticoids in T47D human breast cancer cells. J Endocrinol 130:321–326
Kushner I (1988) The acute phase response: an overview. Methods Enzymol 163:373–382
Kutoh E, Strömstedt P-E, Poellinger L (1992) Functional interference between the ubiquitous and constitutive octamer transcription factor 1 (OTF-1) and the glucocorticoid receptor by direct protein-protein interaction involving the homeo subdomain of OTF-1. Mol Cell Biol 12:4960–4969
Labhart A (1986) Clinical endocrinology. Theory and practice. Springer, Berlin Heidelberg New York
Lam M, Dubyak G, Chen L, Nuñez G, Miesfeld RL, Distelhorst CW (1994) Evidence that BCL-2 represses apoptosis by regulating endoplasmic reticulum-associated Ca2+ fluxes. Proc Natl Acad Sci USA 91:6569–6573
Lamas AM, Leon OG, Schleimer RP (1991) Glucocorticoids inhibit eosinophil responses to granulocyte-macrophage colony-stimulating factor. J Immunol 147:254–259
Lamas M, Sanz E, Martin-Parras L, Espel E, Sperisen P, Collins M, Silva AG (1993) Glucocorticoid hormones upregulate interleukin 2 receptor alpha gene expression. Cell Immunol 151:437–450
Lancz G, Whitaker-Dowling P, Marsh YV, Bradley G, Eppstein DA, Hackney JF, Hulick-Swardstrom B (1990) Inhibition of vesicular stomatitis virus replication in dexamethasone-treated L929 cells. Proc Soc Exp Biol Med 193:190–196
Lange AJ, Espinet C, Hall R, El-Maghrabi MR, Vargas AM, Miksicek RJ, Granner DK, Pilkis SJ (1992) Regulation of gene expression of rat skeletal muscle/liver 6-phosphofructo-2-kinase/fructose-2,6-bisphosphatase. Isolation and characterization of a glucocorticoid response element in the first intron of the gene. J Biol Chem 267:15673–15680
Lange AJ, Argaud D, El-Maghrabi MR, Pan W, Maitra SR, Pilkis SJ (1994) Isolation of a cDNA for the catalytic subunit of rat liver glucose-6-phosphatase: regulation of gene expression in FAO hepatoma cells by insulin, dexamethasone and cAMP. Biochem Biophys Res Commun 201:302–309
Lanier-Smith KL, Currie MG (1991) Glucocorticoid regulation of atrial natriuretic peptide receptors on cultured endothelial cells. Endocrinology 129:2311–2317
Larsson E, Kato N, Cohen M (1989) Human endogenous proviruses. In: Vogt PK (ed) Oncogenes and retroviruses. Selected reviews. Springer, Berlin Heidelberg New York, pp 115–132 (*Current topics in microbiology* and immunology, vol 148)

Laudet V, Vanacker JM, Adelmant G, Begue A, Stehelin D (1993) Characterization of a functional promoter for the human thyroid hormone receptor alpha (c-erbA-1) gene. Oncogene 8:975–982

LeClerc S, Palaniswami R, Xie B, Govindan MV (1991) Molecular cloning and characterization of a factor that binds the human glucocorticoid receptor gene and represses its expression. J Biol Chem 266:17333–17340

Lee J, Tsai SY (1994) Multiple hormone response elements can confer glucocorticoid regulation on the human insulin receptor gene. Mol Endocrinol 8:625–634

Lee JK, Tam JW, Tsai MJ, Tsai SY (1992) Identification of cis- and trans-acting factors regulating the expression of the human insulin receptor gene. J Biol Chem 267:4638–4645

Lee S, Szlachetka M, Christakos S (1991) Effect of glucocorticoids and 1,25-dihydroxyvitamin D3 on the developmental expression of the rat intestinal vitamin D receptor gene. Endocrinology 129:396–401

Lee SW, Tsou A-P, Chan H, Thomas J, Petrie K, Eugui EM, Allison AC (1988) Glucocorticoids selectively inhibit the transcription of the interleukin 1 beta gene and decrease stability of interleukin 1 beta mRNA. Proc Natl Acad Sci USA 85:1204–1208

Lee-Huang S, Lin JJ, Kung HF, Huang PL, Lee L (1993) The human erythropoietin-encoding gene contains a CAAT box, TATA boxes and other transcriptional regulatory elements in its 5' flanking region. Gene 128:227–236

Lefstin JA, Thomas JR, Yamamoto KR (1994) Influence of a steroid receptor DNA-binding domain on transcriptional regulatory functions. Genes Dev 8:2842–2856

Lehoux J-G, Lefebvre A, Bélisle S, Bellabarba D (1989) Hormonal regulation of 3-hydroxy-3-methylglutaryl coenzyme A reductase mRNA in the rat adrenal gland. J Steroid Biochem 34:379–384

Lemaigre FP, Lause P, Rousseau GG (1994) Insulin inhibits glucocorticoid-induced stimulation of liver 6-phosphofructo-2-kinase/fructose-2,6-bisphosphatase gene transcription. FEBS Lett 340:221–225

Lemercier C, Julen N, Coulpier M, Dauchel H, Ozanne D, Fontaine M, Ripoche J (1992) Differential modulation by glucocorticoids of alternative complement protein secretion in cells of the monocyte/macrophage lineage. Eur J Immunol 22:909–915

Levenson CW, Shay NF, Cousins RJ (1994) Cloning and initial characterization of the promoter region of the rat cysteine-rich intestinal protein gene. Biochem J 303:731–736

Li H-L, Kmiec EB (1990) A glucocorticoid response element enhances transcription of a methionine tRNA gene in cis and trans. Mol Endocrinol 4:1173–1182

Li S, Rosen JM (1994) Glucocorticoid regulation of rat whey acidic protein gene expression involves hormone-induced alterations of chromatin structure in the distal promoter region. Mol Endocrinol 8:1328–1335

Lightman SL, Young WS III (1989) Influence of steroids on the hypothalamic corticotropin-releasing factor and preproenkephalin mRNA responses to stress. Proc Natl Acad Sci USA 86:4306–4310

Liley HG, White RT, Benson BJ, Ballard PL (1988) Glucocorticoids both stimulate and inhibit production of pulmonary surfactant protein A in fetal human lung. Proc Natl Acad Sci USA 85:9096–9100

Liu J-L, Papachristou DN, Patel YC (1994) Glucocorticoids activate somatostatin gene transcription through co-operative interaction with the cyclic AMP signalling pathway. Biochem J 301:863–869

Logsdon CD, Perot KJ, McDonald AR (1987) Mechanism of glucocorticoid-induced increase in pancreatic amylase gene transcription. J Biol Chem 262:15765–15769

Lotem J, Sachs L (1993) Hematopoietic cells from mice deficient in wild-type p53 are more resistant to induction of apoptosis by some agents. Blood 82:1092–1096

Lotem J, Sachs L (1994) Control of sensitivity to induction of apoptosis in myeloid leukemic cells by differentiation and bcl-2 dependent and independent pathways. Cell Growth Differ 5:321–327

Lowe WLJ, Meyer T, Karpen CW, Lorentzen LR (1992) Regulation of insulin-like growth factor I production in rat C6 glioma cells: possible role as an autocrine/paracrine growth factor. Endocrinology 130:2683–2691

Lu C, Ikeda K, Deftos LJ, Gazdar AF, Mangin M, Broadus AE (1989) Glucocorticoid regulation of parathyroid hormone-related peptide gene transcription in a human neuroendocrine cell line. Mol Endocrinol 3:2034–2040

Lu SC, Ge J-L, Kuhlenkamp J, Kaplowitz N (1992) Insulin and glucocorticoid dependence of hepatic gamma-glutamylcysteine synthetase and glutathione synthesis in the rat. Studies in cultured hepatocytes and in vivo. J Clin Invest 90:524–532

Lucas PC, O'Brien RM, Mitchell JA, Davis CM, Imai E, Forman BM, Samuels HH, Granner DK (1991) A retinoic acid response element is part of a pleiotropic domain in the phosphoenolpyruvate carboxykinase gene. Proc Natl Acad Sci USA 88:2184–2188 ·

Luedke CE, Cerami A (1990) Interferon-gamma overcomes glucocorticoid suppression of cachectin/tumor necrosis factor biosynthesis by murine macrophages. J Clin Invest 86:1234–1240

Luisi BF, Xu WX, Otwinowski Z, Freedman LP, Yamamoto KR, Sigler PB (1991) Crystallographic analysis of the interaction of the glucocorticoid receptor with DNA. Nature 352:497–505

Luo J, Murphy LJ (1989) Dexamethasone inhibits growth hormone induction of insulin-like growth factor-I (IGF-I) messenger ribonucleic acid (mRNA) in hypophysectomized rats and reduces IGF-I mRNA abundance in the intact rat. Endocrinology 125:165–171

Lyons JG, Birkedal-Hansen B, Pierson MC, Whitelock JM, Birkedal-Hansen H (1993) Interleukin-1 beta and transforming growth factor-alpha/epidermal growth factor induce expression of M(r) 95,000 type IV collagenase/gelatinase and interstitial fibroblast-type collagenase by rat mucosal keratinocytes. J Biol Chem 268:19143–19151

Ma T, Mahajan PB, Thompson EA (1992) Glucocorticoid regulation of c-myc promoter utilization in P1798 T-lymphoma cells. Mol Endocrinol 6:960–968

MacDougald OA, Cornelius P, Lin F-T, Chen SS, Lane MD (1994) Glucocorticoids reciprocally regulate expression of the CCAAT/enhancer-binding protein α and δ genes in 3T3-L1 adipocytes and white adipose tissue. J Biol Chem 269:19041–19047

Mahajan PB, Thompson EA (1990a) Hormonal regulation of transcription of rDNA. Purification and characterization of the hormone-regulated transcription factor IC. J Biol Chem 265:16225–16233

Mahajan PB, Thompson EA Jr (1990b) Glucocorticoid inhibition of transcription from adenovirus major late promoter. Mol Endocrinol 4:1515–1521

Mahajan PB, Thompson EA (1991) Glucocorticoid regulation of rRNA synthesis. Mol Cell Biochem 104:195–200

Mailleux P, Vanderhaeghen J-J (1993) Glucocorticoid regulation of cannabinoid receptor messenger RNA levels in the rat caudate-putamen. An in situ hybridization study. Neurosci Lett 156:51–53

Malbon CC, Hadcock JR (1988) Evidence that glucocorticoid response elements in the 5'-noncoding region of the hamster β2-adrenergic receptor gene are obligate for glucocorticoid regulation of receptor mRNA levels. Biochem Biophys Res Commun 154:676–681

Manyak CL, Tse H, Fischer P, Coker L, Sigal NH, Koo GC (1988) Regulation of class II MHC molecules on human endothelial cells: effects of IFN and dexamethasone. J Immunol 140:3817–3821

Marinkovic S, Baumann H (1990) Structure, hormonal regulation, and identification of the interleukin-6- and dexamethasone-responsive element of the rat haptoglobin gene. Mol Cell Biol 10:1573–1583

Marker AJ, Colosia AD, Tauler A, Solomon DH, Cayre Y, Lange AJ, El-Maghrabi MR, Pilkis SJ (1989) Glucocorticoid regulation of hepatic 6-phosphofructo-2-kinase/fructose-2,6- bisphosphatase gene expression. J Biol Chem 264:7000–7004

Markham PD, Gallo RC (1986) Hydrocortison and some other hormones enhance the expression of HTLV-III. Int J Cancer 37:67

Maroder M, Vacca A, Screpanti I, Petrangeli E, Frati L, Gulino A (1989) Enhancement of c-erbA proto-oncogene expression by glucocorticoid hormones in S49.1 lymphoma cells. Biochim Biophys Acta 1009:188–190

Maroder M, Martinotti S, Vacca A, Screpanti I, Petrangeli E, Frati L, Gulino A (1990) Post-transcriptional control of c-myc proto-oncogene expression by glucocorticoid hormones in human T lymphoblastic leukemic cells. Nucleic Acids Res 18:1153–1157

Masferrer JL, Zweifel BS, Seibert K, Needleman P (1990) Selective regulation of cellular cyclooxygenase by dexamethasone and endotoxin in mice. J Clin Invest 86:1375–1379

Masferrer JL, Seibert K, Zweifel B, Needleman P (1992) Endogenous glucocorticoids regulate an inducible cyclooxygenase enzyme. Proc Natl Acad Sci USA 89:3917–3921

Mason SA, Housley PR (1993) Site-directed mutagenesis of the phosphorylation sites in the mouse glucocorticoid receptor. J Biol Chem 268:21501–21504

Masters JN, Finch CE, Sapolsky RM (1989) Glucocorticoid endangerment of hippocampal neurons does not involve deoxyribonucleic acid cleavage. Endocrinology 124:3083–3088

Mathis JM, Houser WH, Bresnick E, Cidlowski JA, Hines RN, Prough RA, Simpson ER (1989) Glucocorticoid regulation of the rat cytochrome P450c (P450IA1) gene: receptor binding within intron I. Arch Biochem Biophys 269:93–105

Matsumoto K, Tajima H, Okazaki H, Nakamura T (1992) Negative regulation of hepatocyte growth factor gene expression in human lung fibroblasts and leukemic cells by transforming growth factor-β1 and glucocorticoids. J Biol Chem 267:24917–24920

McCarthy TL, Centrella M, Canalis E (1990) Cortisol inhibits the synthesis of insulin-like growth factor-I in skeletal cells. Endocrinology 126:1569–1575

McConkey DJ, Jondal M, Orrenius S (1994) The regulation of apoptosis in thymocytes. Biochem Soc Trans 22:606–610

McDowall IL, Fischer MJ (1995) The immediate 5'-flanking region of the rat phenylalanine hydroxylase-encoding gene. Gene 153:289–290

McGuire JA, Poellinger L, Wikström A-C, Gustafsson J-Å (1992) Cloning and regulation by glucocorticoid receptor ligands of a rat hsp90. J Steroid Biochem Mol Biol 42:813–822

McVicar DW, Merchant RE, Merchant LH, Young HF (1989) Corticosteroids inhibit the generation of lymphokine-activated killer activity in vitro. Cancer Immunol Immunother 29:211–218

Mendelson CR, Boggaram V (1991) Hormonal control of the surfactant system in fetal lung. Annu Rev Physiol 53:415–440

Merino R, Ding L, Veis DJ, Korsmeyer SJ, Nunez G (1994) Developmental regulation of the Bcl-2 protein and susceptibility to cell death in B lymphocytes. EMBO J 13:683–691

Miksicek R, Heber A, Schmid W, Danesch U, Posseckert G, Beato M, Schütz G (1986) Glucocorticoid responsiveness of the transcriptional enhancer of Moloney murine sarcoma virus. Cell 46:283–290

Miller WL (1994) Genetics, diagnosis, and management of 21-hydroxylase deficiency. J Clin Endocrinol Metab 78:241–246

Miner JN, Yamamoto KR (1991) Regulatory crosstalk at composite response elements. TIBS 16:423–426

Misseyanni A, Scheidereit C, Kalff M, Beato M (1991) Neither the endogenous nor a functional steroid hormone receptor binding site transactivate the ribosomal RNA gene promoter in vivo. J Steroid Biochem Mol Biol 39:409–418

Mitchell J, Noisin E, Hall R, O'Brien R, Imai E, Granner D (1994) Integration of multiple signals through a complex hormone response unit in the phosphoenolpyruvate carboxykinase gene promoter. Mol Endocrinol 8:585–594

Mittal R, Pater A, Pater MM (1993) Multiple human papillomavirus type 16 glucocorticoid response elements functional for transformation, transient expression, and DNA-protein interactions. J Virol 67:5656–5659

Mittal R, Kumar KU, Pater A, Pater MM (1994) Differential regulation by c-jun and c-fos protooncogenes of hormone response from composite glucocorticoid response element in human papilloma virus type 16 regulatory region. Mol Endocrinol 8:1701–1708

Miyashita A, Crystal RG, Hay JG (1995) Identification of a 27 bp 5'-flanking region element responsible for the low level constitutive expression of the human cytosolic phospholipase A2 gene. Nucleic Acids Res 23:293–301
Miyashita T, Reed JC (1992) bcl-2 Gene transfer increases relative resistance of S49.1 and WEH17.2 lymphoid cells to cell death and DNA fragmentation induced by glucocorticoids and multiple chemotherapeutic drugs. Cancer Res 52:5407–5411
Miyashita T, Reed JC (1993) Bcl-2 oncoprotein blocks chemotherapy-induced apoptosis in a human leukemia cell line. Blood 81:151–157
Moalli PA, Rosen ST (1994) Glucocorticoid receptors and resistance to glucocorticoids in hematologic malignancies. Leuk Lymph 15:363–374
Moalli PA, Pillay S, Krett NL, Rosen ST (1993) Alternatively spliced glucocorticoid receptor messenger RNAs in glucocorticoid-resistant human multiple myeloma cells. Cancer Res 53:3877–3879
Moffat GJ, Vik DP, Noack D, Tack BF (1992) Complete structure of the murine C4b-binding protein gene and regulation of its expression by dexamethasone. J Biol Chem 267:20400–20406
Moghul A, Lin L, Beedle A, Kanbour-Shakir A, DeFrances MC, Liu Y, Zarnegar R (1994) Modulation of c-MET proto-oncogene (HGF receptor) mRNA abundance by cytokines and hormones: evidence for rapid decay of the 8 kb c-MET transcript. Oncogene 9:2045–2052
Moncada S, Palmer RMJ (1991) Inhibition of the induction of nitric oxide synthase by glucocorticoids: yet another explanation for their anti-inflammatory effects. Trends Pharmacol Sci 12:130–131
Montague JW, Gaido ML, Frye C, Cidlowski JA (1994) A calcium-dependent nuclease from apoptotic rat thymocytes is homologous with cyclophilin. Recombinant cyclophilins A, B, and C have nuclease activity. J Biol Chem 269:18877–18880
Mordacq JC, Linzer DIH (1989) Co-localization of elements required for phorbol ester stimulation and glucocorticoid repression of proliferin gene expression. Genes Dev 3:760–769
Morrison NA, Shine J, Fragonas J-C, Verkest V, McMenemy ML, Eisman JA (1989) 1,25-Dihydroxyvitamin D-responsive element and glucocorticoid repression in the osteocalcin gene. Science 246:1158–1161
Muchardt C, Yaniv M (1993) A human homologue of Saccharomyces cerevisiae SNF2/SWI2 and Drosophila brm genes potentiates transcriptional activation by the glucocorticoid receptor. EMBO J 12:4279–4290
Mudrak I, Ogris E, Rotheneder H, Wintersberger E (1994) Coordinated trans activation of DNA synthesis- and precursor-producing enzymes by polyomavirus large T antigen through interaction with the retinoblastoma protein. Mol Cell Biol 14:1886–1892
Muglia L, Jacobson L, Dikkes P, Majzoub JA (1995) Corticotropin-releasing hormone deficiency reveals major fetal but not adult glucocorticoid need. Nature 373:427–432
Mukaida N, Morita M, Ishikawa Y, Rice N, Okamoto S, Kasahara T, Matsushima K (1994) Novel mechanism of glucocorticoid-mediated gene repression. Nuclear fac-

tor-kappaB is target for glucocorticoid-mediated interleukin 8 gene repression. J Biol Chem 269:13289–13295
Munoz A, Wrighton A, Seliger B, Bernal J, Beug H (1993a) Thyroid hormone receptor/c-erbA: control of commitment and differentiation in the neuronal/chromaffin progenitor line PC12. J Cell Biol 121:423–438
Munoz A, Wrighton C, Seliger B, Bernal J, Beug H (1993b) Thyroid hormone receptor/c-erbA: control of commitment and differentiation in the neuronal/chromaffin progenitor line PC12. J Biol Chem 121:423–438
Mühl H, Geiger T, Pignat W, Märki F, Van den Bosch H, Cerletti N, Cox D, McMaster G, Vosbeck K, Pfeilschifter J (1992) Transforming growth factors type-β and dexamethasone attenuate group II phospholipase A2 gene expression by interleukin-1 and forskolin in rat mesangial cells. FEBS Lett 301:190–194
Nakai T, Endo K, Hosono M, Saga T, Watanabe Y, Sakahara H, Imai K, Yachi A, Kiyozuka Y, Ishiwata I (1991) Drug effects on CA125 antigen expression and antibody binding to cancer cells. Int J Cancer 48:463–467
Nakano T, Ohara O, Teraoka H, Arita H (1990) Glucocorticoids suppress group II phospholipase A2 production by blocking mRNA synthesis and post-transcriptional expression. J Biol Chem 265:12745–12748
Nakano T, Raines EW, Abraham JA, Wenzel FG4, Higashiyama S, Klagsbrun M, Ross R (1993) Glucocorticoid inhibits thrombin-induced expression of platelet-derived growth factor A-chain and heparin-binding epidermal growth factor-like growth factor in human aortic smooth muscle cells. J Biol Chem 268:22941–22947
Naranjo JR (1986) Permissive effect of dexamethasone on the increase of proenkephalin mRNA induced by depolarization of chromaffin cells. Proc Natl Acad Sci USA 83:1513
Naray-Fejes-Toth A, Cornwell GG, Guyre PM (1985) Glucocorticoids inhibit IgE receptor expression on the human monocyte cell line U937. Immunology 56:359–366
Nazareth LV, Harbour DV, Thompson EB (1991) Mapping the human glucocorticoid receptor for leukemic cell death. J Biol Chem 266:12976–12980
Nebert DW, Nelson DR, Coon MJ, Estabrook RW, Feyereisen R, Fujii-Kuriyama Y, Gonzalez FJ, Guengerich FP, Gunsalus IC, Johnson EF, Loper JC, Sato R, Waterman MR, Waxman DJ (1991) The P450 superfamily: update on new sequences, gene mapping, and recommended nomenclature. DNA Cell Biol 10:1–14
Neveu I, Barbot N, Jehan F, Wion D, Brachet P (1991) Antagonistic effects of dexamethasone and 1,25-dihydroxyvitamin D3 on the synthesis of nerve growth factor. Mol Cell Endocrinol 78:R1–R6
Nieto MA, López-Rivas A (1992) Glucocorticoids activate a suicide program in mature T lymphocytes: protective action of interleukin-2. Ann NY Acad Sci 650:115–120
NIH-UC Expert Panel Corticosteroids (1990) Consensus statement on the use of corticosteroids as adjunctive therapy for pneumocystis pneumonia in the acquired immunodeficiency syndrome. N Engl J Med 323:1500–1504
Ning Y-M, Sánchez ER (1995) Evidence for a functional interaction between calmodulin and the glucocorticoid receptor. Biochem Biophys Res Commun 208:48–54

Noda C, Yakiyama M, Nakamura T, Ichihara A (1988) Requirements of both glucocorticoids and glucagon as co-inducers for activation of transcription of the serine dehydratase gene in cultured rat hepatocytes. J Biol Chem 263:14764–14768

Norgren S, Li LS, Luthman H (1994) Regulation of human insulin receptor RNA splicing in HepG2 cells: effects of glucocorticoid and low glucose concentration. Biochem Biophys Res Commun 199:277–284

Northrop JP, Crabtree GR, Mattila PS (1992) Negative regulation of interleukin 2 transcription by the glucocorticoid receptor. J Exp Med 175:1235–1245

O'Banion MK, Levenson RM, Brinckmann UG, Young DA (1992a) Glucocorticoid modulation of transformed cell proliferation is oncogene specific and correlates with effects on c-myc levels. Mol Endocrinol 6:1371–1380

O'Banion MK, Winn VD, Young DA (1992b) cDNA Cloning and functional activity of a glucocorticoid-regulated inflammatory cyclooxygenase. Proc Natl Acad Sci USA 89:4888–4892

O'Brien RM, Noisin EL, Suwanichkul A, Yamasaki T, Lucas RC, Wang J-C, Powell DR, Granner DK (1995) Hepatic nuclear factor 3- and hormone-regulated expression of the phosphoenolpyruvate carboxykinase and insukin-like growth factor-binding protein 1 genes. Mol Cell Biol 15:1747–1758

Oberg KC, Carpenter G (1989) Dexamethasone acts as a negative regulator of epidermal growth factor receptor synthesis in fetal rat lung cells. Mol Endocrinol 3:915–922

O'Callaghan JP, Brinton RE, McEwen BS (1989) Glucocorticoids regulate the concentration of glial fibrillary acidic protein throughout the brain. Brain Res 494:159–161

Oda T (1988) Molecular cloning, complete nucleotide sequence, and gene structure of the provirus genome of a retrovirus produced in a human lymphoblastoid cell line. Virology 167:468

Odenthal M, Neubauer K, Baralle FE, Peters H, Meyer zum Büschenfelde K-H, Ramadori G (1992) Rat hepatocytes in primary culture synthesize and secrete cellular fibronectin. Exp Cell Res 203:289–296

Odom MJ, Snyder JM, Boggaram V, Mendelson CR (1988) Glucocorticoid regulation of the major surfactant associated protein (SP-A) and its messenger ribonucleic acid and of morphological development of human fetal lung in vitro. Endocrinology 123:1712–1720

Offringa R, Smits AMM, Houweling A, Bos JL, Van der Eb AJ (1988) Similar effects of adenovirus E1A and glucocorticoid hormones on the expression of the metalloprotease stromelysin. Nucleic Acids Res 16:10973–10984

Ohtani R, Yayama K, Takano M, Itoh N, Okamoto H (1992) Stimulation of angiotensinogen production in primary cultures of rat hepatocytes by glucocorticoid, cyclic adenosine 3', 5'-monophosphate, and interleukin-6. Endocrinology 130:1331–1338

Ohyama K, Seyer JM, Raghow R, Kang AH (1990) Extracellular matrix phenotype of rat mesangial cells in culture: biosynthesis of collagen types I, III, IV, and V and a low molecular weight collagenous component and their regulation by dexamethasone. J Lab Clin Med 116:219–227

Okazaki R, Riggs BL, Conover CA (1994) Glucocorticoid regulation of insulin-like growth factor-binding protein expression in normal human osteoblast-like cells. Endocrinology 134:126–132

Oltvai ZN, Korsmeyer SJ (1994) Checkpoints of dueling dimers foil death wishes. Cell 79:189–192

Ong JM, Simsolo RB, Saffari B, Kern PA (1992) The regulation of lipoprotein lipase gene expression by dexamethasone in isolated rat adipocytes. Endocrinology 130:2310–2316

Ooi GT, Brown DR, Suh DS, Tseng LY, Rechler MM (1993) Cycloheximide stabilizes insulin-like growth factor-binding protein-1 (IGFBP-1) mRNA and inhibits IGFBP-1 transcription in H4-II-E rat hepatoma cells. J Biol Chem 268:16664–16672

Orchinik M, Murray TF, Moore FL (1991) A corticosteroid receptor in neuronal membranes. Science 252:1848–1851

O'Reilly MA, Clark JC, Whitsett JA (1991) Glucocorticoid enhances pulmonary surfactant protein B gene transcription. Am J Physiol Lung Cell Mol Physiol 260:L37–L43

Orlowski CC, Ooi GT, Brown DR, Yang YW-H, Tseng LY-H, Rechler MM (1991) Insulin rapidly inhibits insulin-like growth factor-binding protein-1 gene expression in H4-II-E rat hepatoma cells. Mol Endocrinol 5:1180–1187

Oro AE, Hollenberg SM, Evans RM (1988) Transcriptional inhibition by a glucocorticoid receptor-β-galactosidase fusion protein. Cell 55:1109–1114

Orson FM, Flagge FP, Cashaw JL (1986) T cell replacing factor for steroids (TRF-S): a 40,000 Dalton protein produced by a T4+ T cell. J Immunol 137:578–584

Ortí E, Bodwell JE, Munck A (1992) Phosphorylation of steroid hormone receptors. Endocr Rev 13:105–128

Orwoll ES, Klein RF (1995) Osteoporosis in men. Endocr Rev 16:87–116

Oursler MJ, Riggs BL, Spelsberg TC (1993) Glucocorticoid-induced activation of latent transforming growth factor-beta by normal human osteoblast-like cells. Endocrinology 133:2187–2196

Pajovic S, Jones VE, Prowse KR, Berger FG, Baumann H (1994) Species-specific changes in regulatory elements of mouse haptoglobin genes. J Biol Chem 269:2215–2224

Palca J (1990) A storm over steroid therapy. Science 250:1196–1198

Paliogianni F, Ahuja SS, Balow JP, Balow JE, Boumpas DT (1993a) Novel mechanism for inhibition of human T cells by glucocorticoids: glucocorticoids inhibit signal transduction through IL-2 receptor. J Immunol 151:4081–4089

Paliogianni F, Raptis A, Ahuja SS, Najjar SM, Boumpas DT (1993b) Negative transcriptional regulation of human interleukin 2 (IL-2) gene by glucocorticoids through interference with nuclear transcription factors AP-1 and NF-AT. J Clin Invest 91:1481–1489

Palmer LA, Hukku B, Harmon JM (1992) Human glucocorticoid receptor gene deletion following exposure to cancer chemotherapeutic drugs and chemical mutagens. Cancer Res 52:6612–6618

Palmiter RD, Findley SD, Whitmore TE, Durnam DM (1992) MT-III, a brain specific member of the metallothionein gene family. Proc Natl Acad Sci USA 89:6333–6337

Pan L, Mendel DB, Zurlo J, Guyre PM (1990) Regulation of the steady state level of Fc gamma RI mRNA by IFN-gamma and dexamethasone in human monocytes, neutrophils, and U-937 cells. J Immunol 145:267–275

Panarelli M, Holloway CD, Mulatero P, Fraser R, Kenyon CJ (1994) Inhibition of lysozyme synthesis by dexamethasone in human mononuclear leukocytes: an index of glucocorticoid sensitivity. J Clin Endocrinol Metab 78:872–877

Parente L, Solito E (1994) Association between glucocorticosteroids and lipocortin 1. Trends Pharmacol Sci 15:362

Parente L, Becherucci C, Perretti M, Solito E, Mugridge KG, Galeotti CL, Raugei G, Melli M, Sansò M (1990) Are the lipocortins the second messengers of the anti-inflammatory action of glucocorticoids. Prog Clin Biol Res 349:55–68

Pash JM, Bailey JM (1988) Inhibition by corticosteroids of epidermal growth factor-induced recovery of cyclooxygenase after aspirin inactivation. FASEB J 2:2613–2618

Pasmanik M, Izhar M, Zolti M, Shemesh M (1991) Dexamethasone inhibition of prostaglandin production in human term placental cells is protein and ribonucleic acid synthesis dependent. Endocrinology 128:2058–2064

Patchev VK, Brady LS, Karl M, Chrousos GP (1994) Regulation of HSP90 and corticosteroid receptor mRNA by corticosterone levels in vivo. Mol Cell Endocrinol 103:57–64

Pater MM, Hughes GA, Hyslop DE, Nakshatri H, Pater A (1988) Glucocorticoid-dependent oncogenic transformation by type 16 but not type 11 human papilloma virus DNA. Nature 335:832–835

Pavé-Preux M, Ferry N, Bouguet J, Hanoune J, Barouki R (1988) Nucleotide sequence and glucocorticoid regulation of the mRNAs for the isoenzymes of rat aspartate aminotransferase. J Biol Chem 263:17459–17466

Peattie DA, Harding MW, Fleming MA, DeCenzo MT, Lippke JA, Livingston DJ, Benasutti M (1992) Expression and characterization of human FKBP52, an immunophilin that associates with the 90-kDa heat shock protein and is a component of steroid receptor complexes. Proc Natl Acad Sci USA 89:10974–10978

Peppel K, Vinci JM, Baglioni C (1991) The AU-rich sequences in the 3' untranslated region mediate the increased turnover of interferon mRNA induced by glucocorticoids. J Exp Med 173:349–355

Peraldi M-N, Rondeau E, Jousset V, El M'Selmi A, Lacave R, Delarue F, Garel J-M, Sraer J-D (1990) Dexamethasone increases preproparathyroid hormone messenger RNA in human hyperplastic parathyroid cells in vitro. Eur J Clin Invest 20:392–397

Perez HD, Kelly E, Holmes R (1992) Regulation of formyl peptide receptor expression and its mRNA levels during differentiation of HL-60 cells. J Biol Chem 267:358–363

Perez JR, Shull S, Gendimenico GJ, Capetola RJ, Mezick JA, Cutroneo KR (1992) Glucocorticoid and retinoid regulation of alpha-2 type I procollagen promoter activity. J Cell Biochem 50:26–34

Perrone-Bizzozero NI, Cansino VV, Kohn DT (1993) Posttranscriptional regulation of GAP-43 gene expression in PC12 cells through protein kinase C-dependent stabilization of the mRNA. J Cell Biol 120:1263–1270

Petersen DD, Koch SR, Granner DK (1989) 3' Noncoding region of phosphoenolpyruvate carboxykinase mRNA contains a glucocorticoid-responsive mRNA-stabilizing element. Proc Natl Acad Sci USA 86:7800–7804

Petroni KC, Shen L, Guyre PM (1988) Modulation of human polymorphonuclear leukocyte IgG Fc receptors and Fc receptor-mediated functions by IFN-gamma and glucocorticoids. J Immunol 140:3467–3472

Picado-Leonard J, Miller WL (1987) Cloning and sequence of the human gene encoding P450c17 (steroid 17α-hydroxylase/17,20-lyase):similarity to the gene for P450c21. DNA 6:439–448

Pierzchalski P, Rokita H, Koj A, Fries E, Åkerström B (1992) Synthesis of α1-microglobulin in cultured rat hepatocytes is stimulated by interleukin-6, leukemia inhibitory factor, dexamethasone and retinoic acid. FEBS Lett 298:165–168

Plisov SY, Nichiporenko MG, Shkapenko AL, Kumarev VP, Baranov LV, Merkulova TI (1995) The immediate vicinity of mouse metallothionein-I gene contains two sites conferring glucocorticoid inducibility to the heterologous promoter. FEBS Lett 358:104

Politis AD, Ozato K, Coligan JE, Vogel SN (1994) Regulation of IFN-gamma-induced nuclear expression of IFN consensus sequence binding protein in murine peritoneal macrophages. J Immunol 152:2270–2278

Poon M, Megyesi J, Green RS, Zhang H, Rollins BJ, Safirstein R, Taubman MB (1991) In vivo and in vitro inhibition of JE gene expression by glucocorticoids. J Biol Chem 266:22375–22379

Powers JH, Hillmann AG, Tang DC, Harmon JM (1993) Cloning and expression of mutant glucocorticoid receptors from glucocorticoid-sensitive and -resistant human leukemic cells. Cancer Res 53:4059–4065

Pratt WB (1993) The role of heat shock proteins in regulating the function, folding, and trafficking of the glucocorticoid receptor. J Biol Chem 268:21455–21458

Pratt WB, Czar MJ, Stancato LF, Owens JK (1993) The hsp56 immunophilin component of steroid receptor heterocomplexes: could this be the elusive nuclear localization signal-binding protein. J Steroid Biochem Mol Biol 46:269–279

Presse F, Hervieu G, Imaki T, Sawchenko PE, Vale W, Nahon J-L (1992) Rat melanin-concentrating hormone messenger ribonucleic acid expression: marked changes during development and after stress and glucocorticoid stimuli. Endocrinology 131:1241–1250

Price WA, Stiles AD, Moats-Staats BM, D'Ercole AJ (1992) Gene expression of insulin-like growth factors (IGFs), the type 1 IGF receptor, and IGF-binding proteins in dexamethasone-induced fetal growth retardation. Endocrinology 130:1424–1432

Pukhalsky AL, Kalashnikova EA, Lyashko VN, Pevnitsky LA (1990) Inhibition of phytohemagglutinin-induced lymphocyte proliferation by dexamethasone: mechanisms of individual susceptibility. Int J Immunopharmacol 12:657–663

Qi M, Hamilton BJ, DeFranco D (1989) v-mos Oncoproteins affect the nuclear retention and reutilization of glucocorticoid receptors. Mol Endocrinol 3:1279–1288

Raab LS, Decker GL, Jonas AJ, Kaetzel MA, Dedman JR (1991) Glucocorticoid regulation of rat liver urate oxidase. J Cell Biochem 47:18–30

Rabek JP, Zhang DE, Torres-Ramos CA, Papaconstantinou J (1994) Analysis of the mechanism of glucocorticoid-mediated down regulation of the mouse alpha-feto-protein gene. Biochim Biophys Acta 1218:136–144

Radomski MW, Palmer RMJ, Moncada S (1990) Glucocorticoids inhibit the expression of an inducible, but not the constitutive, nitric oxide synthase in vascular endothelial cells. Proc Natl Acad Sci USA 87:10043–10047

Raff MC (1992) Social controls on cell survival and cell death. Nature 356:397–400

Raja RH, Paterson AJ, Shin TH, Kudlow JE (1991) Transcriptional regulation of the human transforming growth factor-alpha gene. Mol Endocrinol 5:514–520

Rameh LE, Armelin MCS (1992) Downregulation of JE and KC genes by glucocorticoids does not prevent the Go–G1 transition in BALB/3T3 cells. Mol Cell Biol 12:4612–4621

Rao MV, Rangarajan PN, Padmanaban G (1990) Dexamethasone negatively regulates phenobarbitone-activated transcription but synergistically enhances cytoplasmic levels of cytochrome P-450b/e messenger RNA. J Biol Chem 265:5617–5622

Ratajczak T, Williams PM, DiLorenzo D, Ringold GM (1992) Multiple elements within the glucocorticoid regulatory unit of the rat alpha 1-acid glycoprotein gene are recognition sites for C/EBP. J Biol Chem 267:11111–11119

Ray A, Prefontaine KE (1994) Physical association and functional antagonism between the p65 subunit of transcription factor NF-kappaB and the glucocorticoid receptor. Proc Natl Acad Sci USA 91:752–756

Ray A, LaForge KS, Sehgal PB (1990) On the mechanism for efficient repression of the interleukin-6 promoter by glucocorticoids: enhancer, TATA box, and RNA start site (Inr motif) occlusion. Mol Cell Biol 10:5736–5746

Raz A, Wyche A, Needleman P (1989) Temporal and pharmacological division of fibroblast cyclooxygenase expression into transcriptional and translational phases. Proc Natl Acad Sci USA 86:1657–1661

Re F, Muzio M, DeRossi M, Polentarutti N, Giri JG, Mantovani A, Colotta F (1994) The type II "receptor" as a decoy target for interleukin 1 in polymorphonuclear leukocytes: characterization of induction by dexamethasone and ligand binding properties of the released decoy receptor. J Exp Med 179:739–743

Redondo JM, Fresno M, López-Rivas A (1988) Inhibition of interleukin 2-induced proliferation of cloned murine T cells by glucocorticoids. Possible involvement of an inhibitory protein. Eur J Immunol 18:1555–1559

Reed JC, Nowell PC (1986) Effect of cyclosporin A and dexamethasone on interleukin 2 receptor gene expression. J Immunol 137:150

Reik A, Schütz G, Stewart AF (1991) Glucocorticoids are required for establishment and maintenance of an alteration in chromatin structure: induction leads to a reversible disruption of nucleosomes over an enhancer. EMBO J 10:2569–2576

Reik A, Stewart AF, Schutz G (1994) Cross-talk modulation of signal transduction pathways: two mechanisms are involved in the control of tyrosine aminotransferase gene expression by phorbol esters. Mol Endocrinol 8:490–497

Reisfeld S, Vardimon L (1994) Cell to cell contacts control the transcription activity of the glucocorticoid receptor. Mol Endocrinol 8:1224–1233

Remick DG, Strieter RM, Lynch JP III, Nguyen D, Eskandari M, Kunkel SL (1989) In vivo dynamics of murine tumor necrosis factor-α gene expression: kinetics of dexamethasone-induced suppression. Lab Invest 60:766–771

Rennie PS, Bowden JF, Freeman SN, Bruchovsky N, Cheng H, Lubahn DB, Wilson EM, French FS, Main L (1989) Cortisol alters gene expression during involution of the rat ventral prostate. Mol Endocrinol 3:703–708

Rhee K, Ma T, Thompson EA (1994) The macromolecular state of the transcription factor E2F and glucocorticoid regulation of c-myc transcription. J Biol Chem 269:17035–17042

Rhodes J, Ivanyi J, Cozens P (1986) Antigen presentation by human monocytes: effects of modifying major histocompatibility complex class II antigen expression and interleukin 1 production by using recombinant interferons and corticosteroids. Eur J Immunol 16:370–375

Riegel AT, Lu Y, Remenick J, Wolford RG, Berard DS, Hager GL (1991) Proopiomelanocortin gene promoter elements required for constitutive and glucocorticoid-repressed transcription. Mol Endocrinol 5:1973–1982

Rigaud G, Roux J, Pictet R, Grange T (1991) In vivo footprinting of rat TAT gene: dynamic interplay between the glucocorticoid receptor and a liver-specific factor. Cell 67:977–986

Ringler GE, Kallen CB, Strauss JF III (1989) Regulation of human trophoblast function by glucocorticoids: dexamethasone promotes increased secretion of chorionic gonadotropin. Endocrinology 124:1625–1631

Robertson DG, Marino EM, Thulé PM, Seneviratne CK, Murphy LJ (1994) Insulin and glucocorticoids regulate IGFBP-1 expression via a common promoter region. Biochem Biophys Res Commun 200:226–232

Robin JL, Seldin DS, Austen KF, Lewis RA (1985) Regulation of mediator release from mouse bone marrow-derived mast cells by glucocorticoids. J Immunol 135:2719–2726

Robinson BG, Emanuel RL, Frim DM, Majzoub JA (1988) Glucocorticoid stimulates expression of corticotropin-releasing hormone gene in human placenta. Proc Natl Acad Sci USA 85:5244–5248

Rock D, Lokensgard J, Lewis T, Kutish G (1992) Characterization of dexamethasone-induced reactivation of latent bovine herpesvirus 1. J Virol 66:2484–2490

Rolfe FG, Hughes JM, Armour CL, Sewell WA (1992) Inhibition of interleukin-5 gene expression by dexamethasone. Immunology 77:494–499

Ron D, Brasier AR, Wright KA, Habener JF (1990) The permissive role of glucocorticoids on interleukin-1 stimulation of angiotensinogen gene transcription is mediated by an interaction between inducible enhancers. Mol Cell Biol 10:4389–4395

Rooney SA, Young SL, Mendelson CR (1994) Molecular and cellular processing of lung surfactant. FASEB J 8:957–967

Ros M, Watkins DC, Rapiejko PJ, Malbon CC (1989) Glucocorticoids modulate mRNA levels *for G-protein* β*-subunits*. Biochem J 260:271–275

Rosemblit N, Chen CL (1994) Regulators for the rat clusterin gene: DNA methylation and cis-acting regulatory elements. J Mol Endocrinol 13:69–76
Rosewicz S, Detjen K, Logsdon CD, Chen L-M, Chao J, Riecken EO (1991) Glandular kallikrein gene expression is selectively down-regulated by glucocorticoids in pancreatic AR42J cells. Endocrinology 128:2216–2222
Rosewicz S, Detjen K, Kaiser A, Prosenc N, Cervos-Navarro J, Riecken E-O, Haller H (1994) Bombesin receptor gene expression in rat pancreatic acinar AR42J cells: transcriptional regulation by glucocorticoids. Gastroenterology 107:208–218
Ross ME, Evinger MJ, Hyman SE, Carroll JM, Mucke L, Comb M, Reis DJ, Joh TH, Goodman HM (1990) Identification of a functional glucocorticoid response element in the phenylethanolamine N-methyltransferase promoter using fusion genes introduced into chromaffin cells in primary culture. J Neurosci 10:520–530
Rouleau J, MacLeod AR, Szyf M (1995) Regulation of the DNA methyltransferase by the Ras-AP-1 signaling pathway. J Biol Chem 270:1595–1601
Rozansky DJ, Wu H, Tang K, Parmer RJ, O'Connor DT (1994) Glucocorticoid activation of chromogranin A gene expression. Identification and characterization of a novel glucocorticoid response element. J Clin Invest 94:2357–2368
Russell SB, Trupin JS, Myers JC, Broquist AH, Smith JC, Myles ME, Russell JD (1989) Differential glucocorticoid regulation of collagen mRNAs in human dermal fibroblasts. Keloid-derived and fetal fibroblasts are refractory to down-regulation. J Biol Chem 264:13730–13735
Russell SB, Trupin JS, Kennedy RZ, Russell JD, Davidson JM (1995) Glucocorticoid regulation of elastin synthesis in human fibroblasts: down-regulation in fibroblasts from normal dermis but not from keloids. J Invest Dermatol 104:241–245
Russo AF, Lanigan TM, Sullivan BE (1992) Neuronal properties of a thyroid C-cell line: partial repression by dexamethasone and retinoic acid. Mol Endocrinol 6:207–218
Rüther U, Garber C, Komitowski D, Müller R, Wagner EF (1987) Deregulated c-fos expression interferes with normal bone development in transgenic mice. Nature 325:412–416
Saad MJ, Folli F, Kahn CR (1995) Insulin and dexamethasone regulate insulin receptors, insulin receptor substrate-1, and phosphatidylinositol 3-kinase in Fao hepatoma cells. Endocrinology 136:1579–1588
Saad MJA, Folli F, Araki E, Hashimoto N, Csermely P, Kahn CR (1994) Regulation of insulin receptor, insulin receptor substrate-1 and phosphatidylinositol 3-kinase in 3T3-F442A adipocytes. Effects of differentiation, insulin, and dexamethasone. Mol Endocrinol 8:545–557
Sabol SL, Higuchi H (1990) Transcriptional regulation of the neuropeptide Y gene by nerve growth factor: antagonism by glucocorticoids and potentiation by adenosine 3',5'-monophosphate and phorbol ester. Mol Endocrinol 4:384–392
Sahu A, Phelps CP, White JD, Crowley WR, Kalra SP, Kalra PS (1992) Steroidal regulation of hypothalamic neuropeptide Y release and gene expression. Endocrinology 130:3331–3336

Saito N, Guitart X, Hayward M, Tallman JF, Duman RS, Nestler EJ (1989) Corticosterone differentially regulates the expression of Gsα and Giα messenger RNA and protein in rat cerebral cortex. Proc Natl Acad Sci USA 86:3906–3910

Sakai DD, Helms S, Carlstedt-Duke J, Gustafsson JA, Rottman FM, Yamamoto KR (1988) Hormone-mediated repression: a negative glucocorticoid response element from the bovine prolactin gene. Genes Dev 2:1144–1154

Sakaue M, Hoffman BB (1991) Glucocorticoids induce transcription and expression of the α1B adrenergic receptor gene in DTT1 MF-2 smooth muscle cells. J Clin Invest 88:385–389

Samuelson LC, Keller PR, Darlington GJ, Meisler MH (1988) Glucocorticoid and developmental regulation of amylase mRNAs in mouse liver cells. Mol Cell Biol 8:3857–3863

Sanchez I, Goya L, Vallerga AK, Firestone GL (1993) Glucocorticoids reversibly arrest rat hepatoma cell growth by inducing an early G1 block in cell cycle progression. Cell Growth Differ 4:215–225

Sapi E, Flick MB, Gilmore-Hebert M, Rodov S, Kacinski BM (1995) Transcriptional regulation of the c-fms (CSF-1R) proto-oncogene in human breast carcinoma cells by glucocorticoids. Oncogene 10:529–542

Sapolsky RM, Uno H, Rebert CS, Finch CE (1990) Hippocampal damage associated with prolonged glucocorticoid exposure in primates. J Neurosci 10:2897–2902

Scheinman RI, Gualberto A, Jewell CM, Cidlowski JA, Baldwin AS Jr (1995) Characterization of mechanisms involved in transrepression of NF-kappaB by activated glucocorticoid receptors. Mol Cell Biol 15:943–953

Schlatter LK, Howard KJ, Parker MG, Distelhorst CW (1992) Comparison of the 90-kilodalton heat shock protein interaction with in vitro translated glucocorticoid and estrogen receptors. Mol Endocrinol 6:132–140

Schmidt TJ, Meyer AS (1994) Autoregulation of corticosteroid receptors. How, when, where, and why? Receptor 4:229–257

Schmitt-Ney M, Doppler W, Ball RK, Groner B (1991) β-casein gene promoter activity is regulated by the hormone-mediated relief of transcriptional repression and a mammary-gland-specific nuclear factor. Mol Cell Biol 11:3745–3755

Schrey MP, Patel KV, Tezapsidis N (1992) Bombesin and glucocorticoids stimulate human breast cancer cells to produce endothelin, a paracrine mitogen for breast stromal cells. Cancer Res 52:1786–1790

Schuchard M, Landers JP, Sandhu NP, Spelsberg TC (1993) Steroid hormone regulation of nuclear proto-oncogenes. Endocr Rev 14:659–669

Schuster C, Chasserot-Golaz S, Urier G, Beck G, Sergeant A (1991) Evidence for a functional glucocorticoid responsive element in the Epstein-Barr virus genome. Mol Endocrinol 5:267–272

Schüle R, Muller M, Kaltschmidt C, Renkawitz R (1988a) Many transcription factors interact synergistically with steroid receptors. Science 242:1418–1420

Schüle R, Muller M, Otsuka-Murakami H, Renkawitz R (1988b) Cooperativity of the glucocorticoid receptor and the CACCC-box binding factor. Nature 332:87–90

Schüle R, Rangarajan P, Kliewer S, Ransone LJ, Bolado J, Yang N, Verma IM, Evans RM (1990) Functional antagonism between oncoprotein c-Jun and the glucocorticoid receptor. Cell 62:1217–1226

Sebaldt RJ, Sheller JR, Oates JA, Roberts LJ II, FitzGerald GA (1990) Inhibition of eicosanoid biosynthesis by glucocorticoids in humans. Proc Natl Acad Sci USA 87:6974–6978

Shalhoub V, Conlon D, Tassinari M, Quinn C, Partridge N, Stein GS, Lian JB (1992) Glucocorticoids promote development of the osteoblast phenotype by selectively modulating expression of cell growth and differentiation associated genes. J Cell Biochem 50:425–440

Shapiro SD, Campbell EJ, Kobayashi DK, Welgus HG (1991) Dexamethasone selectively modulates basal and lipopolysaccharide-induced metalloproteinase and tissue inhibitor of metalloproteinase production by human alveolar macrophages. J Immunol 146:2724–2729

Shepherd VL, Cowan HB, Abdolrasulnia R, Vick S (1994) Dexamethasone blocks the interferon-gamma-mediated downregulation of the macrophage mannose receptor. Arch Biochem Biophys 312:367–374

Shields PP, Dixon JE, Glembotski CC (1988) The secretion of atrial natriuretic factor-(99-126) by cultured cardiac myocytes is regulated by glucocorticoids. J Biol Chem 263:12619–12628

Shimizu Y, Kielar D, Masuno H, Minokoshi Y, Shimazu T (1994) Dexamethasone induces the GLUT4 glucose transporter, and responses of glucose transport to norepinephrine and insulin in primary cultures of brown adipocytes. J Biochem (Tokyo) 115:1069–1074

Short NK, Clouthier DE, Schaefer IM, Hammer RE, Magnuson MA, Beale EG (1992) Tissue-specific, developmental, hormonal, and dietary regulation of rat phosphoenolpyruvate carboxykinase-human growth hormone fusion genes in transgenic mice. Mol Cell Biol 12:1007–1020

Sierra-Honigmann MR, Murphy PA (1992) T cell receptor-independent immunosuppression induced by dexamethasone in murine T helper cells. J Clin Invest 89:556–560

Silver G, Reid LM, Krauter KS (1990) Dexamethasone-mediated regulation of 3-methylcholanthrene-induced cytochrome P-450d mRNA accumulation in primary rat hepatocyte cultures. J Biol Chem 265:3134–3138

Simard J, De Launoit Y, Haagensen DE, Labrie F (1992) Additive stimulatory action of glucocorticoids and androgens on basal and estrogen-repressed apolipoprotein-D messenger ribonucleic acid levels and secretion in human breast cancer cells. Endocrinology 130:1115–1121

Simmons DL, Nell BG, Stevens R, Evett G, Erikson RL (1992) Identification of an early-growth-response gene encoding a novel putative protein kinase. Mol Cell Biol 12:4164–4169

Simonet WS, Ness GC (1989) Post-transcriptional regulation of 3-hydroxy-3-methylglutaryl-CoA reductase mRNA in rat liver. Glucocorticoids block the stabilization caused by thyroid hormones. J Biol Chem 264:569–573

Simonson MS, Wolfe JA, Konieczkowiski M, Sedor JR, Dunn MJ (1991) Regulation of prostaglandin endoperoxide synthase gene expression in cultured rat mesangial cells: induction by serum via a protein kinase-C-dependent mechanism. Mol Endocrinol 5:441–451

Singh P, Coe J, Hong W (1995) A role for retinoblastoma protein in potentiating transcriptional activation by the glucocorticoid receptor. Nature 374:562–565

Sirois J, Simmons DL, Richards JS (1992) Hormonal regulation of messenger ribonucleic acid encoding a novel isoform of prostaglandin endoperoxide H synthase in rat preovulatory follicles. Induction in vivo and in vitro. J Biol Chem 267:11586–11592

Sivitz WI, Pasley AM (1995) Glucocorticoid effects on glucose transport and transporter gene expression in L6 muscle cells. Biochem Biophys Res Commun 206:51–56

Slater EP, Rabenau O, Karin M, Baxter JD, Beato M (1985) Glucocorticoid receptor binding and activation of a heterologous promoter by dexamethasone by the first intron of the human growth hormone gene. Mol Cell Biol 5:2984–2992

Slater EP, Redeuilh G, Beato M (1991) Hormonal regulation of vitellogenin genes: an estrogen-responsive element in the Xenopus A2 gene and a multihormonal regulatory region in the chicken II gene. Mol Endocrinol 5:386–396

Slater EP, Hesse H, Müller JM, Beato M (1993) Glucocorticoid receptor binding site in the mouse α-amylase 2 gene mediates response to the hormone. Mol Endocrinol 7:907–914

Slavin BG, Ong JM, Kern PA (1994) Hormonal regulation of hormone-sensitive lipase activity and mRNA levels in isolated rat adipocytes. J Lipid Res 35:1535–1541

Smith MA, Makino S, Kvetnansky R, Post RM (1995) Stress and glucocorticoids affect the expression of brain-derived neurotrophic factor and neurotrophin-3 mRNAs in the hippocampus. J Neurosci 15:1768–1777

Snyder DS, Unanue ER (1982) Corticosteroids inhibit murine macrophage Ia expression and interleukin 1 production. J Immunol 129:1803–1805

Somasekhar MB, Gorski J (1988) Two elements of the rat prolactin 5' flanking region are required for its regulation by estrogen and glucocorticoids. Gene 69:13–21

Soudeyns H, Geleziunas R, Shyamala G, Hiscott J, Wainberg MA (1993) Identification of a novel glucocorticoid response element within the genome of the human immunodeficiency virus type 1. Virology 194:758–768

Soury B, Hentzen D, Vignal M, Christeff N, Doly J (1995) Induction of interferon-β gene expression by dexamethasone in murine L929 cells. Mol Endocrinol 9:199–207

Spiegelman BM, Lowell B, Napolitano A, Dubuc P, Barton D, Francke U, Groves DL, Cook KS, Flier JS (1989) Adrenal glucocorticoids regulate adipsin gene expression in genetically obese mice. J Biol Chem 264:1811–1815

Srinivasan G, Patel NT, Thompson EB (1994) Heat shock protein is tightly associated with the recombinant human glucocorticoid receptor: glucocorticoid response element complex. Mol Endocrinol 8:189–196

Srivastava G, Kwong SKY, Lam KSL, May BK (1992) Effect of dexamethasone on mRNA levels for 5-aminolevulinate synthase in different rat tissues. Eur J Biochem 203:59–63
Staels B, Van Tol A, Chan L, Verhoeven G, Auwerx J (1991) Variable effects of different corticosteroids on plasma lipids, apolipoproteins, and hepatic apolipoprotein mRNA levels in rats. Arteriosclerosis 11:760–769
Standke GJ, Meier VS, Groner B (1994) Mammary gland factor activated by prolactin on mammary epithelial cells and acute-phase response factor activated by interleukin-6 in liver cells share DNA binding and transactivation potential. Mol Endocrinol 8:469–477
Stanimirovic DB, McCarron RM, Spatz M (1994) Dexamethasone down-regulates endothelin receptors in human cerebromicrovascular endothelial cells. Neuropeptides 26:145–152
Stefanovic V, Vlahovic P, Ardaillou N, Ardaillou R (1991) Receptor-mediated induction of aminopeptidase A (APA) of human glomerular epithelial cells (HGEC) by glucocorticoids. FEBS Lett 294:171–174
Stein R, Orit S, Anderson DJ (1988) The induction of a neural-specific gene, SCG10, by nerve growth factor in PC12 cells is transcriptional, protein synthesis dependent, and glucocorticoid inhibitable. Dev Biol 127:316–325
Steineger HH, Sorensen HN, Tugwood JD, Skrede S, Spydevold O, Gautvik KM (1994) Dexamethasone and insulin demonstrate marked and opposite regulation of the steady-state mRNA level of the peroxisomal proliferator-activated receptor (PPAR) in hepatic cells-Hormonal modulation of fatty-acid-induced transcription. Eur J Biochem 225:967–974
Stern L, Kunos G (1988) Synergistic regulation of pulmonary β-adrenergic receptors by glucocorticoids and interleukin-1. J Biol Chem 263:15876–15879
Strasser-Wozak EMC, Hattmannstorfer R, Hála M, Hartmann BL, Fiegl M, Geley S, Kofler R (1995) Splice site mutation in the glucocorticoid receptor gene causes resistance to glucocorticoid-induced apoptosis in a human acute leukemic cell line. Cancer Res 55:348–353
Strawhecker JM, Betz NA, Neades RY, Houser W, Pelling JC (1989) Binding of the 97 kDa glucocorticoid receptor to the 5' upstream flanking region of the mouse c-Ha-ras oncogene. Oncogene 4:1317–1322
Strähle U, Schmid W, Schütz G (1988) Synergistic action of the glucocorticoid receptor with transcription factors. EMBO J 7:3389–3395
Strähle U, Schmidt A, Kelsey G, Stewart AF, Cole TJ, Schmid W, Schütz G (1992) At least three promoters direct expression of the mouse glucocorticoid receptor gene. Proc Natl Acad Sci USA 89:6731–6735
Strobl JS, Van Eys GJJM, Thompson EB (1989) Dexamethasone control of growth hormone mRNA levels in GH3 pituitary cells is cycloheximide-sensitive and primarily posttranscriptional. Mol Cell Endocrinol 66:71–82
Strömstedt P-E, Poellinger L, Gustafsson J-Å, Carlstedt-Duke J (1991) The glucocorticoid receptor binds to a sequence overlapping the TATA box of the human osteocalcin promoter: a potential mechanism for negative regulation. Mol Cell Biol 11:3379–3383

Su Y, Pitot HC (1992) Location and characterization of multiple glucocorticoid-responsive elements in the rat serine dehydratase gene. Arch Biochem Biophys 297:239–243

Subramaniam M, Colvard D, Keeting PE, Rasmussen K, Riggs BL, Spelsberg TC (1992) Glucocorticoid regulation of alkaline phosphatase, osteocalcin, and proto-oncogenes in normal human osteoblast-like cells. J Cell Biochem 50:411–424

Suh D-S, Ooi GT, Rechler MM (1994) Identification of cis-elements mediating the stimulation of rat insulin-like growth factor-binding protein-1 promoter activity by dexamethasone, cyclic adenosine 3',5'-monophosphate, and phorbol esters, and inhibition by insulin. Mol Endocrinol 8:794–805

Suto N, Ikura K, Sasaki R (1993) Expression induced by interleukin-6 of tissue-type transglutaminase in human hepatoblastoma HepG2 cells. J Biol Chem 268:7469–7473

Suwanichkul A, Allander SV, Morris SL, Powell DR (1994) Glucocorticoids and insulin regulate expression of the human gene for insulin-like growth factor-binding protein-1 through proximal promoter elements. J Biol Chem 269:30835–30841

Suzuki K, Tatsumi H, Satoh S, Senda T, Nakata T, Fujii J, Taniguchi N (1993) Manganese-superoxide dismutase in endothelial cells: localization and mechanism of induction. Am J Physiol 265:H1173–H1178

Suzuki N, Kobayashi M, Sakata K, Suzuki T, Hosoya T (1991) Synergistic stimulatory effect of glucocorticoid, EGF and insulin on the synthesis of ribosomal RNA and phosphorylation of nucleolin in primary cultured rat hepatocytes. Biochim Biophys Acta Mol Cell Res 1092:367–375

Suzuki N, Suzuki T, Uchida A, Thompson EA, Hosoya T (1992) Effect of dexamethasone on nuclear casein kinase II activity and phosphorylation of nucleolin in lymphosarcoma P1798 cells. J Steroid Biochem Mol Biol 42:305–312

Szapary D, Oshima H, Simons SS Jr (1993) A new cis-acting element involved in tissue-selective glucocorticoid inducibility of tyrosine aminotransferase gene expression. Mol Endocrinol 7:941–952

Tai P-KK, Albers MW, Chang H, Faber LE, Schreiber SL (1992) Association of a 59-kilodalton immunophilin with the glucocorticoid receptor complex. Science 256:1315–1318

Takahashi H, Iizuka H (1991) Regulation of β2-adrenergic receptors in keratinocytes: glucocorticoids increase steady-state levels of receptor mRNA in foetal rat keratinizing epidermal cells (FRSK cells). Br J Dermatol 124:341–347

Takimoto K, Fomina AF, Gealy R, Trimmer JS, Levitan ES (1993) Dexamethasone rapidly induces Kv1.5 K+ channel gene transcription and expression in clonal pituitary cells. Neuron 11:359–369

Tata JR, Baker BS, Machuca I, Rabelo EML, Yamauchi K (1993) Autoinduction of nuclear receptor genes and its significance. J Steroid Biochem Mol Biol 46:105–119

Thompson EB (1994) Apoptosis and steroid hormones. Mol Endocrinol 8:665–673

Thorens B, Mermod J-J, Vasalli P (1987) Phagocytosis and inflammatory stimuli induce GM-CSF mRNA in macrophages through posttranscriptional regulation. Cell 48:671–679

Thulasi R, Harbour DV, Thompson EB (1993) Suppression of c-myc is a critical step in glucocorticoid-induced human leukemic cell lysis. J Biol Chem 268:18306–18312

Tobler A, Meier R, Seitz M, Dewald B, Baggiolini M, Fey MF (1992) Glucocorticoids downregulate gene expression of GM-CSF, NAP-1/IL-8, and IL-6, but not of M-CSF in human fibroblasts. Blood 79:45–51

Touray M, Ryan F, Saurer S, Martin F, Jaggi R (1991) Mos-induced inhibition of glucocorticoid receptor function is mediated by Fos. Oncogene 6:211–217

Trapp T, Rupprecht R, Castren M, Reul JM, Holsboer F (1994) Heterodimerization between mineralocorticoid and glucocorticoid receptor: a new principle of glucocorticoid action in the CNS. Neuron 13:1457–1462

Tremblay Y, Tretjakoff I, Peterson A, Antakly T, Zhang CX, Drouin J (1988) Pituitary-specific expression and glucocorticoid regulation of a proopiomelanocortin fusion gene in transgenic mice. Proc Natl Acad Sci USA 85:8890–8894

Trudel C, Couët J, Martel C, Labrie C, Labrie F (1991) Regulation of adrenal 3β-hydroxysteroid dehydrogenase/Delta5-Delta4-isomerase expression and activity by adrenocorticotropin and corticosterone in the rat. Endocrinology 129:2077–2084

Tsai SY, Carlstedt-Duke J, Weigel NL, Dahlman K, Gustafsson J-Å, Tsai M-J, O'Malley BW (1988) Molecular interactions of steroid hormone receptor with its enhancer element: evidence for receptor dimer formation. Cell 55:361–369

Tsai SY, Tsai M-J, O'Malley BW (1989) Cooperative binding of steroid hormone receptors contributes to transcriptional synergism at target enhancer elements. Cell 57:443–448

Tuncer AM, Ozsoylu S, Ersoy F, Yeniay I (1991) The effects of glucocorticoids on lymphocyte numbers. Immunol Today 12:207

Tuomela T, Miettinen P, Pesonen K, Viinikka L, Perheentupa J (1990) Epidermal growth factor in mice: effects of estradiol, testosterone and dexamethasone. Acta Endocrinol (Copenh) 123:211–217

Tur-Kaspa R (1986) Hepatitis B virus DNA contains a glucocorticoid-responsive element. Proc Natl Acad Sci USA 83:1627–1631

Turnbow MA, Keller SR, Rice KM, Garner CW (1994) Dexamethasone down-regulation of insulin receptor substrate-1 in 3T3-L1 adipocytes. J Biol Chem 269:2516–2520

Tverberg LA, Russo AF (1992) Cell-specific glucocorticoid repression of calcitonin/calcitonin gene-related peptide transcription. Localization to an 18-base pair basal enhancer element. J Biol Chem 267:17567–17573

Uchida C, Funai T, Oda T, Ohbayashi K, Ichiyama A (1994) Regulation by glucagon of serine: pyruvate/alanine: glyoxylate aminotransferase gene expression of cultured rat hepatocytes. J Biol Chem 269:8849–8856

Umlauf SW, Beverly B, Kang S-M, Brorson K, Tran A-C, Schwartz RH (1993) Molecular regulation of the IL-2 gene: rheostatic control of the immune system. Immunol Rev 133:177–197

Uno S, Guo D-F, Nakajima M, Ohi H, Imada T, Hiramatsu R, Nakakubo H, Nakamura N, Inagami T (1994) Glucocorticoid induction of rat angiotensin II type 1A receptor gene promoter. Biochem Biophys Res Commun 204:210–215

Unterman TG, Lacson RG, McGary E, Whalen C, Purple C, Goswami RG (1992) Cloning of the rat insulin-like growth factor binding protein-1 gene and analysis of its 5' promoter region. Biochem Biophys Res Commun 185:993–999

Vacca A, Felli MP, Farina AR, Martinotti S, Maroder M, Screpanti I, Meco D, Petrangeli E, Frati L, Gulino A (1992) Glucocorticoid receptor-mediated suppression of the interleukin 2 gene expression through impairment of the cooperativity between nuclear factor of activated T cells and AP-1 enhancer elements. J Exp Med 175:637–646

Vamvakopoulos NC, Chrousos GP (1993) Regulated activity of the distal promoter-like element of the human corticotropin-releasing hormone gene and secondary structural features of its corresponding transcripts. Mol Cell Endocrinol 94:73–78

Van de Stolpe A, Caldenhoven E, Stade BG, Koenderman L, Raaijmakers JAM, Johnson JP, Van der Saag PT (1994) 12-O-tetradecanoylphorbol-13-acetate- and tumor necrosis factor α-mediated induction of intercellular adhesion molecule-1 is inhibited by dexamethasone. Functional analysis of the human intercellular adhesion molecule-1 promoter. J Biol Chem 269:6185–6192

Van Itallie CM (1990) Thyroid hormone and dexamethasone increase the levels of a messenger ribonucleic acid for a mitochondrially encoded subunit but not for a nuclear-encoded subunit of cytochrome c oxidase. Endocrinology 127:55–62

Van Zonneveld A-J, Curriden SA, Loskutoff DJ (1988) Type 1 plasminogen activator inhibitor gene: functional analysis and glucocorticoid regulation of its promoter. Proc Natl Acad Sci USA 85:5525–5529

Van LP, Spengler DH, Holsboer F (1990) Glucocorticoid repression of 3',5'-cyclic-adenosine monophosphate-dependent human corticotropin-releasing-hormone gene promoter activity in a transfected mouse anterior pituitary cell line. Endocrinology 127:1412–1418

Vannice JL, Taylor JM, Ringold GM (1984) Glucocorticoid-mediated induction of alpha1-acid glycoprotein: evidence for hormone-regulated RNA processing. Proc Natl Acad Sci USA 81:4241–4245

Vardimon L, Fox LL, Degenstein L, Moscona AA (1988) Cell contacts are required for induction by cortisol of glutamine synthetase gene transcription in the retina. Proc Natl Acad Sci USA 85:5981–5985

Vaux DL, Haecker G, Strasser A (1994) An evolutionary perspective on apoptosis. Cell 76:777–779

Velasco S, Tarlow M, Olsen K, Shay JW, McCracken GH Jr, Nisen PD (1991) Temperature-dependent modulation of lipopolysaccharide-induced interleukin-1β and tumor necrosis factor α expression in cultured human astroglial cells by dexamethasone and indomethacin. J Clin Invest 87:1674–1680

Verbeeck MAE, Sutanto W, Burbach JPH (1991) Regulation of vasopressin messenger RNA levels in the small cell lung carcinoma cell line GLC-8: interactions between glucocorticoids and second messengers. Mol Endocrinol 5:795–801

Verdi JM, Campagnoni AT (1990) Translational regulation by steroids. Identification of a steroid modulatory element in the 5'-untranslated region of the myelin basic protein messenger RNA. J Biol Chem 265:20314–20320

Verdi JM, Kampf K, Campagnoni AT (1989) Translational regulation of myelin protein synthesis by steroids. J Neurochem 52:321–324

Villiger PM, Terkeltaub R, Lotz M (1992) Monocyte chemoattractant protein-1 (MCP-1) expression in human articular cartilage. Induction by peptide regulatory factors and differential effects of dexamethasone and retinoic acid. J Clin Invest 90:488–496

Vivanco MDM, Johnson R, Galante PE, Hanahan D, Yamamoto KR (1995) A transition in transcriptional activation by the glucocorticoid and retinoic acid receptors at the tumor stage of dermal fibrosarcoma development. EMBO J 14:2217–2228

Von Knebel Doeberitz M, Koch S, Drzonek H, Hausen HZ (1990) Glucocorticoid hormones reduce the expression of major histocompatibility class I antigens on human epithelial cells. Eur J Immunol 20:35–40

Von Knebel Doeberitz M, Bauknecht T, Bartsch D, Zur Hausen H (1991) Influence of chromosomal integration on glucocorticoid-regulated transcription of growth-stimulating papillomavirus genes E6 and E7 in cervical carcinoma cells. Proc Natl Acad Sci USA 88:1411–1415

Von Knebel Doeberitz M, Rittmuller C, Aengeneyndt F, Jansen-Durr P, Spitkovsky D (1994) Reversible repression of papillomavirus oncogene expression in cervical carcinoma cells: consequences for the phenotype and E6-p53 and E7-pRB interactions. J Virol 68:2811–2821

Vrtovsnik F, Jourdain M, Cherqui G, Lefebvre J, Friedlander G (1994) Glucocorticoid inhibition of Na-Pi cotransport in renal epithelial cells is mediated by protein kinase C. J Biol Chem 269:8872–8877

Wan Y-JY, Wang L, Wu T-CJ (1994) Dexamethasone increases the expression of retinoid X receptor genes in rat hepatoma cell lines. Lab Invest 70:547–552

Wand GS, Takiyyuddin M, O'Connor DT, Levine MA (1991) A proposed role for chromogranin A as a glucocorticoid-responsive autocrine inhibitor of proopiomelanocortin secretion. Endocrinology 128:1345–1351

Wang C, Leung A (1989) Glucocorticoids stimulate plasminogen activator production by rat granulosa cells. Endocrinology 124:1595–1601

Wang WW, Howells RD (1994) Sequence of the 5'-flancing region of the rat c-fos proto-oncogene. Gene 143:261–264

Wang XC, Smith TJ, Lau JTY (1990) Transcriptional regulation of the liver β-galactoside α2, 6-sialyltransferase by glucocorticoids. J Biol Chem 265:17849–17853

Warren MK, Vogel SN (1985) Opposing effects of glucocorticoids on interferon-gamma-induced murine macrophage Fc receptor and Ia antigen expression. J Immunol 134:2462

Watanabe Y, Lee S, Allison AC (1990) Control of the expression of a class II major histocompatibility gene (HLA-DR) in various human cell types: down-regulation by IL-1 but not by IL-6, prostaglandin E2, or glucocorticoids. Scand J Immunol 32:601–609

Webb ML, Mealey-Cavender JF, Jacob ST (1989) Glucocorticoid-induced stimulation of ribosomal gene transcription in rat hepatoma cells is mediated by modification of RNA polymerase I or an associated factor. Mol Endocrinol 3:1861–1868

Weber CM, Eke BC, Maines MD (1994) Corticosterone regulates heme oxygenase-2 and NO synthase transcription and protein expression in rat brain. J Neurochem 63:953–962

Webster MK, Goya L, Ge Y, Maiyar AC, Firestone GL (1993) Characterization of sgk, a novel member of the serine/threonine protein kinase gene family which is transcriptionally induced by glucocorticoids and serum. Mol Cell Biol 13:2031–2040

Webster NJG, Green S, Jin JR, Chambon P (1988) The hormone-binding domains of the estrogen and glucocorticoid receptors contain an inducible transcription activation function. Cell 54:199–207

Wehrenberg WB, Janowski BA, Piering AW, Culler F, Jones KL (1990) Glucocorticoids: potent inhibitors and stimulators of growth hormone secretion. Endocrinology 126:3200–3203

Welte T, Philipp S, Cairns C, Gustafsson J-Å, Doppler W (1993) Glucocorticoid receptor binding sites in the promoter region of milk protein genes. J Steroid Biochem Mol Biol 47:75–81

Wen L, Huang JK, Blackshear PJ (1989) Rat ornithine decarboxylase gene, nucleotide sequence, potential regulatory elements, and comparison to the mouse gene. J Biol Chem 264:9016–9021

Wertheim WA, Kunkel SL, Standiford TJ, Burdick MD, Becker FS, Wilke CA, Gilbert AR, Strieter RM (1993) Regulation of neutrophil-derived IL-8: the role of prostaglandin E2, dexamethasone, and IL-4. J Immunol 151:2166–2175

Westermann J, Pabst R (1990) Lymphocyte subsets in the blood: a diagnostic window on the lymphoid system. Immunol Today 11:406–410

Wheeler TT, Sadowski HB, Young DA (1994) Glucocorticoid and phorbol ester effects in 3T3-L1 fibroblasts suggest multiple and previously undescribed mechanisms of glucocorticoid receptor-AP-1 interaction. Mol Cell Endocrinol 104:29–38

Whitehouse BJ (1989) Lipocortins, mediators of the anti-inflammatory actions of corticosteroids. J Endocrinol 123:363–366

Williams P, Ratajczak T, Lee SC, Ringold GM (1991) AGP/EBP(LAP) expressed in rat hepatoma cells interacts with multiple promoter sites and is necessary for maximal glucocorticoid induction of the rat alpha-1 acid glycoprotein gene. Mol Cell Biol 11:4959–4965

Winter LA, Stewart MJ, Shean ML, Dong Y, Poellinger L, Okret S, Gustafsson J-Å, Duester G (1990) A hormone response element upstream from the human alcohol dehydrogenase gene ADH2 consists of three tandem glucocorticoid receptor binding sites. Gene 91:233–240

Wira CR, Rossoll RM (1991) Glucocorticoid regulation of the humoral immune system. Dexamethasone stimulation of secretory component in serum, saliva, and bile. Endocrinology 128:835–842

Wira CR, Sandoe CP, Steele MG (1990) Glucocorticoid regulation of the humoral immune system. I. In vivo effects of dexamethasone on IgA and IgG in serum and at mucosal surfaces. J Immunol 144:142–146

Wong DL, Lesage A, Siddall B, Funder JW (1992) Glucocorticoid regulation of phenylethanolamine N-methyltransferase in vivo. FASEB J 6:3310–3315

Wood AC, Waters CM, Garner A, Hickman JA (1994) Changes in c-myc expression and the kinetics of dexamethasone-induced programmed cell death (apoptosis) in human lymphoid leukaemia cells. Br J Cancer 69:663–669

Wright APH, McEwan IJ, Dahlman-Wright K, Gustafsson J-Å (1991) High level expression of the major transactivation domain of the human glucocorticoid receptor in yeast cells inhibits endogenous gene expression and cell growth. Mol Endocrinol 5:1366–1372

Wright K, Morgan ET (1991) Regulation of cytochrome P450IIC12 expression by interleukin-1α, interleukin-6, and dexamethasone. Mol Pharmacol 39:468–474

Wu CY, Fargeas C, Nakajima T, Delespesse G (1991a) Glucocorticoids suppress the production of interleukin 4 by human lymphocytes. Eur J Immunol 21:2645–2647

Wu CY, Sarfati M, Heusser C, Fournier S, Rubio-Trujillo M, Peleman R, Delespesse G (1991b) Glucocorticoids increase the synthesis of immunoglobulin E by interleukin 4-stimulated human lymphocytes. J Clin Invest 87:870–877

Wu T, Ikezono T, Angus CW, Shelhamer JH (1994) Characterization of the promoter for the human 85 kDa cytosolic phospholipase A2 gene. Nucleic Acids Res 22:5093–5098

Wyllie AH, Kerr JFR, Currie AR (1980) Cell death: the significance of apoptosis. Int Rev Cytol 68:251–307

Xing L, Quinn PG (1993) Involvement of 3',5'-cyclic adenosine monophosphate regulatory element binding protein (CREB) in both basal and hormone-mediated expression of the phosphoenolpyruvate carboxykinase (PEPCK) gene. Mol Endocrinol 7:1484–1494

Xu Z-X, Stenzel W, Sasic SM, Smart DA, Rooney SA (1993) Glucocorticoid regulation of fatty acid synthase gene expression in fetal rat lung. Am J Physiol Lung Cell Mol Physiol 265:L140–L147

Xun C-Q, Ensor CM, Tai H-H (1991) Regulation of synthesis and activity of NAD+-dependent 15-hydroxy-prostaglandin dehydrogenase (15-PGDH) by dexamethasone and phorbol ester in human erythroleukemia (HEL) cells. Biochem Biophys Res Commun 177:1258–1265

Yakovlev AG, De Bernardi MA, Fabrazzo M, Brooker G, Costa E, Mocchetti I (1990) Regulation of nerve growth factor receptor mRNA content by dexamethasone: in vitro and in vivo studies. Neurosci Lett 116:216–220

Yamamoto KR (1985) Steroid receptor regulated transcription of specific genes and gene networks. Annu Rev Genet 19:209–252

Yamanishi H, Tanaka A, Miyamoto K, Takeda M, Nishizawa Y, Koga M, Miyata M, Matsumoto K (1995) An essential role of androgen-induced growth factor in glu-

cocorticoid-dependent autocrine loop in Shionogi carcinoma 115 cells. J Steroid Biochem Mol Biol 52:49–53
Yanagimoto T, Itoh S, Muller-Enoch D, Kamataki T (1992) Mouse liver cytochrome P-450 (P-450IIIAM1): its cDNA cloning and inducibility by dexamethasone. Biochim Biophys Acta Gene Struct Expr 1130:329–332
Yang J, DeFranco DB (1994) Differential roles of heat shock protein 70 in the in vitro nuclear import of glucocorticoid receptor and simian virus 40 large tumor antigen. Mol Cell Biol 14:5088–5098
Yang J, Tashjian AH Jr (1993) Transcriptional regulation by dexamethasone of endogenous thyrotropin-releasing hormone receptor messenger ribonucleic acid in rat pituitary GH4C1 cells. Endocrinology 133:487–490
Yang-Yen H-F, Chambard J-C, Sun Y-L, Smeal T, Schmidt TJ, Drouin J, Karin M (1990) Transcriptional interference between c-Jun and the glucocorticoid receptor: mutual inhibition of DNA binding due to direct protein-protein interaction. Cell 62:1205–1215
Yewdell JW, Bennink JR (1990) The binary logic of antigen processing and presentation to T cells. Cell 62:203–206
Yoganathan T, Sells BH (1992) A downstream sequence of the rpL32 promoter competes with the glucocorticoid responsive element for a protein factor. Biochem Cell Biol 70:787–791
Yoshinaga SK, Peterson CL, Herskowitz I, Yamamoto KR (1992) Roles of SWI1, SWI2, and SWI3 proteins for transcriptional enhancement by steroid receptors. Science 258:1598–1604
Yuh Y-S, Thompson EB (1989) Glucocorticoid effect on oncogene/growth gene expression in human T lymphoblastic leukemic cell line CCRF-CEM. Specific c-myc mRNA suppression by dexamethasone. J Biol Chem 264:10904–10910
Yun CHC, Gurubhagavatula S, Levine SA, Montgomery JLM, Brant SR, Cohen ME, Cragoe E Jr, Pouyssegur J, Tse C-M, Donowitz M (1993) Glucocorticoid stimulation of ileal Na+ absorptive cell brush border Na+/H+ exchange and association with an increase in message for NHE-3, an epithelial Na+/H+ exchanger isoform. J Biol Chem 268:206–211
Zach TL, Hill LD, Herrman VA, Leuschen MP, Hostetter MK (1992) Effect of glucocorticoids on C3 gene expression by the A549 human pulmonary epithelial cell line. J Immunol 148:3964–3969
Zakar T, Hirst JJ, Mijovic JE, Olson DM (1995) Glucocorticoids stimulate the expression of prostaglandin endoperoxide H synthase-2 in amnion cells. Endocrinology 136:1610–1619
Zernik J, Kream B, Twarog K (1991) Tissue-specific and dexamethasone-inducible expression of alkaline phosphatase from alternative promoters of the rat bone/liver/kidney/placenta gene. Biochem Biophys Res Commun 176:1149–1156
Zhang H, Young AP (1991) A single upstream glucocorticoid response element juxtaposed to an AP1/ATF/CRE-like site renders the chicken glutamine synthetase gene hormonally inducible in transfected retina. J Biol Chem 266:24332–24338

Zhang H, Young AP (1993) Exogenous, but not endogenous, glucocorticoid receptor induces glutamine synthetase gene expression in early stage embryonic retina. J Biol Chem 268:2850–2856

Zhang J-X, Fasciotto BH, Cohn DV (1993) Dexamethasone and calcium interact in the regulation of parathormone and chromogranin-A secretion and messenger ribonucleic acid levels in parathyroid cells. Endocrinology 133:152–158

Zhang Z, Tarone G, Turner DC (1993) Expression of integrin alpha 1 beta 1 is regulated by nerve growth factor and dexamethasone in PC12 cells. Functional consequences for adhesion and neurite outgrowth. J Biol Chem 268:5557–5565

Zhao J-Y, Ikeguchi M, Eckersberg T, Kuo MT (1993) Modulation of multidrug resistance gene expression by dexamethasone in cultured hepatoma cells. Endocrinology 133:521–528

Zhao Y, Mendelson CR, Simpson ER (1995) Characterization of the sequences of the human CYP19 (aromatase) gene that mediate regulation by glucocorticoids in adipose stromal cells and fetal hepatocytes. Mol Endocrinol 9:340–349

Zhivotovsky B, Wade D, Nicotera P, Orrenius S (1994) Role of nucleases in apoptosis. Int Arch Allergy Immunol 105:333–338

Zhong H, Minneman KP (1993) Close reciprocal regulation of beta 1- and beta 2-adrenergic receptors by dexamethasone in C6 glioma cells: effects on catecholamine responsiveness. Mol Pharmacol 44:1085–1093

Zhu YS, Branch AD, Robertson HD, Inturrisi CE (1994) Cloning and characterization of hamster proenkephalin gene. DNA Cell Biol 13:25–35

Zong J, Ashraf J, Thompson EB (1990) The promoter and first, untranslated exon of the human glucocorticoid receptor gene are GC rich but lack consensus glucocorticoid receptor element sites. Mol Cell Biol 10:5580–5585

Editor-in charge: Professor H. Grunicke

Regulation of Glucose Transport into Skeletal Muscle

J.O. Holloszy and P.A. Hansen

Washington University School of Medicine, Department of Internal Medicine, 4566 Scott Ave., Campus Box 8113, St. Louis, MO 63110 USA

Contents

1
Brief Overview

Glucose transport across the plasma membrane is a major rate limiting step in glucose utilization and an important regulatory point in muscle carbohydrate metabolism (Ziel et al. 1988; Furler et al. 1991). Glucose entry into muscle occurs by facilitated diffusion; this process exhibits saturation kinetics and, in contrast to active transport, does not require ATP (Morgan and Whitfield 1973). Stimulation of glucose transport by insulin is largely mediated by translocation of glucose transporters from intracellular sites into the plasma membrane (Cushman and Wardzala 1980; Suzuki and Kono 1980; Wardzala and Jeanrenaud 1981; Simpson and Cushman 1986; Klip et al. 1987; Holman et al. 1990; Slot et al. 1991a; Lund et al. 1993). In striated muscle, glucose transport is also stimulated by contractile activity; this effect is independent of insulin action (Holloszy and Narahara 1965; Wallberg-Henriksson and Holloszy 1984, 1985; Ploug et al. 1984; Nesher et al. 1985). As with insulin, muscle contractions induce translocation of glucose transporters into the plasma membrane (Hirshman et al. 1988; Douen et al. 1989, 1990b; Gao et al. 1994b). The maximal effects of insulin and muscle contractions on glucose transport are additive (Nesher et al. 1985; Zorzano et al. 1986a; Wallberg-Henriksson et al. 1988; Constable et al. 1988; Henriksen et al. 1990). This finding, together with the evidence that it is possible to inhibit the effect of insulin without inhibiting that of contractions (Lee et al. 1995b; Yeh et al. 1995), indicates that these stimuli activate glucose transport by different pathways.

Two isoforms of the facilitative glucose transporter family, GLUT4 and GLUT1, are expressed in skeletal muscle (Douen et al. 1990b; Marette et al. 1992). In the unstimulated state, the GLUT4 isoform is located intracellularly, and is translocated to the sarcolemma by the actions of insulin and contractions (Douen et al. 1990b; Marette et al. 1992). The much less abundant GLUT1 isoform appears to reside primarily in the plasma membrane where it likely plays a role in basal, but not in insulin or contraction stimulated glucose transport (Douen et al. 1990b; Marette et al. 1992; Mueckler 1994). The different types of skeletal muscle fiber vary considerably in their content of GLUT4, and there is a close correlation between GLUT4 concentration and maximally stimulated glucose transport (Henriksen et al. 1990; Kern et al. 1990). Furthermore, skeletal muscle GLUT4 protein concentration and maximal glucose transport activity generally change in parallel in response to adaptive stimuli (Rodnick et al. 1992a;

Ren et al. 1993c, 1994), providing evidence that GLUT4 concentration determines the capacity for stimulated transport when the signaling pathways that activate glucose transport are intact. A major adaptive stimulus that induces an increase in muscle GLUT4 is exercise training (Rodnick et al. 1990, 1992a; Friedman et al. 1990; Ploug et al. 1990; Goodyear et al. 1992; Brozinick et al. 1993; Ren et al. 1994).

As the acute, insulin-independent, increase in glucose transport in response to exercise wears off, a second effect becomes evident. This effect of exercise consists of an increased sensitivity of the glucose transport process to insulin (Richter et al. 1982; Garetto et al. 1984; Cartee et al. 1989; Cartee and Holloszy 1990; Gulve et al. 1990) as well as to stimulation by other agents, including vanadate and hypoxia (Cartee and Holloszy 1990). Reversal of the increase in muscle insulin sensitivity after exercise occurs concomitantly with muscle glycogen repletion and can be speeded by carbohydrate feeding and slowed by keeping muscle glycogen low (Fell et al. 1982; Cartee et al. 1989; Gulve et al. 1990).

A number of other reviews of glucose transport in muscle, or relevant to this topic, that present different perspectives, may also be of interest to the readers of this article (Klip 1987; Carruthers 1990; Klip and Paquet 1990; Barnard and Youngren 1992).

2 Measurement of Glucose Transport

In numerous studies, information about glucose transport activity has been obtained using measurement of glucose uptake. In studies on perfused skeletal muscles, such as the perfused rat hindlimb, rates of glucose uptake have been determined by measuring arteriovenous glucose differences and perfusate flow rate. Since glucose uptake can theoretically be limited by either transport across the plasma membrane or by phosphorylation in the hexokinase reaction, it is essential to determine whether or not free glucose has accumulated in the muscle during the period of measurement. If it has, this indicates that phosphorylation rather than transport is rate limiting. In the great majority of muscle perfusion studies free glucose has not been measured. Although the assumption that transport is rate limiting under most physiological circumstances is probably correct, the results of such studies should be interpreted with caution, in view of the uncertainty regarding the rate limiting step. Glucose has also been used as the substrate in some studies of glucose transport activity in isolated, incubated muscle preparations. One approach has been to use glucose that is tritiated in the five position, because this tritium ends up in water during glycolysis, making it possible to determine the rate of glycolytic disposal by measuring production of tritiated water (Nesher et al. 1980). Some of the glucose entering the muscle is incorporated into glycogen, particularly if the muscles are well oxygenated. It is, therefore, also necessary to measure accumulation of tritiated glucose in glycogen. The sum of the glucose undergoing glycolysis, determined from production of tritiated water, plus the glucose converted to glycogen is a measure of the amount of glucose transported into the muscle (Nesher et al. 1980; Nesher et al. 1985). However, to determine whether transport rather than phosphorylation is rate limiting under the experimental conditions used, it is also necessary to measure free glucose, as has been done in the elegant studies of Nesher and coworkers (Nesher et al. 1980, 1985).

2.1 Glucose Analogs

Measurement of tritiated water, of tritium incorporation into glycogen and of free intracellular glucose is cumbersome and time consuming, and most investigators have used the simpler approach of measuring the intracellular accumulation of nonmetabolizable glucose analogs as a measure of glucose transport activity (Fig. 1). 3-O-Methylglucose (3MG) has generally been considered the gold standard for measurement of glucose transport, because, it is transported across the plasma membrane by the same glucose transport process as glucose but is not further metabolized, accumulating as free 3MG in the cell (Narahara and Özand 1963). There is, however, a major problem with the use of 3MG for measurement of glucose transport activity, because as the concentration of 3MG in the cytoplasm increases, the rate of its transport out of the cell progressively rises. If a muscle is incubated with 3MG for a sufficiently long time, a steady state is attained in which the rate of efflux of 3MG out of the muscle equals the rate of entry. This problem makes it difficult, and in some cases impossible, to obtain an accurate estimate of the rate of transport in muscles with a high capacity for glucose uptake under conditions in which transport activity is stimulated maximally. The reason for this is that it takes a few minutes, even in very thin muscles, for the extracellular space to equilibrate with the medium (Henriksen and Hol-

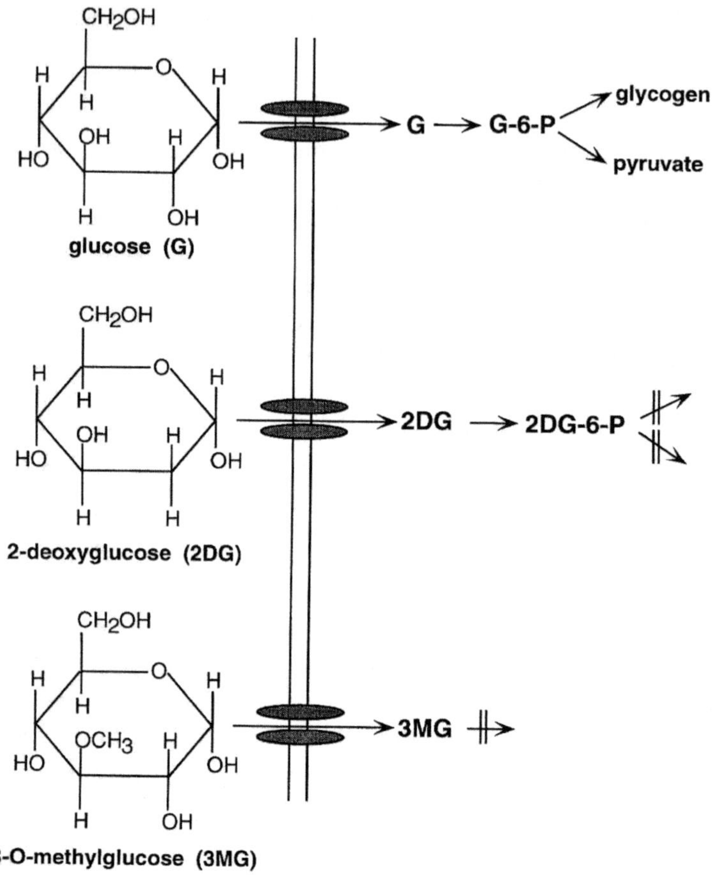

Fig. 1. Structure of glucose (*G*), 3-methylglucose (*3MG*), and 2-deoxyglucose (*2DG*). Glucose enters muscle cells by a facilitated diffusion process mediated by specific transporter proteins. After transport into skeletal muscle glucose is phosphorylated by hexokinase to form glucose-6-phosphate (*G-6-P*). G-6-P can then enter either the glycogen synthetic pathway or the glycolytic pathway. 2DG and 3MG enter muscle cells via the same glucose transporters as glucose. 3MG undergoes no intracellular metabolism. 2DG is phosphorylated by hexokinase to form 2-deoxyglucose-6-phosphate (*2DG-6-P*). 2DG-6-P is not metabolized to a significant extent over the time course employed in most glucose transport measurements

loszy 1991); at high rates of transport activity, the intracellular 3 MG concentration in the superficial fibers can increase sufficiently so that efflux becomes a problem at a time when the interstitial space around the deep fibers has not yet equilibrated with the medium. This problem can be circumvented with the use of the other commonly used nonmetabolizable glucose analog, 2-deoxyglucose (2DG). 2DG is transported into the cell via the glucose transporters and is phosphorylated by hexokinase, but not metabolized to any significant extent within the time frame of most glucose transport studies (Kipnis and Cori 1959). Since phosphorylated sugars cannot be transported out of the cell, efflux is not a problem, making it possible to measure 2DG-6-P accumulation for long periods of time. This has the major additional advantage that the initial few minutes, during which the interstitial space is equilibrating with the medium, represents a relatively small portion of the total period of sugar uptake measurement. This makes it possible to approximate initial rates of transport. Some investigators have viewed the use of 2DG with suspicion, because hexokinase is exquisitely sensitive to inhibition by glucose-6-P and it has been assumed that 2DG-6-P has a similar inhibitory effect. However, as clearly demonstrated by Kipnis and Cori (1959), 2DG-6-P is an extremely weak inhibitor of hexokinase. This finding was recently confirmed by Hansen et al. (1994), who showed that 2DG uptake in isolated rat skeletal muscles is linear until an extremely high intracellular 2DG-6-P concentration, in the range of 30 m*M*, is attained. Similarly, hexokinase activity in muscle homogenates is not significantly inhibited until the concentration of 2DG-6-P attains approx. 30 m*M* (Hansen et al. 1994). This is in contrast to glucose-6-P, which causes an approx. 50% inhibition of hexokinase when its concentration is only about 300 μ*M*. This very weak inhibitory effect of 2DG-6-P makes it possible to measure initial rates of transport for prolonged periods of time in skeletal muscle using 2DG.

2.2 Preparations For Studying Regulation of Muscle Glucose Transport

The two preparations that are in general use for studying glucose transport and its regulation in skeletal muscle are perfused rat or mouse hindlimbs and isolated rat or mouse skeletal muscles incubated in vitro. These preparations have been reviewed in detail by Bonen et al. (1994) and are therefore dealt with only briefly here. Another preparation that is used in a number of laboratories consists of skeletal muscle plasma membrane vesicles prepared by differential and sucrose gradient centrifugations of homogenates of disrupted muscles. Glucose transport can, of course, no longer be regulated in the sarcolemmal vesicles, and their use is based on the assumption that they retain the effects on transport induced by experimental treatments of the muscles prior to their disruption. As we have no experience with this preparation, it is not reviewed here; the reader is referred to papers by investigators who have expertise in its use (Grimditch et al. 1986; Hirshman et al. 1988; King et al. 1989; Ploug et al. 1992).

Another approach that has been used to evaluate the effects of exercise and insulin on muscle glucose transport, and to compare the responses of the different types of muscle fiber in rats, is the euglycemic clamp procedure combined with a bolus injection of radioactive 2DG (Kraegen et al. 1985; James et al. 1985). This procedure is suitable for evaluating the regulation of glucose uptake under physiological conditions by muscle in vivo in rats that are not anesthetized. However, it does not provide information regarding the regulation of muscle glucose transport per se, because the rate of glucose uptake in vivo is determined by glucose delivery, i.e., blood flow, number of capillaries per muscle fiber and degree of capillary dilation, as well as by permeability of the sarcolemma to glucose (Schultz et al. 1977; Baron et al. 1994; Hespel et al. 1995). It is, therefore, not possible to distinguish between effects of interventions on glucose delivery and glucose transport.

2.2.1
Perfused Muscle Preparations

Much information regarding glucose transport activity has been obtained using perfused muscle preparations. Perfused rat hindquarter or rat hindlimb, and to a lesser extent mouse hindquarter preparations, have been utilized in numerous studies and have provided much information regarding the effects of insulin, contractions and other agents on the regulation of glucose transport activity. The procedure commonly used for perfusing these preparations is one developed by Ruderman et al. (1971), and various modifications thereof. These preparations have also provided insights regarding changes in insulin sensitivity and responsiveness induced by exercise. However, the perfused hindlimb is less useful for comparative studies of glucose transport in different types of muscle fiber, for accurately quantifying adaptive changes in glucose transport capacity, or for studies of glucose transport kinetics. The reason for this is that at high rates of glucose uptake, delivery of glucose to the muscles becomes rate limiting (see Hespel et al. 1995). Glucose transport is a saturable process, and over a wide range of concentrations the rate of transport varies with the concentration of sugar to which the muscle fibers are exposed. At high rates of glucose uptake the concentration gradient from the blood to the sarcolemma increases, i.e., the concentration of sugar to which the muscle cells are exposed decreases if perfusate flow cannot be increased sufficiently to keep up with increases in glucose transport rate induced by insulin and/or contractions (Schultz et al. 1977; Hespel et al. 1995).

2.2.2 Isolated Muscles Incubated In Vitro

To accurately measure glucose transport activity, one should, ideally, determine the initial rate of 3MG or 2DG uptake, i.e., during the period when the rate of intracellular sugar accumulation is linear. While it is possible to measure initial rates of glucose transport in suspensions of adipocytes or cultured cells, this is not possible in intact skeletal muscles because of the time required for sugar to diffuse into the extracellular space. It is only after the lag phase, during which the extracellular space equilibrates with the medium and the concentration of the sugar around all of the muscle fibers increases to its final steady-state level, that the rate of sugar uptake becomes linear and accurately reflects transport activity (Young et al. 1986).

In the case of isolated muscles incubated in vitro, the thicker a muscle is, the greater is the diffusion distance to the deepest fibers, and the longer it takes for the extracellular space to equilibrate with the incubation medium. Oxygen diffusion can also become a problem with thick muscles incubated in vitro because of development of hypoxia of the fibers in the center of the muscle. Even in muscles from very young and small rats (60–70 g) the time required for the extracellular space to equilibrate with the incubation medium is approx. 12 min for the intact extensor digitorum longus and about 9 min for the intact soleus, compared to only about 4 min in the much thinner epitrochlearis muscle (Henriksen and Holloszy 1991). As a consequence, when intracellular 3MG accumulation is measured for 10 min in intact soleus and epitrochlearis muscles that are incubated with a maximally effective insulin concentration, it appears that these two muscles have similar rates of 3MG uptake, because diffusion is limiting in the thick soleus (Henriksen and Holloszy 1991). On the other hand, when the soleus muscle is split so as to give strips of approximately the same thinness as the epitrochlearis, it becomes evident that maximally insulin-stimulated 3MG transport is 2.5- to 3-fold higher in the soleus than in the epitrochlearis (Henriksen et al. 1990; Henriksen and Holloszy 1991). Muscle preparations that are suitable for measuring sugar transport in vitro include the epitrochlearis, flexor digitorum brevis, diaphragm, split soleus, and split extensor digitorum brevis, from rats in the 60–130 g weight range (Kipnis and Cori 1959; Henriksen et al. 1990). The same muscles from mice can also be used.

2.2.3
Muscle Cells in Culture

Our understanding of the regulation of glucose transport in skeletal muscle would undoubtedly be increasing more rapidly if a suitable cultured muscle cell preparation were available. Numerous studies of sugar transport have been conducted on a wide range of muscle cells ranging from BC3H-1 myocytes, to L6 and L8 cells, to primary cultures of human and rat skeletal muscle (Klip et al. 1982; Standaert et al. 1988; Walker et al. 1989; Calderhead et al. 1990; Mitsumoto and Klip 1992; Greco-Perotto et al. 1992; Sarabia et al. 1992; Bashan et al. 1993). These studies have provided much information regarding glucose transport that is of interest in itself, but often has limited relevance to the regulation of glucose transport in adult skeletal muscle. One major reason why muscle cells in culture are not representative of skeletal muscle is that the glucose transporter expressed in most cultured muscle cells is exclusively, or predominantly, GLUT1 (Calderhead et al. 1990; Greco-Perotto et al. 1992; Sarabia et al. 1992). As a consequence, these myocytes are only modestly responsive to insulin, with extremely high insulin concentrations resulting in 50%–100% increases in glucose transport activity.

The BC3H-1 myocyte has the characteristics of smooth muscle, in which the regulation of glucose transport is quite different than in skeletal muscle in a number of respects. Apparently BC3H-1 cells contain only the GLUT1 form of glucose transporter (Calderhead et al. 1990). In L8 myocytes, glucose transport also appears to be mediated entirely by the GLUT1 type of transporter, with GLUT4 concentration being too low to quantify (Greco-Perotto et al. 1992). In primary cultures of human muscle cells, the myotube stage does contain some GLUT4 protein, but at only about

14% as high a concentration as GLUT1 protein (Sarabia et al. 1992). Apparently, the only muscle cell line that expresses GLUT4 is the L6 myocyte (Mitsumoto and Klip 1992; Bashan et al. 1993). L6 myocytes appear to contain roughly equal amounts of GLUT1 and GLUT4 protein (Mitsumoto and Klip 1992). However, even the L6 myocyte differs from skeletal muscle in some aspects of glucose transport regulation. An interesting example of such a difference is the response to hypoxia. As with skeletal muscles (Cartee et al. 1991), L6 cells respond to conditions that require them to depend on anaerobic metabolism with an increase in glucose transport activity (Bashan et al. 1993). In skeletal muscle the increase in glucose transport activity in response to hypoxia is mediated by an increase in GLUT4 in the plasma membrane (Cartee et al. 1991). In contrast, the increase in glucose transport in response to hypoxia or inhibitors of respiration in L6 cells is mediated by increases in GLUT1 in the plasma membrane, as well as by increases in the total quantity of GLUT1 due to a stimulation of synthesis of GLUT1 protein (Bashan et al. 1993).

3 The Glucose Transport System

Glucose transport across the cell membrane is, under most physiological conditions, the major rate-limiting step in glucose utilization and an important site of metabolic regulation in striated muscle (Ziel et al. 1988; Furler et al. 1991).

3.1 Physiological Characteristics of the Glucose Transport Process

Glucose entry into muscle and various other cells occurs by a process that has been variously called mediated transport, passive transport, or facilitated diffusion. This process is distinguished from simple diffusion in that it exhibits saturation kinetics, substrate stereospecificity, competitive inhibition, and countertransport. The difference between passive and active transport is that the passive process does not use ATP (Morgan and Whitfield 1973).

Glucose entry into striated muscle follows saturation kinetics of the Michaelis-Menten type. This is evidenced by the finding in skeletal (Narahara et al. 1960; Narahara and Özand 1963; Holloszy and Narahara 1965;

Nesher et al. 1985) and heart (Morgan et al. 1961; Post et al. 1961) muscle that the rate of entry of glucose approaches a maximum as the concentration of sugar to which the muscle is exposed is increased. This phenomenon provides evidence that glucose must bind to some membrane component that serves as a transporter. The transport process is highly selective for certain sugars. The transport site has a poor affinity for L-glucose, which does not penetrate muscle in significant amounts despite the fact that it is extremely close to D-glucose in its chemical and physical properties (Morgan et al. 1964). This stereospecificity of the glucose transporter, which led to design of the glucose analogues 3MG and 2DG, greatly strengthened the concept of a glucose binding site. Glucose and its analogues compete for transport (Park et al. 1959; Narahara and Özand 1963).

An early finding that was interpreted as evidence for a glucose transporter came from studies demonstrating countertransport of the nonmetabolized glucose analogue 3MG (Park et al. 1959; Morgan et al. 1964). In these experiments tissues were exposed to medium containing 3MG until a considerable rise in the intracellular concentration of this sugar occurred. At this point, a high concentration of glucose was added to the medium; this caused a rapid drop in intracellular 3MG concentration. Thus, 3MG was transported out of the cell despite the fact that extracellular sugar concentration was higher than the intracellular concentration (Park et al. 1959; Morgan et al. 1964). These investigators explained their findings by hypothesizing that the glucose transporter is mobile, and that a gradient for glucose-loaded transporters is established within the membrane from outside to inside. They further hypothesized that the transporter is accessible to sugars at only one side of the membrane at a time, and that as glucose is transported into the cell, empty transporters become available at the inner side of the membrane for 3MG to bind to, creating a gradient of 3MG-loaded carriers from inside to outside. They argued that if the transporter was an immobile channel or pore spanning the membrane and lined with glucose binding sites, glucose on the outside would compete with 3MG on the inside for transport, and inhibit rather than stimulate efflux of the nonmetabolizable sugar. Although this explanation seemed reasonable at that time, when nothing was known regarding the structure of transporter molecules in the plasma membrane, it is not tenable in light of current knowledge. As is discussed below, it is now necessary to explain the countertransport phenomenon in terms of pores through the plasma membrane with glucose binding sites on their inner surface, that are formed by the glucose transporters.

3.2
The Glucose Transporters

There has never been any real doubt that the plasma membrane component that transports glucose is a protein, as only proteins have the complexity of structure needed for distinguishing between such closely related molecules as D- and L-glucose. Early evidence that the glucose transporter is, in fact, a protein was provided by the finding that cytochalasin B, which interacts with certain proteins, inhibits glucose transport when present in low concentrations (Kohn and Clausen 1971). The binding of cytochalasin B was used extensively to quantify glucose transporters prior to the availability of specific antibodies.

The first glucose transporter to be isolated was obtained from human erythrocytes. Kasahara and Hinkle (1977) reconstituted purified erythrocyte membrane proteins into artificial lipid bilayers and measured glucose transport activity, while Baldwin et al. (1979) used cytochalasin B binding to identify the erythrocyte glucose transporter in their purification procedure. The next major advance, made possible by the preparation of antibodies against the erythrocyte glucose transporter, was the expression cloning of the transporter from human HepG2 cells and deduction of its amino acid sequence (Mueckler et al. 1985). The erythrocyte-type glucose transporter, which has now been termed GLUT1, is present in high concentrations in cultured cells and brain tissue, but is low in striated muscle and adipocytes, the tissues in which glucose transport is regulated by insulin. This suggested that insulin stimulated glucose transport is mediated by another glucose transporter.

The existence of an insulin regulated glucose transporter that is specific to striated muscle and adipose tissue was demonstrated by James et al. (1988). These investigators partially purified intracellular vesicles that contain the glucose transporters that are translocated to the cell surface by insulin. They used vesicle preparations from rat adipocytes to immunize mice and were able to obtain a monoclonal antibody specific for the predominant glucose transporter in striated muscle and adipocytes. This isoform of the glucose transporter has been named GLUT4. Studies by several groups, who screened adipocyte, muscle and liver cDNA libraries for clones that cross-hybridized with a GLUT1 cDNA probe, confirmed that GLUT4 is expressed only in adipocytes and striated muscle (Birnbaum 1989; Charron et al. 1989; Fukumoto et al. 1989; James et al. 1989). The cross-hybridization approach has led to the isolation of other members of the facilitative sugar transporter protein family. Of the six isoforms that have been identified, four (GLUT1, GLUT2, GLUT3 and GLUT4) are glucose transporters that are expressed in a tissue-specific manner and differ in their kinetic and regulatory properties. GLUT2 is the primary glucose transporter in liver, absorptive epithelial cells and pancreatic cells, while GLUT3 is the major glucose transporter of neurons (Bell et al. 1993; Mueckler 1994). Based on the amino acid sequence for GLUT1 deduced by Mueckler et al. (1985), this protein consists of 13 segments that are hydrophilic, which alternate with 12 segments that are hydrophobic. Based on this analysis, Mueckler et al. (1985) proposed a model for GLUT1 in the plasma membrane of 12 membrane spanning helical, hydrophobic segments connected by intracytoplasmic and external surface loops, with the NH_2 and COOH termini located intracellularly. They proposed that each transmembrane segment is coiled into a helix and that in some of the transmembrane segments, in which hydrophilic and hydrophobic amino acid groups are appropriately arranged, the helices form pores whose inner surface binds glucose (Mueckler et al. 1985; Mueckler 1994). Available evidence suggests that, when exposed to glucose, the transporter binds the sugar on an outward facing binding site in the pore. This results in a change in conformation with closing of the end of the pore facing the extracellular space, transfer of the glucose to a binding site at the cytoplasmic end of the pore, which opens and releases glucose into the cytoplasm. The transporter then changes its conformation so that the cytoplasmic end closes, the outer end opens, with the glucose binding site again facing outward (Lienhard et al. 1992). If this scenario is correct, countertransport of 3MG out of muscles preloaded with this nonmetabolizable glucose analog could be explained by a reversal of the process made possible by release of glucose into the cytoplasm, with freeing-up of the binding site at the open cytoplasmic end of the pore to which 3MG can then bind. Most of the studies relating structures to function have involved GLUT1. However, GLUT4 and GLUT1 have a 68% sequence identity and GLUT4 is though to have a structure similar to that of GLUT1 (Bell et al. 1993; Mueckler 1994).

4
Regulation of Glucose Transport

Glucose transport into skeletal muscle under basal conditions is mediated primarily by the GLUT1 isoform of the glucose transporter (Mueckler 1994). GLUT1 is the predominant isoform in the plasma membrane under basal conditions, and the number of GLUT1 in the membrane does not change significantly in response to hormonal or other stimuli in skeletal muscle (Douen et al. 1990b; Mueckler 1994). Therefore, the major factor that determines glucose uptake under basal conditions is the glucose concentration in the interstitial fluid.

4.1
Acute Stimulation of Glucose Transport by Insulin

Glucose transport in striated muscles and adipocytes is rapidly stimulated by insulin and insulin mimetic agents. Insulin increases the maximal velocity of glucose transport (Narahara and Özand 1963; Rennie et al. 1983; Nesher et al. 1985; Hansen et al. 1995a). The V_{max} of glucose transport is determined by both the catatylic turnover number, i.e., the intrinsic activity of the glucose transporters, and by the number of glucose transporters in the plasma membrane. The effect of insulin on glucose transport could, therefore, be mediated by either an increase in glucose transporter intrinsic activity or by an increased number of glucose transporters at the cell surface. Although some investigators have reported that insulin also has an effect on the K_{m} of glucose transport, this finding has not been confirmed, and if changes in K_{m} do occur, they are probably small.

To determine whether insulin increases the number of glucose transporters in the plasma membrane, Wardzala et al. (1978) developed a cytochalasin B binding assay to quantify the number of glucose transporters and showed that insulin induces a large increase in glucose transporters in the plasma membrane of adipocytes. The next major advance in our understanding of the regulation of glucose transport by insulin came from studies of the subcellular distribution of glucose transporters. Using different methodologies, Cushman and coworkers (Cushman and Wardzala 1980; Karnieli et al. 1981) and Kono and coworkers (Suzuki and Kono 1980; Kono et al. 1982) showed that insulin treatment causes translocation of glucose transporters from a large intracellular pool into the plasma membrane (Simpson and Cushman 1986). These initial studies were performed on rat adipocytes, but subsequent studies have shown that insulin also stimulates glucose transporter translocation into the sarcolemma in striated muscle (Wardzala and Jeanrenaud 1981; Klip et al. 1987; Fushiki et al. 1989; Hirshman et al. 1990).

Cushman's group used D-glucose inhibitable, cytochalasin B binding to cellular fractions to quantify glucose transporters (Cushman and Wardzala 1980; Karnieli et al. 1981), while Kono's group (Suzuki and Kono 1980; Kono et al. 1982) extracted membrane proteins, reconstituted them into artificial liposomes and measured their glucose transport activity. Both groups showed that the increase in glucose transporters in the plasma membrane is paralleled by a decrease in transporter content of an intracellular "low-density microsomal" fraction. Studies using the cytochalasin B binding methodology have consistently shown a large discrepancy between the magnitude of the stimulatory effect of insulin on glucose transport activity and the increase in the number of glucose transporters in the plasma membrane. In skeletal muscle, insulin stimulates glucose transport 4- to 10-fold, while the increase in cytochalasin B-binding in the plasma membrane has been in the range of only 1.5- to 2-fold (Wardzala and Jeanrenaud 1981; Klip et al. 1987; Fushiki et al. 1989; Hirshman et al. 1990). Now that specific antibodies against the GLUT1 and GLUT4 isoforms are available, it has become evident that both GLUT1 and GLUT4 are expressed in skeletal muscle (Douen et al. 1990b; Marette et al. 1992). GLUT4 is the major isoform in muscle, and is responsible for the acute increase in glucose transport in response to insulin. The much less abundant GLUT1 isoform appears to reside constitutively in the plasma membrane and to play a role in basal, but not stimulated, glucose transport in skeletal muscle (Douen et al. 1990b; Mueckler 1994). GLUT4 transporters may have a higher intrinsic activity than GLUT1, so relating increases in cytochalasin B binding to the plasma membrane fraction to the increase in glucose transport activity following insulin treatment is not appropriate. However, even when the insulin-induced increase in glucose transport activity is compared to the increase in GLUT4 protein there is still a large discrepancy between the approx. fivefold increase in glucose transport activity and the only 50%–100% increase in GLUT4 in the plasma membrane (Douen et al. 1990b; Hirshman et al. 1990; Marette et al. 1992; Goodyear et al. 1992; Etgen et al. 1993b; Rosholt et al. 1994). Taken at face value these modest increases in the GLUT4 content of the sarcolemmal fraction could account for 20% or less of the increase in glucose transport activity induced by the maximal effect of insulin. This discrepancy has led some investigators to conclude that insulin also induces an increase in GLUT4 intrinsic activity.

Thus, there is currently some controversy regarding whether or not the increase in glucose transport activity induced by insulin is mediated entirely by translocation of GLUT4 into the sarcolemma or whether an increase in the intrinsic activity of the glucose transporters also plays a role. In the subcellular fractionation studies, the relatively small increase in GLUT4 in the plasma membrane fraction of skeletal muscle in response to insulin is due to the large amount of GLUT4 in the plasma membrane fraction isolated from the control, noninsulinized muscles. In fact, using this separation procedure, the GLUT4 content of the plasma membrane fraction in the basal state usually exceeds the increase in GLUT4 content induced by a maximal insulin stimulus (Douen et al. 1990b; Hirshman et al. 1990; Marette et al. 1992; Goodyear et al. 1992; Etgen et al. 1993b; Rosholt et al. 1994). In contrast, sophisticated immunocytochemical studies have shown that there are few GLUT4 in the sarcolemma of muscle in the basal state, and that insulin causes a marked increase in the number of GLUT4 in the sarcolemma (Rodnick et al. 1992b; Bornemann et al. 1992). It has also been demonstrated using immunocytochemistry that GLUT4 is essentially absent from the plasma membrane in adipocytes in the basal state, and that treatment with insulin causes about 40% of the total number of GLUT4 in the cell to move to the cell surface (Slot et al. 1991b). Furthermore, in contrast to the findings obtained by subcellular fractionation, experiments using a cell surface glucose transporter label have shown a close correlation between the increase in GLUT4 in the plasma membrane and the increase in glucose transport activity induced by insulin (Fig. 2). Using the exofacial GLUT photoaffinity label ATB-[2-^{3}H]-BMPA (Holman et al. 1990), Lund et al. (1993) and Wilson and Cushman (1994) were able to entirely account for a sixfold increase in 3MG transport in rat soleus muscle in response to insulin by a six- to eightfold rise in GLUT4 protein in the sarcolemma. In these studies, the labeled GLUT4 protein was immunoprecipitated using a GLUT4 specific antibody prior to quantitation (Lund et al. 1993; Wilson and Cushman 1994). This discrepancy between the subcellular fractionation and exofacial labeling studies will have to be resolved by further research. However, we suspect that the finding of an insulin-induced increase in intrinsic activity, i.e., a larger increase in transport than in transporter number in the subcellular fractionation studies, is an artifact caused by trapping of cytosolic GLUT4 in plasma

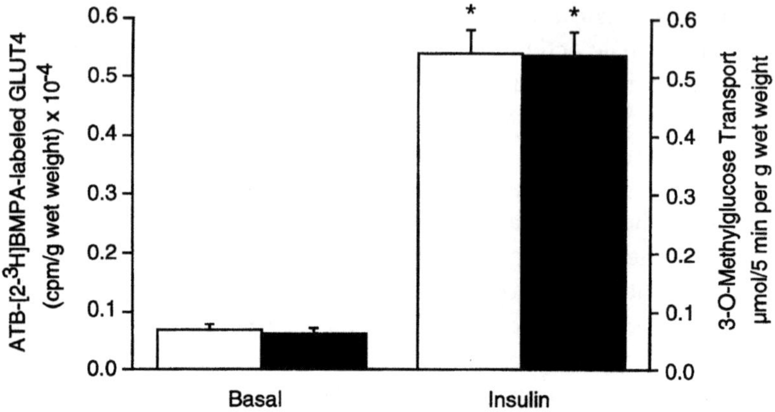

Fig. 2. Comparison of surface-accessible GLUT4 with glucose tranpsort activity. Isolated soleus muscles were equilibrated and incubated as basal or insulin-stimulated (20 mU/ml insulin). They were then used for transport measurements or were photolabeled and immunoprecipitated, and GLUT4 was quantified as described (Wilson and Cushman 1994). Surface-accessible GLUT4 (*open bars*) determined by photolabeling was eightfold greater in insulin-stimulated compared with basal (n=30 for each). Glucose transport activity (*solid bars*), determined using 3-*O*-[^{14}C]methylglucose, showed an 8.6-fold greater transport activity in insulin-stimulated than in basal muscles (n=24 for each condition). *, $p<0.001$ versus insulin-stimulated. (From Wilson and Cushman 1994)

membrane vesicles. These contaminating GLUT4 contribute to the total amount of GLUT4 in the vesicles without contributing to transport activity.

4.2 The Insulin Signaling Pathway

The first step in the insulin signaling pathway involves the binding of insulin to the extracellular α-subunits of insulin receptors on the surface of adipocytes and muscle cells. This insulin binding activates a tyrosine (Tyr) kinase in the intracellular tails of the β-subunits, which results in autophosphorylation of Tyr residues in the insulin receptors (for detailed reviews, see White and Kahn 1994; Lee and Pilch 1994). The principal insulin receptor substrate, IRS-1, is phosphorylated on multiple tyrosines by the activated insulin receptor (White et al. 1985; Thies et al. 1990; Rothenberg et al. 1991; Sun et al. 1991, 1992). The phosphorylated IRS-1 binds the Src homology-2 (SH2) domains of a number of signal transducing proteins through its approx. eight tyrosine phosphorylation sites. These include phosphatidylinositol (PI) 3-kinase, the GRB2/mSOS complex and SH-PTP-2. Stimulation of glucose transport is only one of the many effects of insulin, which also include regulation of gene expression and protein synthesis, activation of glycogen synthase, regulation of Na^+/K^+ATPase, amino acid uptake and fatty acid synthesis. Treatment of striated muscle and adipocytes with insulin results in activation of a serine/threonine protein kinase cascade that is now thought to be initiated by the IRS-1 bound GRB-2/mSOS complex which activates $p21^{ras}$. The activated Ras binds to Raf-1 and this activates the cascade which includes MAP kinase kinase and MAP kinase. As a consequence, in addition to tyrosine phosphorylation of insulin receptor substrates, a considerable number of proteins undergo serine/threonine phosphorylation in response to insulin. Treatment with insulin also brings about activation of protein serine/threonine phosphatases, resulting in dephosphorylation of other proteins (Lawrence 1992).

These multiple protein phosphorylations and dephosphorylations (Lawrence 1992) have greatly complicated elucidation of the signal transduction pathways that mediate the pleiotropic effects of insulin. However, methodological advances have recently made possible more rapid progress in this area. It has been shown that the serine/threonine kinase cascade initiated by the binding of the GRB-2/mSOS complex to IRS-1 is involved in regulating gene expression and protein synthesis, not activation of glucose transport (White and Kahn 1994). A major breakthrough in our understanding of the regulation of glucose transport occurred recently when it was shown that activation of PI 3-kinase is involved in the stimulation of glucose transport by insulin (see Fig. 3). The initial evidence that PI 3-kinase is involved in insulin action came from the finding that this enzyme is activated by insulin resulting in an increase in the concentrations of its products, PI 3-P and PI 3,4-P. PI 3-kinase exists as a 110-kDa catalytic subunit and an 85-kDa regulatory subunit that contains two SH2 domains. The binding of the SH2 domains of the 85-kDa subunit of PI 3-kinase to IRS-1, which occurs when IRS-1 is phosphorylated by the activated insulin receptor, is responsible for PI 3-kinase activation by insulin (Sun et al. 1991, 1992; Folli et al. 1992). The evidence that activation of PI 3-kinase is involved in the stimulation of glucose transport was provided by the finding that inhibition of PI 3-kinase blocks the stimulation of glucose transport by insulin in adipocytes (Okada et al. 1994; Lam et al. 1994) and skeletal muscle (Lee

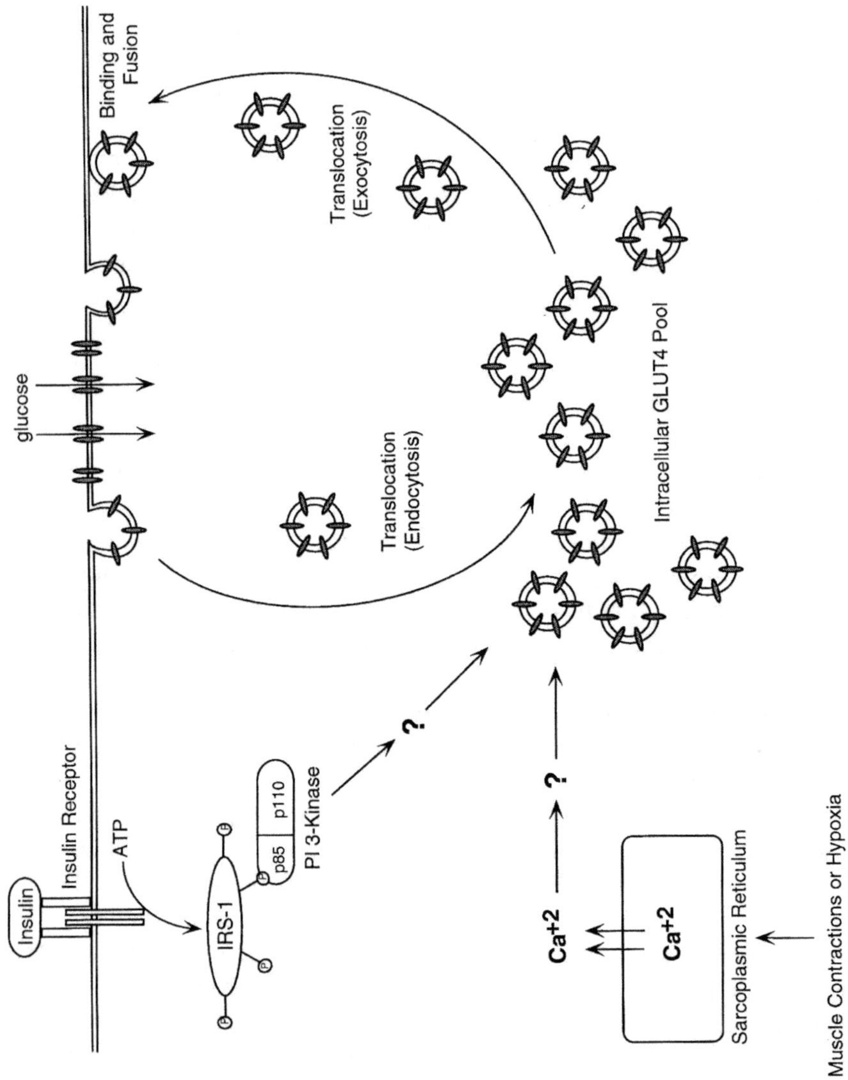

Fig. 3. Insulin and contractions or hypoxia stimulate GLUT4 translocation and glucose transport in skeletal muscle via distinct pathways. Detailed descriptions of the two pathways for stimulation of muscle glucose transport are provided in the text

et al. 1995b; Yeh et al. 1995). At the time this is written, the mechanism by which PI 3-kinase activation leads to GLUT4 translocation is not known.

4.3 Acute Stimulation of Glucose Transport by Contractile Activity

4.3.1 Early Studies of Contraction-Activated Sugar Uptake

In studies on normal and diabetic rats, Ingle et al. (1950) found that stimulation of leg muscles to contract caused a rapid fall in blood glucose levels. Goldstein et al. (1953) showed that muscle contraction lowers the concentration of injected D-galactose, D-xylose, and L-arabinose, but not of sorbitol (to which the muscle is not normally permeable), in the blood of eviscerated rats. Helmreich and Cori (1957) studied this phenomenon more directly by measuring the concentration of pentoses in plasma and muscles of nephrectomized rats and found that accumulation of pentoses within the muscle cells was increased when muscles were stimulated to contract. These results were confirmed by Dulin and Clark (1961). It is also well documented that exercise increases glucose removal from the blood in exercising humans and dogs. It was not clear from these studies whether the increased uptake of sugar was due to an insulinlike effect of muscle contractions on glucose transport or to increased delivery of sugar and insulin as a result of effects on the capillary bed and blood flow. A detailed study of the effects of muscle contractions on glucose transport was, therefore, undertaken using isolated frog sartorius muscles.

4.3.2 Initial Studies of the Effect of Contractions in Frog Muscle

The isolated frog sartorius muscle, was first used to study the effects of insulin on sugar transport by Narahara et al. (1960); it is thin enough to be oxygenated by diffusion, and is viable in vitro for long periods. It was found that contractile activity results in a large increase in 3MG transport into the frog sartorius, with a maximal stimulus resulting in an approx. 20-fold increase (Holloszy and Narahara 1965). When muscles were dissected out with the nerve intact and stimulated via the nerve, the effect on 3MG transport was the same as when the muscle was stimulated directly. At 19°C the temperature at which these studies were conducted, the increase in permeability of frog muscle to 3MG persists essentially unchanged for a number of hours after contractile activity stops. This makes it possible to stimulate a muscle to contract for as long as necessary to attain a full effect on glucose transport activity and then measure 3MG transport after stimulation is stopped. It was found that glucose transport rate increases progressively during contractile activity until a plateau is attained. The greater the number of contractions per minute, the greater is the increase in permeability, and the more rapidly the plateau level is attained. For frequencies in the range of 2.5–20 contractions/min, the increase in permeability is proportional to the frequency of contraction. If contractile activity is very brief and intense, permeability continues to increase for a period after stimulation is stopped. At 19°C the increase in permeability persists essentially unchanged for 5 h after stimulation and then progressively returns to very low baseline levels. However, if frog muscles are kept at 4°C, the increase in permeability persists unchanged for at least 24 h. Hypoxia also stimulates 3MG transport in the frog sartorius; however, the increase in transport activity does not begin until after about 60 min of hypoxia (Holloszy and Narahara 1965). In contrast, there is no lag period with contractile activity.

The transport of 3MG into frog sartorius muscles in which permeability has been increased by contractile activity follows saturation kinetics. When rates of sugar penetration at different concentrations of sugar are plotted according to the method of Lineweaver and Burke, the apparent K_m of sugar transport in stimulated muscles is not significantly different than in resting muscle, while the V_{max} (i.e., the theoretical maximum rate of transport when the transport process is completely saturated) is markedly increased (Holloszy and Narahara 1965). These findings were interpreted as indicating that the affinity of the transporters for sugar is unchanged, and that the greater rate of sugar transport after contractile activity is mediated by an increase in the number of transport sites that are operative in the plasma membrane (Holloszy and Narahara 1965).

To evaluate the effect of the amount of work performed on permeability, frog muscles were made to contract isotonically against either a 0.3- or 10-g load for various periods of time at the same contraction frequency. Although the muscles contracting against the 10-g load performed approximately seven times more work, produced significantly larger amounts of lactic acid and broke down more creatine phosphate than the lightly loaded muscles, the same changes in permeability to 3MG occurred in the two groups. Thus, it appears that changes in the amount of work performed at a given frequency of contraction do not modify the effect of stimulation on permeability. Furthermore, the change in permeability was not correlated with the rate of formation of lactate or with the intracellular concentrations of lactate or creatine phosphate (Holloszy and Narahara 1965).

4.3.3 Discrepancies between Certain Findings in Frog and Rat Muscles

Although the studies on frog sartorius muscle provided the first clear demonstration that contractile activity stimulates glucose transport, they also resulted in some long-lasting confusion. In the study on frog muscle, the maximal effect of insulin and contractions in combination was not significantly greater than that of either stimulus alone (Holloszy and Narahara 1965). This lack of additivity was interpreted as indicating that insulin and contractions act upon the same system and increase the same operative sites in the plasma membrane but by different mechanisms. However, subsequent studies on rat skeletal muscles have shown that the maximal effects of insulin and contractions on glucose transport are additive. It is not clear whether this discrepancy is due to a species difference or to the design of the experiment, as this question has not been reexamined in frog muscle using a different protocol. In a study on the rat epitrochlearis muscle, it was found that incubating the muscles with an extremely high concentration of insulin, 20 mU/ml, for 60 min prior to stimulating them to contract prevented the additive effect (Constable et al. 1988). This finding suggests that prior incubation with an unphysiologically high insulin concentration for a prolonged period can induce changes that interfere with subsequent translocation of GLUT4 into the plasma membrane in response to contractions. In the study on frog sartorius, the muscles were incubated with 2 mU/ml insulin for 3 h prior to electrical stimulation (Holloszy and Narahara 1965). Although only one-tenth as high as the insulin concentration used in the study on rat muscle (Constable et al. 1988) this is still a very high insulin level. Whether or not it was responsible for the lack of additivity is not clear, as the additive effect of insulin and contractions has not been reinvestigated in frog muscle using a lower insulin concentration or with a reversal of the sequence in which contractions and insulin are applied.

Another source of confusion that appears to be due to a species difference, was the finding that inhibition of protein synthesis with cycloheximide or puromycin completely blocked reversal of the increase in 3MG transport in frog sartorius muscle following contractions (Garthwaite and Holloszy 1982). This finding was taken as evidence that contractions activate glucose transport by a process that requires protein synthesis for its reversal. However, in a subsequent study on rat epitrochlearis muscle, inhibition of protein synthesis had no effect at all on the reversal of glucose transport activity after exercise (Young et al. 1987).

4.4 Regulation of Glucose Transport by Contractile Activity in Rat Skeletal Muscle

Both normal exercise, such as running or swimming, and stimulation of muscles to contract in situ or in vitro, result in an insulin-independent increase in glucose transport activity. This effect persists sufficiently long for it to be measured after exercise is stopped (Ivy and Holloszy 1981; Young et al. 1987; Constable et al. 1988; Gulve et al. 1990). Studies on muscles that had recovered in vivo for various time periods following a bout of exercise gave the impression that the increase in glucose transport activity induced by contractions wears off slowly (Young et al. 1983; Cartee et al. 1989). In these studies about 25%–50% of the increase in glucose transport activity seen shortly after exercise was still present 18 h postexercise in muscles of rats in which glycogen concentration was kept low by means of fasting or a carbohydrate-free diet (Young et al. 1983; Cartee et al. 1989).

In contrast to recovery in vivo, muscles incubated without insulin in vitro following exercise or electrical stimulation undergo a rapid reversal of the increase in glucose transport activity, which is complete within 1–3 h (Young et al. 1987; Ploug et al. 1987; Gulve et al. 1990). This rapid reversal occurs regardless of whether protein synthesis is inhibited or whether or not glycogen resynthesis occurs (Young et al. 1987). Inclusion of a very low concentration of insulin in the incubation medium results in a persistent increase in glucose transport activity in muscles incubated in vitro after exercise (Young et al. 1987; Gulve et al. 1990). During the first 60 min this persistent increase in glucose transport appears to be due to prevention by insulin of reversal of the exercise-induced increase in glucose transport activity. The persistent increase in permeability to glucose thereafter is due to an exercise-induced increase in muscle insulin sensitivity (Young et al. 1987; Gulve et al. 1990). Thus, the persistent effect of exercise in muscles allowed to recover in vivo and then studied in the absence of insulin in vitro (Young et al. 1983; Cartee et al. 1989) appears to be due to persistent effects of the insulin present in vivo on muscles in which insulin sensitivity is markedly increased, i.e., insulin still bound to the insulin receptors. In contrast, the acute stimulation of glucose transport by contractions occurs in the absence of insulin (Wallberg-Henriksson and Holloszy 1984, 1985; Ploug et al. 1984; Richter et al. 1985).

As with insulin, muscle contractions induce translocation of glucose transporters into the sarcolemma (Hirshman et al. 1988; Douen et al. 1989, 1990b; Goodyear et al. 1990a, b, 1991; Etgen et al. 1993b), and it is the GLUT4 isoform that moves to the cell surface in response to contractile activity (Douen et al. 1990b; Goodyear et al. 1991; Etgen et al. 1993b). Exercise does not appear to have any significant effect on the distribution of the GLUT1 isoform in skeletal muscle (Douen et al. 1990b). As in the case of insulin, there is currently some disagreement regarding whether contractile activity stimulates glucose transport entirely by means of translocation of GLUT4 into the sarcolemma, or whether there is also an increase in GLUT4 intrinsic activity. As discussed above in the section on insulin action, it appears likely that the smaller increase in GLUT4 content of the plasma membrane than in glucose transport activity in the subcellular fractionation studies is an artifact that is due to extensive contamination of the plasma membrane fraction with intracellular GLUT4 containing vesicles.

As shown in Fig. 4, the maximal effects of muscle contractions and insulin on glucose transport are additive (Nesher et al. 1985; Zorzano et al. 1986a; Wallberg-Henriksson et al. 1988; Constable et al. 1988; Henriksen et al. 1990; Ploug et al. 1992), providing evidence that these stimuli activate glucose transport by separate pathways. There are also two other lines of evidence that separate pathways are involved. One is that exercise-stimulated glucose transport is not impaired in muscles of obese Zucker rats (Brozinick et al. 1993; King et al. 1993) or in fat fed rats (Kusunoki et al. 1993) whose muscles are severely insulin resistant. The other is that it is possible to selectively inhibit stimulation of glucose transport by insulin without inhibiting the effect of contractions (Lee et al. 1995b; Yeh et al. 1995). One possible explanation for the finding that the maximal effects of contractions and insulin on glucose transport are additive is that there are two separate pools of GLUT4 in skeletal muscle, and that the GLUT4 from one pool are translocated into the sarcolemma by the action of insulin but not by contractions, while those in the other pool are susceptible to translocation by contractions but not insulin. Some support of

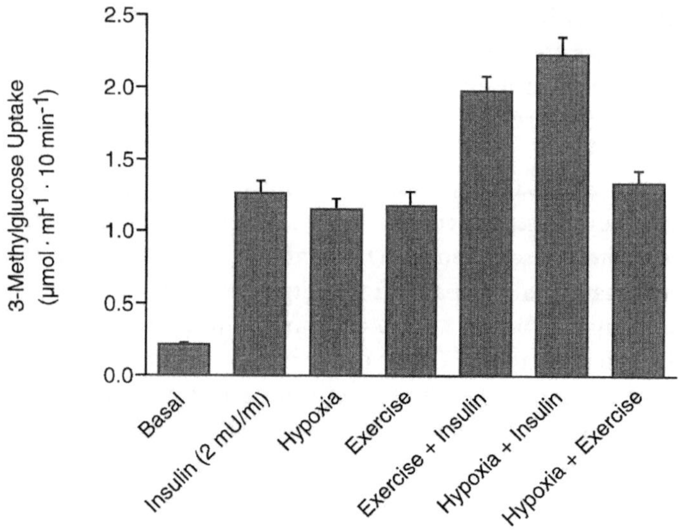

Fig. 4. Effects of insulin, hypoxia, or exercise, alone or in combination, on 3-*O*-[^{3}H]methylglucose (*3MG*) transport in the rat epitrochlearis muscle. Muscles were incubated at 35°C under oxygenated or hypoxic conditions for 60 min with or without insulin (2 mU/ml). In one experiment animals were exercised by swimming before their muscles were incubated with or without insulin. Muscles were then transferred to oxygenated medium containing 40 m*M* mannitol and insulin, if present previously, for 15 min at 29°C to remove glucose from the extracellular space and to reoxygenate hypoxic muscles. Glucose transport activity was then measured using 3MG. Values are means ± SE for 6–12 muscles/group. (From Cartee et al. 1991)

this possibility is provided by subcellular fractionation studies. Depending on the procedures used for disruption and subcellular fractionation of muscle, exercise either does, or does not, cause a decrease in GLUT4 content of the intracellular fraction that is partially depleted of GLUT4 by the action of insulin. In a number of studies, insulin caused a decrease in glucose transporters in the "intracellular fraction," while exercise did not, even though both stimuli caused an increase in GLUT4 in the plasma membrane fraction (Douen et al. 1989, 1990b; Broznick et al. 1994b), while in other studies, using a different fractionation procedure, decreases in the GLUT4 content of the intracellular fraction occurred in response to exercise (Hirshman et al. 1988; Fushiki et al. 1989; Gao et al. 1994b). These findings seem compatible with the existence of separate glucose transporter pools, one sensitive to insulin, the other to contractions.

If there are distinct glucose transporter pools recruitable by insulin or contractions, one would expect the maximal effects of insulin and contractions on GLUT4 translocation into the sarcolemma to be additive. This possibility has been investigated in at least four studies using subcellular fractionation procedures. The results have been conflicting, with three studies showing no additivity (Douen et al. 1990a; Goodyear et al. 1990b; Broznick et al. 1994b), and one showing an additive effect of contractions and insulin on GLUT4 translocation (Gao et al. 1994b). Resolution of this disagreement will probably require the use of a more quantitative procedure, such as photoaffinity labeling of the GLUT4 in the plasma membranes with ATB-[^{3}H]BMPA (Lund et al. 1993; Wilson and Cushman 1994).

The mechanism by which contractions activate GLUT4 translocation bypasses the steps in the insulin signaling pathway known to be involved in stimulation of glucose transport. This is evidenced by the finding that wortmannin, an inhibitor of PI 3-kinase, completely blocks stimulation of glucose transport by insulin without inhibiting the stimulation of glucose transport by contractions (Fig. 5; Lee et al. 1995b; Yeh et al. 1995).

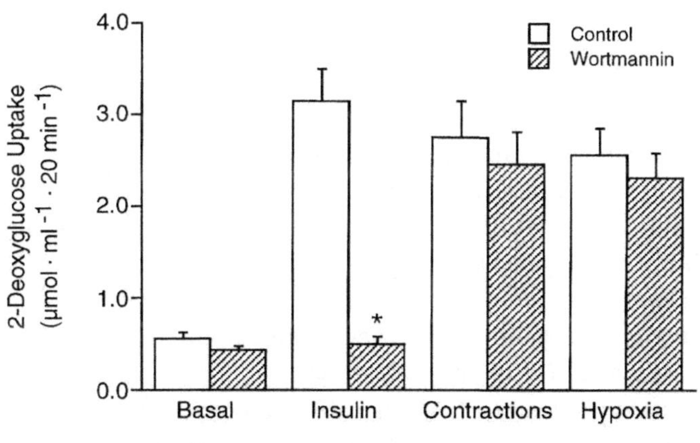

Fig. 5. Effect of wortmannin (2µM) on basal, insulin-, contraction- and hypoxia-stimulated 2-deoxy-[^{3}H]glucose (*2DG*) transport in skeletal muscle. Epitrochlearis muscles were incubated for 30 min at 30°C with or without 2 µ*M* wortmannin. Muscles were then incubated for 30 min with or without 2 mU/ml insulin, stimulated to contract in vitro (10 min), or incubated in hypoxic medium (90 min), all in the presence or absence of wortmannin. Glucose transport activity was then assessed by measuring the intracellular accumulation of 2DG at 30°C for 20 min. Values are means ± SE for 6–11 muscles/group. **, $p<0.001$ versus control. (From Lee et al. 1995b)

Furthermore, contractions do not cause phosphorylation of IRS-1 or activation of PI 3-kinase (Goodyear et al. 1995a). Some inhibitors, including polymyxin B (Henriksen et al. 1989b) and sphingosine (Gulve and Holloszy 1993) inhibit stimulation of glucose transport by both contractions and insulin, indicating that the two pathways have a late step in common. From what is known about the action of polymyxin B, this step may be the fusion of the GLUT4 containing vesicles with the plasma membrane (Cormont et al. 1992).

4.5 Stimulation of Glucose Transport in Muscle by Hypoxia

Hypoxia stimulates glucose transport in striated muscle (Randle and Smith 1958; Özand et al. 1962; Holloszy and Narahara 1965; Cartee et al. 1991). The increase in glucose transport activity occurs slowly in response to hypoxia; in rat epitrochlearis incubated in vitro at 35°C a maximal effect is usually attained in approx. 60 min (Cartee et al. 1991). This is in contrast to contractile activity which results in an almost immediate increase in glucose transport with a maximal effect occurring within 10 min in response to some stimulation protocols (Constable et al. 1988; Henriksen et al. 1990). As with insulin and exercise, hypoxia causes a translocation of GLUT4 into the sarcolemma (Cartee et al. 1991). The magnitude of the maximal effect of hypoxia on glucose transport is similar to that induced by exercise or in vitro muscle contractions. Furthermore, the effects of a maximal hypoxic stimulus and in vitro contractile activity in combination are no greater than that of either stimulus alone (Cartee et al. 1991). This lack of additivity suggests that hypoxia and muscle contractions stimulate glucose transport by the same mechanism. On the other hand, as shown in Fig. 4, the maximal effects of insulin and hypoxia on 3MG transport are completely additive (Cartee et al. 1991).

It has been hypothesized that the rate of glucose transport is somehow coupled to the requirement for ATP, as reflected in the decrease in phosphocreatine and the increases in inorganic phosphate and ADP concentrations under hypoxic conditions or during exercise (Walker et al. 1982; Katz et al. 1986). The decrease in high energy phosphates may initiate the events that lead to a stimulation of glucose transport by hypoxia. However, a direct linkage between energy demand and glucose transport is ruled out by the finding that the increase in 3MG transport induced by hypoxia is still present in muscles that have recovered for 15 min in oxygenated medium; by this time, phosphocreatine concentration has returned to the oxygenated control level (Cartee et al. 1991). Complete reversal of the increase in glucose transport activity occurs within 3 h in muscles incubated in oxygenated medium following 80 min of hypoxia.

4.6 Possible Role of Ca^{2+} in Contraction- and Hypoxia-Stimulated Glucose Transport

Early studies on frog sartorius muscle showed that caffeine contractures cause a large increase in glucose transport activity (Holloszy and Narahara 1967a). Caffeine causes contraction by directly releasing Ca^{2+} from the sarcoplasmic reticulum. Furthermore, increasing the work done per isotonic contraction does not influence the magnitude of the increase in glucose transport activity (Holloszy and Narahara 1965). These findings provided evidence that the increase in glucose transport activity is not directly dependent on plasma membrane depolarization or on the number of interactions between actin and myosin (i.e., the rate of ATP utilization). Rather, they suggest that the other fundamental event in the stimulation of muscle contraction, the increase in cytoplasmic Ca^{2+}, might mediate the increase in glucose transport. In studies designed to test this possibility it was found that, within limits, the greater the amount of Ca^{2+} entering a muscle during a potassium contracture, the greater the increase in glucose transport activity (Holloszy and Narahara 1967b). Valant and Erlij (1983) have confirmed the stimulation of glucose transport by potassium contractures. NO_3^- impairs the uptake of Ca^{2+} by the sarcoplasmic reticulum and, therefore, potentiates twitch tension (Ebashi 1976). Additional indirect evidence for a role of Ca^{2+} came from a study in which frog sartorius muscles were stimulated to contract while immersed in frog Ringer's solution in which NO_3^- was substituted for Cl^-. NO_3^- caused a potentiation of a submaximal effect of muscle contractions on glucose transport activity, even when the potentiation of twitch tension was prevented (Holloszy and Narahara 1967a).

Clausen and coworkers (Clausen et al. 1975; Sorenson et al. 1980) followed up these studies on frog muscle with a series of experiments in which they raised cytosolic Ca^{2+} concentration in rat soleus muscles in vitro using a variety of agents. Increases in the cytoplasmic Ca^{2+} concentration were documented by measuring the rate of $^{45}Ca^{2+}$ efflux across the plasma membrane from preloaded muscles. Increases in glucose transport activity were determined by measuring the rate of 3MG efflux from preloaded muscles. Treatment of muscles with the metabolic poisons 2,4-dinitrophenol, cyanide, or salicylate resulted in accelerated washout of $^{45}Ca^{2+}$ and 3MG from preloaded rat soleus muscles (Clausen et al. 1975; Sorenson et al. 1980). Other agents that produced a stimulation of 3MG washout that was preceded by, or coincided with, a rise in the washout of $^{45}Ca^{2+}$ from soleus muscles, were trypsin, *p*-chloromercuriphenylsulfonic acid, veratrine, H_2O_2, hyperosmolarity, caffeine, K^+-substituted buffer, and exposure to 0°C followed by warming to 30°C (Clausen et al. 1975; Sorenson et al. 1980). Many of these agents also caused muscle contraction (Clausen et al. 1975; Sorenson et al. 1980).

Further evidence that an increase in cytosolic Ca^{2+} results in stimulation of glucose transport in muscle was obtained fortuitously during a study involving phospholipase C. It had been proposed by Standaert et al. (1988), on the basis of studies on BC3H-1 myocytes, that phospholipase C stimulates glucose transport by generation of diacylglycerol from membrane phospholipids with activation of protein kinase C. In a study conducted to reevaluate this hypothesis it was found that the broadly active phospholipase C from *Clostridium perfringens* (PLC-Cp) stimulates glucose transport in a dose-dependent and cytochalasin B inhibitable manner (Henriksen et al. 1989a). Studies of the interactions among PLC-Cp, insulin, muscle contractions and hypoxia provided evidence that PLC-Cp stimulates glucose transport by the same pathway as exercise and hypoxia, not by the insulin activated pathway (Henriksen et al. 1989a). Additional experiments were, therefore, conducted to evaluate whether PLC-Cp causes an increase in cytosolic Ca^{2+} concentration, and whether an increase in cytosolic Ca^{2+} is involved in the stimulation of glucose transport by PLC-Cp. PLC-Cp caused a significant increase in the proportion of phosphorylase in the *a* form, providing evidence for an increase in cytosolic Ca^{2+} concentration in epitrochlearis muscle (Henriksen et al. 1989a). Additional evidence for an increase in cytoplasmic Ca^{2+} in muscle treated with PLC-Cp was provided by the finding of an increase in twitch tension. Omission of Ca^{2+} from the incubation medium resulted in inhibition of the stimulation of glucose transport by PLC-Cp (Henriksen et al. 1989a). This finding is in keeping with the evidence that PLC-Cp causes an influx of Ca^{2+} into sarcolemmal vesicles (Philipson et al. 1983). The Ca^{2+} channel blockers verapamil, diltiazem and nifedipine had little effect on the stimulation of 3MG transport by PLC-Cp, providing evidence that this agent increases Ca^{2+} influx independently of the voltage sensitive Ca^{2+} channels. Additional evidence that Ca^{2+} mediates the effect of PLC-Cp on glucose transport came from the finding that dantrolene, which inhibits Ca^{2+} release from the sarcoplasmic reticulum, completely blocked both the activation of phosphorylase and the stimulation of glucose transport activity by PLC-Cp (Henriksen et al. 1989a). An explanation for these findings could be that an influx of Ca^{2+} mediated by PLC-Cp induces Ca^{2+}-activated Ca^{2+} release from the SR (Ebashi 1976) resulting in an increase in cytoplasmic Ca^{2+} which initiates the events leading to an increase in glucose transport activity via the same pathway that is activated during stimulation of muscle contractions.

Hypoxia appears to stimulate muscle glucose transport via the same pathway as contractile activity (Cartee et al. 1991). Furthermore, hypoxia has been shown to increase Ca^{2+} concentration in heart muscle (Steenbergen et al. 1987). Hypoxia has, therefore, been used as another model to evaluate the possibility that an increase in cytoplasmic Ca^{2+} stimulates glucose transport in the absence of muscle contraction. Caffeine at a concentration that is too low (1.0 m*M*) to cause muscle contraction or an increase in glucose transport activity markedly potentiated the effect of a submaximal hypoxic stimulus on glucose transport activity in rat epitrochlearis muscle (Cartee et al. 1991). Similarly, substitution of $NaNO_3$ which impairs uptake of Ca^{2+} by the SR (Ebashi 1976), for NaCl in the incubation medium caused a large potentiation of the effect of a submaximal hypoxic stimulus. This substitution had no effect on glucose transport in oxygenated medium. A low concentration of dantrolene partially inhibited the effect of hypoxia on glucose transport. These findings provide additional support for the hypotheses that an increase in cytosolic Ca^{2+} concentration induces an increase in glucose transport activity in skeletal muscle and that Ca^{2+} plays a role in the stimulation of glucose transport by hypoxia. Another agent that causes release of Ca^{2+} from skeletal muscle sarcoplasmic reticulum is alkaline pH (Dettbarn and Palade 1991). Incubation of epitrochlearis at pH 8.6, which raised cytosolic pH to 8.2, induced an approx. threefold increase in glucose transport activity, which was completely blocked by 25 μ*M* dantrolene (Ren et al. 1993b).

The concept that the increase in sarcoplasmic Ca^{2+} during stimulation of muscle contraction activates a separate series of events that cause the increase in glucose transport activity, independent of contractions, has been difficult to test. One difficulty is that it is not possible to distinguish between the consequences of muscle contraction, such as changes in the concentrations of high energy phosphates (~P) and metabolites, and independent effects of the rise in cytosolic Ca^{2+} during contractile activity. Another problem is that altering the magnitude of the increase in cytoplasmic Ca^{2+} during muscle contractions also affects contractile force development or work output. To circumvent these problems, the effects of increases in cytoplasmic Ca^{2+} that are too low to cause muscle contractions were examined by Youn et al. (1991) using agents that release Ca^{2+} from the sarcoplasmic reticulum. The compound *N*-(6-aminohexyl)-5-chloro-1-napthalenesulfonamide (W-7), which has been shown to release Ca^{2+} from the SR (Palade 1987), caused a dose-dependent increase in tension in rat epitrochlearis muscle at concentrations of 200 μ*M* or greater, providing evidence for an increase in cytoplasmic Ca^{2+} (Youn et al. 1991). Concentrations of W-7 in a range that did not increase muscle tension (50–100 μ*M*), speeded the loss of $^{45}Ca^{2+}$ from preloaded muscles, providing evidence that these low concentrations of W-7 must have raised cytoplasmic Ca^{2+} to levels that were below the contraction threshold. Glucose transport activity increased sixfold in muscles incubated for 100 min with 50 μ*M* W-7, and eightfold in muscles incubated with 100 μ*M* W-7, despite no increase in muscle tension (Fig. 6).

ATP and creatine phosphate concentrations were unchanged in muscles incubated with 50 μ*M* W-7. Dantrolene, which inhibits Ca^{2+} release from the SR, prevented the effects of W-7 on both Ca^{2+} release and glucose transport activity (Youn et al. 1991). This effect of dantrolene occurred at a concentration of 5 μ*M*, which had no effect on insulin action (Youn et al. 1994a). 9-Aminoacridine, another inhibitor of Ca^{2+} release from the SR (Palade 1987), also blocked the stimulation of glucose transport activity by W-7 (Fig. 6; Youn et al. 1991, 1994a). W-7 is a calmodulin antagonist (Hidaka et al. 1981) and, in addition to inducing release of Ca^{2+} from the SR, inhibits a number of calcium-dependent protein kinases (Tanaka et al. 1982; Schatzman et al. 1983). The finding that dantrolene blocks both the increase in the rate of $^{45}Ca^{2+}$ efflux from preloaded muscles and the stimulation of sugar transport induced by W-7 provides evidence that W-7 stimulates glucose transport by releasing Ca^{2+} from the SR, not by inhibiting calmodulin-dependent enzymes. If stimulation of glucose transport by W-7 was mediated by inhibition of Ca^{2+}-calmodulin dependent enzymes rather than by Ca^{2+} release from the SR, dantrolene and 9-aminoacridine should, by lowering cytoplasmic Ca^{2+} concentration, potentiate the effect of W-7 on glucose transport instead of inhibiting it. The results of these experiments using W-7 provide strong evidence that an increase in cytoplasmic Ca^{2+} concentration that is too low to cause muscle con-

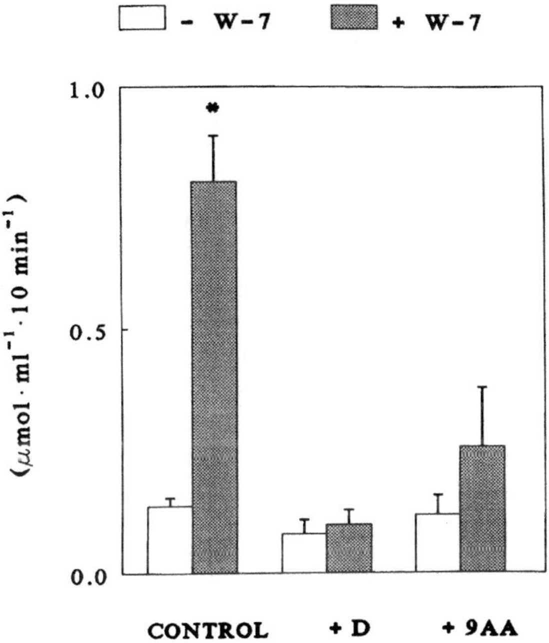

Fig. 6. Effects of dantrolene (*D*) and 9-aminoacridine (*9AA*) on 3-*O*-[^{3}H]methylglucose (3MG) transport without and with W-7. Muscles were incubated at 35°C for 100 min without or with W-7 (100 μ*M*) in the presence or absence of dantrolene (25 μ*M*) or 9-aminoacridine (100 μ*M*). Rate of 3-MG transport was then measured in absence of dantrolene, 9-aminoacridine, and W-7. Values are means ± SE for 5–11 muscles. *, p<0.001 versus other groups. (From Youn et al. 1991)

traction or a decrease in high energy phosphate concentration can bring about an increase in glucose transport activity in skeletal muscle. Support for this interpretation is provided by studies of the effect of caffeine. As with W-7 caffeine causes release of Ca^{2+} from the sarcoplasmic reticulum (Palade 1987). As might be expected, glucose transport activity is markedly increased by exposure to concentrations of caffeine that cause muscle contraction (Holloszy and Narahara 1967b; Youn et al. 1991). However, lower concentrations of caffeine that do not result in muscle contraction or a decrease in ~P also cause a significant stimulation of glucose transport activity (Youn et al. 1991). As with W-7, the increase in glucose transport induced by caffeine is completely blocked by dantrolene.

Studies of the interaction between Ca^{2+} and insulin in adipocytes have shown that a sustained elevation of cytosolic Ca^{2+} has no effect on basal glucose transport but significantly inhibits stimulation of glucose transport by insulin (Draznin et al. 1987, 1989; Reusch et al. 1991). In these studies cytosolic Ca^{2+} was raised by various means including the Ca^{2+} ionophore ionomycin (Draznin et al. 1987), plasma membrane depolarization with K^+ (Draznin et al. 1989; Reusch et al. 1991), and parathyroid hormone (Draznin et al. 1989; Reusch et al. 1991). Studies using the Ca^{2+} ionophore A-23187 (Bihler et al. 1980) and the Ca^{2+}-channel agonist BAY K 8644 (Westfall and Sayeed 1990) have provided evidence that an increase in cytoplasmic Ca^{2+} also inhibits insulin-stimulated glucose transport in skeletal muscle. These findings seem to be in conflict with the evidence that the effects of insulin and contractile activity on glucose transport are additive in contracting muscle, in which Ca^{2+} is being released into the cytoplasm from the sarcoplasmic reticulum with each stimulus (Nesher et al. 1985; Zorzano et al. 1986a). They also appear to argue against the concept that an increase in cytoplasmic Ca^{2+} stimulates glucose transport in skeletal muscle.

In a study undertaken to explain this discrepancy, Lee et al. (1995a) used the relatively specific Ca^{2+} ionophore ionomycin (Liu and Hermann 1978) to raise cytosolic Ca^{2+}. They found that continuous incubation of rat epitrochlearis and soleus muscles with ionomycin induces a significant inhibition of insulin stimulated glucose transport. The influx of Ca^{2+} induced by ionomycin was sufficiently great to result in an approx. 70% increase in % phosphorylase *a*, which is a good, physiological indicator of increases in cytosolic Ca^{2+} (Wilde et al. 1986). This inhibitory effect of ionomycin was prevented by lowering the Ca^{2+} concentration in the medium to a very low level, providing evidence that the inhibition is mediated by the Ca^{2+} influx induced by this ionophore. These findings are in keeping with the results of previous studies showing that sustained increases in cytosolic Ca^{2+} inhibit insulin stimulated glucose transport (Bihler et al. 1980; Draznin et al. 1987, 1989; Westfall and Sayeed 1990; Reusch et al. 1991). Lee et al. (1995a) also found that when basal glucose transport was measured in the presence of ionomycin, no stimulation was observed. However, when the measurement of glucose transport was preceded by a 15 min washout period without ionomycin, a significant stimulation of glucose transport activity by prior exposure to ionomycin was evident, providing further support for the concept that an increase in sarcoplasmic Ca^{2+} induces an increase in glucose transport activity in skeletal muscle (Lee et al. 1995a). These findings, together with a similar observation in an earlier study involving potassium contractures (Holloszy and Narahara 1967b), provide evidence that increased influx of Ca^{2+} into the sarcoplasm can, in addition to stimulating glucose transport, cause a temporary inhibition of both insulin and Ca^{2+} stimulated glucose transport.

In contrast to the finding with ionomycin and K^{+} that a recovery period is needed before the stimulation of glucose transport becomes evident, contractile activity results in an almost immediate increase in glucose transport that is readily measurable and additive to the effect of insulin during sustained contractile activity (Wallberg-Henriksson and Holloszy 1984; Nesher et al. 1985). A likely explanation for this difference is that the brief increases in cytoplasmic Ca^{2+} that trigger contraction are localized to a region adjacent to the sarcoplasmic reticulum (SR) so that the action of the Ca^{2+} released from the SR is limited to those Ca^{2+}-activated enzymes that are located on, or close to, the SR, such as actomyosin adenosinetriphosphatase and phosphorylase kinase. Apparently the Ca^{2+}-activated enzyme that initiates the process responsible for the inhibitory effect of Ca^{2+} on insulin-stimulated glucose transport is sufficiently distant from the SR that the Ca^{2+} concentration to which it is exposed during exercise does not reach the critical level. Along the same line, if the concept that Ca^{2+} mediates the stimulation of glucose transport by exercise is correct, the Ca^{2+}-binding protein responsible for activating the process by which exercise stimulates glucose transport must be located close to the SR.

An increase in the concentration of cytosolic Ca^{2+} is the signal for the regulation of a wide range of cellular processes (Rasmussen and Barrett 1984). Many of the actions of Ca^{2+} are mediated by activation of the Ca^{2+} binding regulatory protein calmodulin, which binds to and activates a number of enzymes in a Ca^{2+} dependent manner (Rasmussen and Barrett 1984; Klee et al. 1986). However, the finding that W-7, a potent calmodulin antagonist (Hidaka et al. 1981), stimulates glucose transport by raising cytosolic Ca^{2+}, provides evidence that stimulation of glucose transport by Ca^{2+} in skeletal muscle does not involve a calmodulin-dependent process. Among the actions of Ca^{2+} that do not appear to involve calmodulin are (a) the regulation of cytoplasmic structure and coupling of the cytomatrix to the plasma membrane by regulation of actin filament organization (Mooseker et al. 1986), (b) Ca^{2+}-dependent proteolysis by Ca^{2+}-dependent proteases, the calpains (DeMartino and Croall 1987; Mellgren 1987), and (c) the stimulation of exocytosis (Knight 1986). Any one of these processes could possibly be responsible for initiating the events leading to an increase in glucose transport. However, the possibility that an increase in Ca^{2+} acts by triggering exocytosis of glucose transporters is of particular interest because stimulation of glucose transport by contractions and hypoxia in striated muscle is, at least in part, due to translocation of GLUT4 glucose transporters (Douen et al. 1990b; Klip and Paquet 1990; Cartee et al. 1991), a process in which internal glucose transporter-containing vesicles are exocytosed to the plasma membrane (Cushman and Wardzala 1980; Suzuki and Kono 1980; Simpson and Cushman 1986; James et al. 1987; Jhun et al. 1992).

5 GLUT4 Protein Translocation is an Exocytosis-like Process

Morphological studies employing immunocytochemical techniques have shown that the intracellular GLUT4 are components of small vesicles that are located in the trans Golgi reticulum, scattered throughout the cytosol in clusters, and adjacent to the cell membrane (Slot et al. 1991a; Rodnick et al. 1992b). Kinetic studies using photolabeling of GLUT4 in adipocytes have provided evidence that GLUT4 cycles between the cytosol and the plasma membrane in the basal state, i.e., constitutively, and in response to insulin (Jhun et al. 1992; Yang and Holman 1993). During this process, the intracellular GLUT4 containing vesicles undergo exocytosis, moving to and fusing with the plasma membrane, and then eventually cycle back into the cell via an endocytic pathway. Insulin induces an increase in the number of GLUT4 at the cell surface both by stimulating exocytosis and decreasing the rate of endocytosis (Fig. 3; Jhun et al. 1992; Yang and Holman 1993).

5.1 Exocytosis

In the process of exocytosis, secretory vesicles in the interior of the cell are captured by molecular motors that transport them to the cell surface where they fuse with the plasma membrane (Almers 1990). The vesicle membrane thus becomes a component of the cell membrane, and material contained in the vesicle diffuses out of the cell. Exocytosis functions either to insert new components, such as GLUT4, into the cell membrane, or to export material contained in the vesicles. In regulated exocytosis, a cytosolic signaling factor causes fusion of the vesicles and cell membrane, probably by causing conformational changes in proteins on the vesicles and on the cell membrane that enable fusion to occur. Agents that have been implicated in the triggering of exocytosis in different cell types and in response to various stimuli include Ca^{2+}, guanine nucleotides, and cyclic AMP (Almers 1990; Gomperts 1990). In some tissues, including neurons and eggs, an increase in Ca^{2+} is sufficient to cause exocytosis, while in others it just serves as a cofactor.

5.2
Vesicle Membrane Proteins that may be Involved in Exocytosis

The cycle of exocytosis followed by endocytosis that the GLUT4 containing vesicles undergo is reminiscent of the process that synaptic vesicles in nerve terminals and secretory vesicles in endocrine cells undergo during secretion of neurotransmitters and hormones (De Camilli and Jahn 1990; Sudhof and Jahn 1991; Jahn and Sudhof 1994). A number of the membrane proteins of synaptic and secretory vesicles that are thought to be involved in exocytosis have been characterized. Some of these proteins have also been identified as components of the GLUT4-containing vesicles. Among these are the vesicle-associated membrane proteins or VAMPs also termed synaptobrevins (Cain et al. 1992; Ralston et al. 1994). VAMPs are located on the cytoplasmic surface of the vesicles, and appear to be involved in the docking/fusion of the vesicles with the cell membrane (Jahn and Sudhof 1994). Treatment with insulin causes translocation of VAMPs to the plasma membrane along with the GLUT4 (Cain et al. 1992). Secretory carrier membrane proteins (SCAMPs), which are generally present in regulated secretory vesicles, are also found in the GLUT4 containing vesicles (Laurie et al. 1993; Thoidis et al. 1993). A potentially important component of the pathway by which insulin stimulates glucose transport is suggested by the finding that the nonhydrolyzable GTP analog, GTPγS, induces GLUT4 translocation to the same extent as insulin in permeabilized fat cells (Baldini et al. 1991; Robinson et al. 1992). The maximal effects of insulin and GTPγS are not additive (Robinson et al. 1992). It is interesting in this context that low molecular weight GTP-binding proteins are components of the GLUT4 containing vesicles (Cormont et al. 1991, 1993; Etgen et al. 1993b). In adipocytes, the low molecular weight GTP binding protein associated with the GLUT4 vesicles appears to be Rab4 (Cormont et al. 1993). It has been reported that, in adipocytes, insulin treatment results in a 50% decrease in Rab4 content of the microsomal fraction that occurs concomitantly with the departure of the GLUT4 to the plasma membrane (Cormont et al. 1993). However, instead of appearing in the plasma membrane together with the GLUT4, Rab4 is redistributed to the cytosol (Cormont et al. 1993). In contrast, it has been reported that in skeletal muscle parallel increases in GLUT4 and low molecular weight GTP binding proteins occur in the plasma membrane in response to insulin and to muscle contractions (Etgen et al. 1993b)

While the molecular mechanisms responsible for GLUT4 translocation are still not known, it seems probable that the above proteins, as well as others, that are components of the GLUT4 containing vesicles play key roles in the movement of GLUT4 vesicles to, and fusion with, the sarcolemma in response to insulin and contractile activity.

6 Factors that Determine Sensitivity and Responsiveness of Glucose Transport in Muscle

The glucose transport process is usually characterized in terms of its sensitivity and responsiveness to insulin (Kahn 1978). Insulin sensitivity is generally quantified by determining the concentration of insulin required to induce 50% of its maximal effect on glucose transport activity. Insulin responsiveness of glucose transport is defined as the increase in glucose transport induced by a maximally effective insulin concentration. Insulin resistance can be due to a decrease in insulin sensitivity, a decrease in insulin responsiveness or a combination of the two. It is important to determine whether a change in insulin action is due to an alteration in sensitivity or in responsiveness, because the underlying mechanisms can be different. Factors that could be responsible for a change in insulin sensitivity include a) the magnitude of the GLUT4 translocating signal generated by a submaximal insulin concentration, b) the susceptibility of the GLUT4 containing vesicles to a submaximally effective insulin signal; for example, a change in the affinity of a GLUT4 vesicle protein for a product of the insulin signaling pathway that triggers the translocation process, or c) an change in the activity of the GLUT4 vesicle and/or plasma membrane docking and/or fusion proteins.

A change in insulin sensitivity does not affect maximally insulin stimulated glucose transport. It seems logical, therefore, that the end result of

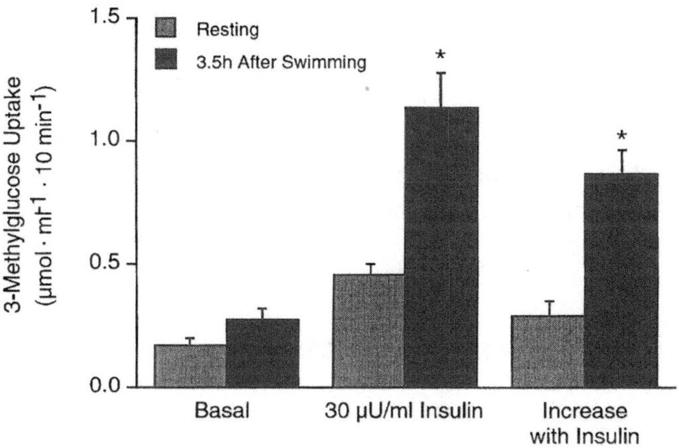

Fig. 7. Basal and insulin-stimulated 3-*O*-[^{3}H]methylglucose (*3MG*) transport in rat epitrochlearis muscles 3.5 h after a bout of swimming. Muscles were dissected out immediately after exercise, and incubated for 3 h at 35°C in buffer containing 8 m*M* glucose and 32 m*M* mannitol, in the absence of insulin. After 3 h the muscles were transferred to buffer containing 2 m*M* pyruvate and 36 m*M* mannitol, with or without 30 μU/ml insulin, and incubated at 29°C for 30 min before measurement of 3MG transport. *Each bar*, mean ± SE for 7–10 muscles. *, p<0.01 versus rested. (From Gao et al. 1994a)

a change in insulin sensitivity is the translocation of a greater number (increased sensitivity) or a smaller number (decreased sensitivity) of GLUT4 into the plasma membrane in response to a given submaximal insulin concentration, without a change in the total number of transporters available for translocation by insulin action. A change in insulin responsiveness of glucose transport can be due to an increase or a decrease in the number of GLUT4 available for translocation in response to a maximally effective insulin stimulus. If the intrinsic activity of the glucose transporter molecule can vary, as some studies suggest, another mechanism for a change in responsiveness would be an increase or decrease in GLUT4 intrinsic activity. A third mechanism for a decrease in insulin responsiveness of glucose transport is a reduction in the magnitude of the signal generated in response to a maximally effective insulin stimulus to such an extent that it results in translocation of fewer GLUT4 into the plasma membrane.

Because there is a second pathway for stimulating glucose transport in skeletal muscle that is activated by contractions or hypoxia, it is necessary to also characterize the sensitivity and responsiveness of the glucose transport process to these stimuli. This is particularly true in view of the evidence that what was thought to be an increase in sensitivity to insulin in response to exercise or to lithium is, in fact, due to an increase in sensitivity to stimulation via both pathways. There is also evidence that muscle glucose transport can be resistant to stimulation by one pathway but not via the other.

6.1 Agents that Increase the Sensitivity of Glucose Transport

6.1.1 Exercise

Exercise, such as running or swimming, and stimulation of muscles to contract in vitro, induce an increase in glucose transport activity in skeletal muscle (Holloszy and Narahara 1965; Ivy and Holloszy 1981; Garetto et al. 1984; Nesher et al. 1985; Constable et al. 1988). This acute effect of muscle contractions on glucose transport is independent of the effect of insulin (Wallberg-Henriksson and Holloszy 1984, 1985; Ploug et al. 1984), and reverses rapidly after cessation of exercise (Young et al. 1987; Gulve et al. 1990). Richter et al. (1982) found that, following exercise, a second effect that consists of a large increase in the sensitivity of glucose transport to insulin becomes evident. This increase in insulin sensitivity manifests itself after the insulin-independent effect of exercise on glucose transport has partially or completely reversed (Garetto et al. 1984; Wallberg-Henriksson et al. 1988). The magnitude of this effect is remarkable, with an insulin concentration of 30 μU/ml inducing a maximal effect on glucose transport in the rat epitrochlearis muscle studied 3 h after a bout of swimming (Fig. 7; Gao et al. 1994a); this concentration of insulin produces less than 50% of the maximal effect in epitrochlearis muscles from non-exercised animals.

A number of studies have provided evidence that feeding a high-carbohydrate diet speeds reversal of the enhanced glucose transport after exercise. The possibility that carbohydrate feeding might speed reversal of the increase in insulin sensitivity after exercise was suggested by studies showing that (a) 18 h postexercise, glucose uptake measured in the absence of added insulin had returned to resting baseline in muscles of carbohydrate-fed rats, but was still twofold above baseline in muscles of fasting animals, presumably due to the effect of endogenous insulin still bound to muscle (Young et al. 1983, 1987), and (b) muscle glucose uptake in the presence of submaximally effective insulin levels was higher in rats fed a carbohydrate-free diet than in carbohydrate-fed rats 18 h after exercise (Fell et al. 1982).

In a study designed to determine the effect of carbohydrate restriction on the reversal of the increase in muscle insulin sensitivity after exercise, Cartee et al. (1989) found that the effect of 60 μU/ml insulin on 3MG transport was approx. twofold greater in muscles studied 3 h after exercise than in nonexercised controls, regardless of the rats' carbohydrate intake. This concentration of insulin, which normally induces roughly a half-maximal effect on 3MG transport in epitrochlearis muscles from young rats, causes a maximal effect 3 h postexercise. The increase in insulin sensitivity was completely lost within 18 h in carbohydrate-fed rats, and the reversal of the increase in insulin sensitivity appeared to coincide with the increase in muscle glycogen above the usual fed level, i.e., with the muscle glycogen "supercompensation" that occurs in response to carbohydrate feeding after exercise (Bergström and Hultman 1966). In contrast, the increase in muscle insulin sensitivity was still present 48 h after exercise in animals fed a carbohydrate-free diet (Cartee et al. 1989).

It is of interest that the increase in insulin sensitivity was still present in carbohydrate-fed rats 3 h postexercise, even though muscle glycogen had returned to the level seen in fed resting rats. This finding suggests that the increase in insulin sensitivity may play a role in the glycogen supercompensation phenomenon seen with carbohydrate feeding after exercise (Bergström and Hultman 1966). When muscles are incubated with a high concentration of glucose and a low insulin concentration for 3 h after exercise, the exercise-induced increase in insulin sensitivity is reduced by approx. 50% (Gulve et al. 1990). In contrast, incubation of muscles with a similarly high concentration of 2-deoxyglucose (2DG) did not alter insulin sensitivity, even though a large amount of 2DG entered the muscles and was phosphorylated (Gulve et al. 1990). As 2DG is not further metabolized to a significant extent in this time frame, this finding provides evidence that glucose regulates the enhancement of insulin sensitivity by means of a signal generated beyond the transport and phosphorylation steps. For a brief period after exercise there is also a significant increase in responsiveness of muscle glucose transport to insulin (Richter et al. 1984; Zorzano et al. 1986b; Cartee et al. 1989). The increase in responsiveness is completely prevented by feeding rats carbohydrate immediately after exercise (Cartee et al. 1989) and does not occur in muscles allowed to recover for 3 h in vitro instead of in situ (Gulve et al. 1990).

The increase in muscle insulin sensitivity induced by exercise is probably one of the most important health benefits of physical activity, and there has been much interest in the mechanisms responsible for this adaptation. Much of this interest has focused on the possibility that exercise: (a) increases the affinity of muscle insulin receptors for insulin, (b) increases the number of functional insulin receptors, (c) enhances the degree to which the tyrosine kinase activity of the insulin receptor is increased by a given submaximal insulin concentration, and/or (d) amplifies the activation of one of the insulin signaling steps beyond the insulin receptor. The majority of studies have shown no change in insulin binding to intact muscles or to sarcolemmal preparations after exercise (Horton 1983; Bonen et al. 1984, 1985; Zorzano et al. 1985; Treadway et al. 1989), although one study reported an increase (Webster et al. 1986). When the insulin binding step was bypassed in a study that used the insulin-mimetic agent vanadate, a tyrosine phosphatase inhibitor (Green 1986), it was found that stimulation of glucose transport is markedly enhanced in rat epitrochlearis muscles 18 h after a bout of swimming (Cartee and Holloszy 1990). This finding clearly shows that the increased sensitivity of muscle glucose transport to insulin is not mediated by increased binding of insulin to its receptor.

The increase in sensitivity is also not due to an increase in insulin receptor tyrosine kinase activity. Treadway et al. (1989) have reported that insulin-stimulated insulin receptor autophosphorylation, as well as phosphorylation of an exogenous substrate by insulin-stimulated receptor kinase, are unchanged after exercise. In a more recent study, Goodyear et al. (1995a) found that tyrosine phosphorylation of the insulin receptor and of IRS-1 in response to insulin were actually decreased by 20% after exercise. They also found that insulin-stimulated activation of PI 3-kinase was blunted by prior exercise (Goodyear et al. 1995a). The finding that the increase in muscle insulin sensitivity induced by exercise is not due to an enhancement of the activity of the steps (that have been identified) in the insulin-signaling pathway for glucose transport activation is not too surprising, in view of the evidence that this phenomenon is not specific to insulin action. It has been shown that the effect of a submaximal hypoxic stimulus on glucose transport is greatly amplified in muscles studied 3 h after exercise, by which time the effect of contractions per se on glucose transport has worn off (Cartee and Holloszy 1990).

The increased sensitivity of glucose transport to stimulation by hypoxia following exercise provides evidence that the increase in insulin sensitivity is mediated by an effect of exercise on a late step that is common to the two pathways for stimulating glucose transport in skeletal muscle. The concept that the two pathways for stimulating glucose transport in muscle converge at a late step was suggested previously by the findings that sphingosine (Gulve and Holloszy 1993) and polymyxin B (Henriksen et al. 1989b) inhibit the stimulation of glucose transport both by insulin and contractile activity/hypoxia. Thus the persistent effect of exercise that has been thought of specifically as an increase in the insulin sensitivity of muscle appears to actually be a nonspecific increase in the susceptibility of glucose transport to stimulation by a variety of agents and to involve both pathways for stimulating glucose transport in muscle.

In a study to determine whether the increase in insulin sensitivity is due to local or systemic factors, Richter et al. (1984) stimulated the hindlimb muscles of rats to contract via the sciatic nerve. They found that in situ contractile activity resulted in the same pattern of response as treadmill running, i.e., a relatively brief stimulation of insulin-independent glucose transport and a more prolonged increase in insulin sensitivity. These results show that the increase in the insulin sensitivity of glucose transport in muscle after exercise is mediated by local contraction-induced effects rather than by systemic factors. However, rather surprisingly, muscles that are washed after being dissected out and then stimulated to contract in vitro show no enhancement of insulin sensitivity (Cartee and Holloszy 1990; Gao et al. 1994a). The reason for this is that an interaction between a serum factor and an effect of muscle contractions is necessary for development of the increase in the susceptibility of the glucose transport process in muscle to stimulation by insulin. Muscles stimulated to contract in vitro while immersed in rat or human serum subsequently develop an increase in insulin sensitivity that is similar in magnitude to that which occurs in response to exercise such as swimming (Gao et al. 1994a). Contrary to what one might expect, the serum factor is not something that is produced during exercise but is present in serum obtained from sedentary rats. The serum factor has not been identified, but appears to be a protein.

6.1.2 Lithium

The results of a study by Tabata et al. (1994) have shown that lithium markedly enhances the sensitivity of the glucose transport process in skeletal muscle to stimulation by both the insulin-activated and the exercise/hypoxia-activated pathways. The results of previous studies had suggested that the effects of lithium are insulin-mimetic (Clausen 1968; Haugaard et al. 1974; Cheng et al. 1983; Rossetti 1989). Lithium does appear to be as effective as insulin in activating glycogen synthase in skeletal muscle (Haugaard et al. 1974); however, the insulinlike effect of lithium on glucose transport activity is small in muscle, with an increase in 3MG transport in response to 10 m*M* lithium that is only 10% as great as that induced by a maximal insulin stimulus (Tabata et al. 1994).

The effects of lithium on the regulation of glucose metabolism are strikingly similar to those seen following a bout of exercise. As with exercise, lithium induces not only a large increase in insulin sensitivity but also a significant increase in insulin responsiveness (Tabata et al. 1994). Also as with exercise, lithium increases the sensitivity of the glucose transport process to stimulation via the muscle contraction/hypoxia-activated pathway. Glycogen synthase is activated in skeletal muscle following exercise (Bergström et al. 1972; Conlee et al. 1978) and is similarly increased by lithium (Horn et al. 1973; Haugaard et al. 1974). The synthesis of glycogen is markedly enhanced in skeletal muscle after exercise, resulting in the glycogen "supercompensation" phenomenon (Bergström and Hultman 1966; Conlee et al. 1978). Lithium also greatly enhances glycogen synthesis (Haugaard et al. 1974). As is the case for exercise, the mechanism by which lithium enhances the sensitivity of the glucose transport process to stimulation is still not known. However, the finding that the susceptibility of the glucose transport system in skeletal muscle to stimulation by submaximally effective hypoxia or contractile activity is increased by treatment with lithium provides evidence that a distal step common to the insulin and contraction/hypoxia-mediated pathways is involved.

6.1.3 Catecholamines

Lupien et al. (1990) infused norepinephrine (NE) into young rats over a 10 day period. Insulin sensitivity was evaluated using a multistage euglycemic-hyperinsulinemic clamp. They found that at submaximal plasma insulin concentrations, glucose disposal rate was approx. 60% higher in NE-treated than in control animals, while at a maximally effective insulin concentration glucose disposal was about 13% higher in the NE group. The increase in insulin sensitivity was lost within 36 h after cessation of the NE infusion. The effect of catecholamines on muscle insulin sensitivity does not depend on prolonged treatment, as Nolte et al. (1994) found that muscles of rats that had been given an injection of epinephrine showed a significant increase in insulin sensitivity. The epinephrine injection resulted in a moderate (approx. 40%) decrease in muscle glycogen concentration. Muscles were taken 2 h after the epinephrine injection and incubated in vitro in the absence of epinephrine for 60 min before being exposed to insulin. The increase in insulin sensitivity following epinephrine injection was roughly 50% as great as that normally seen following a bout of exercise. At this time, there is no information regarding the mechanisms responsible for the catecholamine-induced increase in insulin sensitivity.

6.1.4 Fasting

Goodman and Ruderman (1979) found that fasting rats for 48 h markedly enhances insulin action on glucose transport in skeletal muscle. In experiments using the perfused rat hindquarter preparation, they found that glucose uptake was higher at submaximally, but not at maximally, effective insulin concentrations. This finding led them to conclude that insulin sensitivity was enhanced. However, in an experiment in which glucose uptake was measured in isolated soleus and extensor digitorum longus muscles in vitro, the increase in glucose uptake induced by a maximally effective insulin concentration was clearly higher in the fasting group (Goodman and Ruderman 1979). Similarly, LeMarchand-Brustel and Freychet (1979) found that 2DG transport was higher at a maximally effective insulin concentration in soleus muscles of mice fasted 48 h than in those of fed control mice. Thus, although it is generally stated that fasting increases muscle insulin sensitivity, it appears from the experiments performed on muscles incubated in vitro that insulin responsiveness is also increased. The reason for this detailed evaluation of these old studies is that two more recent studies have shown that fasting induces a large increase in GLUT4 protein concentration in skeletal muscle of rats (Bourey et al. 1990;

Charron and Kahn 1990). Skeletal muscle glucose transport activity was not measured in these studies. However, as reviewed in detail in the next section, other studies have shown a close correlation between muscle GLUT4 content and maximally stimulated glucose transport activity. One would, therefore, expect fasting to induce an increase in insulin responsiveness. Further studies are needed to clarify the effects of fasting on skeletal muscle glucose transport, particularly the relationship between GLUT4 protein content and maximally insulin-stimulated and contraction-stimulated glucose transport.

6.2 Role of GLUT4 Content in Determining Responsiveness of Glucose Transport

When the signaling pathways that mediate the stimulation of glucose transport by insulin and by exercise are intact, the primary factor that determines a muscle's maximally stimulated glucose transport capacity appears to be its content of GLUT4.

6.2.1 Comparison of Different Muscles

There are large differences among skeletal muscle fiber types in their content of GLUT4 protein (James et al. 1989; Henriksen et al. 1990; Kern et al. 1990; Marette et al. 1992; Megeney et al. 1993; Kong et al. 1994). Fig. 8 compares the GLUT4 protein concentrations in four rat muscles (Henriksen et al. 1990). These muscles were chosen because they are suitable for measurement of glucose transport in vitro, and because they represent a wide range of muscle fiber type compositions: epitrochlearis, 11% type I, 12% type IIa, and 77% type IIb (Rodnick et al. 1992a); soleus, 84% type I, and 16% type IIa (Ariano et al. 1973); extensor digitorum longus (EDL), 3% type I, 57% type IIa, and 40% type IIb (Ariano et al. 1973); and flexor digitorum brevis (FDB), 7% type I, 92% type IIa and 1% type IIb (Carlsen et al. 1985). For those not familiar with this muscle fiber type terminology, type I fibers are slow contracting and have a high oxidative capacity, the type IIa fibers are fast contracting and the majority of them have both high oxidative and high glycolytic capacities, while the type IIb fibers are fast contracting and the majority have a high glycolytic capacity and a

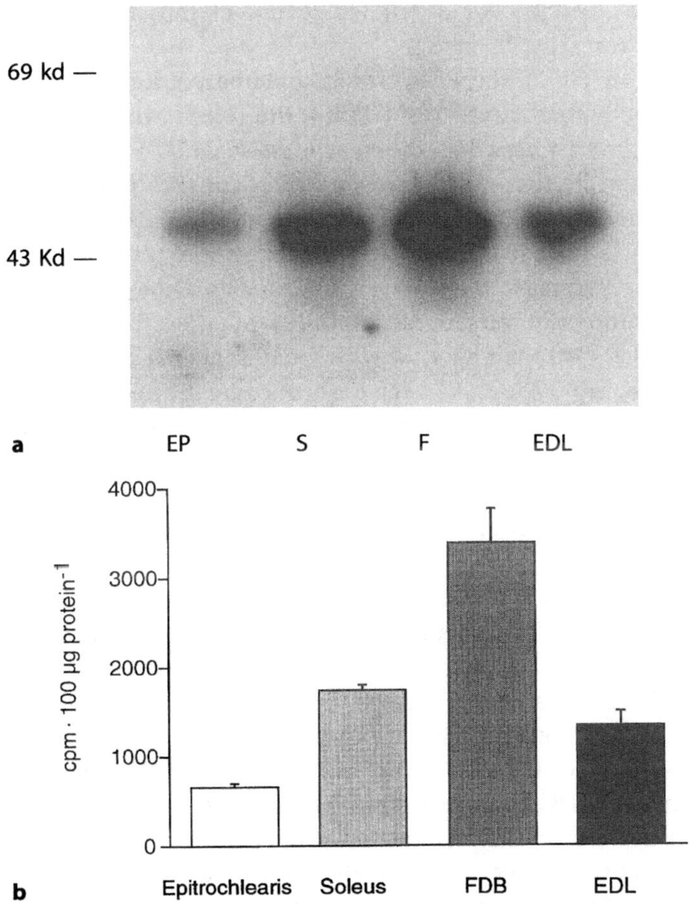

Fig. 8a,b. Glucose transporter (GLUT4) protein content in epitrochlearis (*EP*), soleus (*S*), extensor digitorum longus (*EDL*), and flexor digitorum brevis (*FDB*) muscles. **a** Protein (100 µg) from each muscle was analyzed by SDS-polyacrylamide gel electrophoresis followed by immunoblotting with a polyclonal antiserum (R820) against GLUT4 protein. Immunolabeled bands were then visualized with ^{125}I-labeled protein A. **b** Bands from 5–6 individual immunoblots were counted for radioactivity. Each bar represents the mean ± SE. (From Henriksen et al. 1990)

low oxidative capacity. For a detailed review of muscle fiber types see Pette and Staron (1990).

As shown in Fig. 8, the FDB, which contains predominantly type IIa fibers, has the highest content of GLUT4; the soleus, which contains predominantly type I fibers, has the next highest GLUT4 content followed by the EDL; while the epitrochlearis, which consists predominantly of type IIb fibers, has the lowest GLUT4 content. The FDB is not a typical type IIa muscle in that it has a considerably lower respiratory capacity than other fast red muscles. However, Megeny et al. (1993), who compared the GLUT4 content of various rat hindlimb muscles, found that the red portion of the gastrocnemius, which contains primarily type IIa fibers (68% type IIa, 21% type I), had the highest content of GLUT4, the soleus had the next highest and the white portion of the gastrocnemius, which contains almost entirely type IIb fibers had the lowest content of GLUT4. Generally, the GLUT4 content of muscles parallel their respiratory capacity, i.e., content of mitochondria, and also their hexokinase activity, although there are exceptions.

As shown in Figs. 9 and 10, there was an almost perfect correlation between muscle GLUT4 content and 2DG transport stimulated maximally by both insulin and contractile activity, providing evidence that, in the absence of insulin resistance, maximal glucose transport activity is determined by GLUT4 content. On the other hand, responsiveness to insulin is not as well correlated with muscle GLUT4 content. The maximally insulin stimulated 2DG transport rate was higher in the soleus than in the FDB, even though GLUT4 content was higher in the FDB. On the other hand, contractile activity induced a threefold greater stimulation of glucose transport activity in the FDB than in the soleus, and a similar increase in the EDL as in the soleus (Fig. 9), even though the EDL has a considerably lower GLUT4 content. Of particular interest are the findings that, although maximally insulin-stimulated and maximally contraction-stimulated glucose transport activities are similar in the epitrochlearis and the EDL, sugar transport stimulated maximally by insulin was more than twofold greater than contraction stimulated transport in the soleus. In contrast, the maximal effect of contractions on glucose transport activity was 2.5-fold greater than that of insulin in the FDB (Fig. 9).

As discussed above, it has been hypothesized that there are two separate pools of GLUT4 in muscle, with one pool being available for translocation by the action of insulin but not contractions, while the other is translocated in response to contractions but not insulin (Douen et al. 1990b; Henriksen

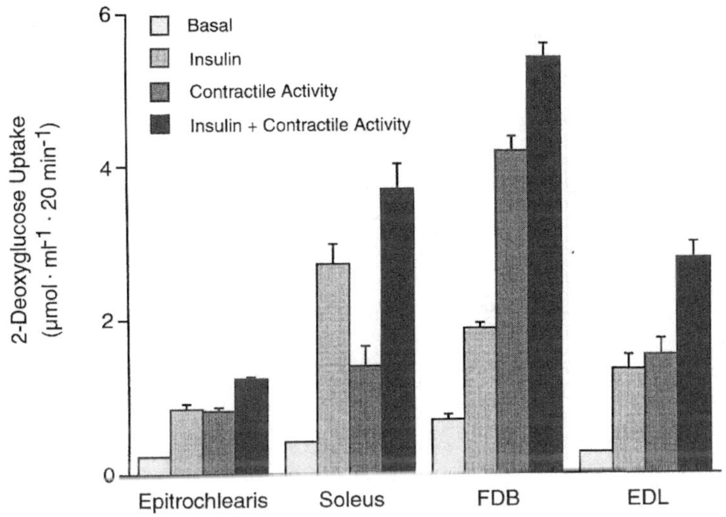

Fig. 9. Effects of insulin, contractions, or insulin and contractions in combination on 2-deoxy-[^{3}H]glucose (*2DG*) uptake in epitrochlearis, soleus, extensor digitorum longus (*EDL*), and flexor digitorum brevis (*FDB*) muscles. Muscles were initially incubated at 35°C for 60 min in Krebs-Henseleit bicarbonate buffer containing 8 m*M* glucose and 32 m*M* mannitol before being treated with or without insulin for 30 min. Muscles were then either washed immediately for 10 min at 29°C in glucose-free medium containing 40 m*M* mannitol and insulin, if present previously, or were first stimulated electrically to contract before subsequent incubations. Glucose transport activity was then assessed by measuring the intracellular accumulation of 2DG at 29°C for 20 min. Values are means ± SE for 6–12 muscles/group. (From Henriksen et al. 1990)

et al. 1990; Gao et al. 1994b). The differences in response of the soleus and FDB to insulin and contractions (Fig. 9) could be explained in terms of this hypothesis if the relative sizes of these pools are different, with a large insulin-regulatable, and a small contraction-regulatable, pool of GLUT4 in the soleus, and with the reverse situation in the FDB. An alternative possibility is that there is a single pool of glucose transporters, with differential recruitment in response to insulin or contractile activity in various muscles as a result of differences in the magnitudes of the signals generated.

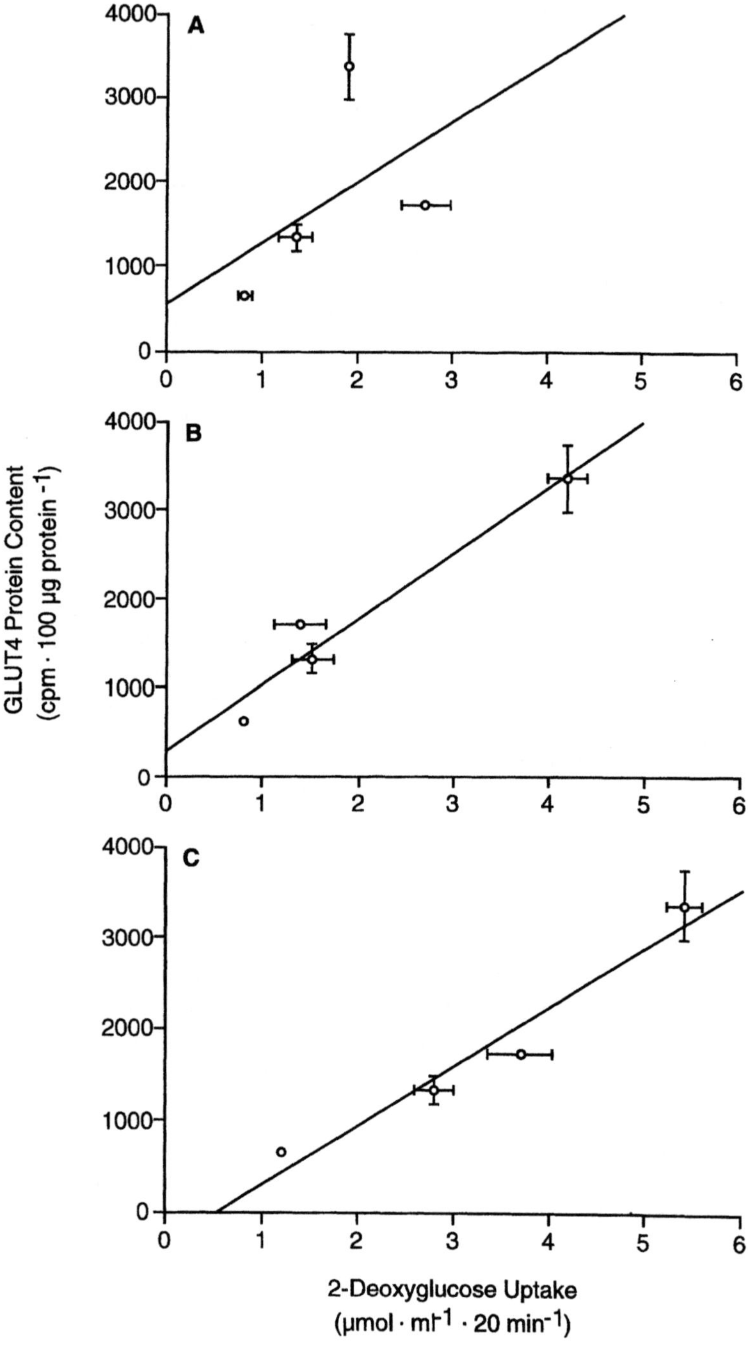
A
B
C
4000
3000
2000
1000
0
0 1 2 3 4 5 6
GLUT4 Protein Content
(cpm · 100 µg protein -1)
2-Deoxyglucose Uptake
(µmol · ml-1 · 20 min-1)

6.2.2 Role of Physical Activity in Determining GLUT4 Concentration and Glucose Transport Capacity in Muscle

Endurance exercise results in a remarkable adaptive increase in the capacity of skeletal muscle to oxidize carbohydrate (Holloszy 1967) and fat (Molé et al. 1971) as the result of a proliferation of mitochrondria (Holloszy and Booth 1976; Holloszy and Coyle 1984). Exercise-training also induces an increase in muscle hexokinase activity (Holloszy and Booth 1976). In view of the finding that GLUT4, mitochondrial proteins, and hexokinase generally vary in parallel in skeletal muscles, it seemed probable that endurance exercise training might also induce an increase in muscle GLUT4 content. Studies on normal rats (Rodnick et al. 1990, 1992a; Ploug et al. 1990; Goodyear et al. 1992; Slentz et al. 1992; Ren et al. 1994), as well as on obese Zucker rats (Friedman et al. 1990; Brozinick et al. 1993; Torgan et al. 1995), old, fat rats (Ezaki et al. 1992), young healthy humans (Dela et al. 1993), previously sedentary middle-aged men (Houmard et al. 1993), and patients with impaired glucose tolerance (Hughes et al. 1993) or non insulin dependent diabetes (Dela et al. 1994) have demonstrated that exercise training does indeed induce an increase in muscle GLUT4. In contrast to GLUT4, there is no significant change in GLUT1 concentration in skeletal muscle in response to exercise training (Rodnick et al. 1992a). In both the epitrochlearis (Rodnick et al. 1992a; Ren et al. 1994) and the split soleus preparation (Slentz et al. 1992) glucose transport activity, stimulated maximally with insulin and/or contractions, is increased roughly in proportion to the increase in GLUT4.

There is considerable confusion in the literature regarding the adaptation of GLUT4 protein and glucose transport activity in rat soleus muscle to endurance exercise training. In a study in which rats were exercised by means of running in voluntary running wheels (Rodnick et al. 1992a),

Fig. 10A–C. Relationship between stimulated glucose transport and GLUT4 protein content in epitrochlearis, soleus, extensor digitorum longus, and flexor digitorum brevis muscles. Data shown in Figs. 8 and 9 for each muscle were analyzed using linear regression to determine the correlation between mean values for GLUT4 protein content and 2DG uptake stimulated by insulin alone (**A**), contractions alone (**B**), and insulin and contractions in combination (**C**). Regression equations and correlation coefficients: **A** $y=910x+86$, $r=0.720$; **B** $y=59x+537$, $r=0.950$; **C** $y=551x-74$, $r=0.992$. (From Henriksen et al. 1990).

the exercise induced increases in GLUT4 concentration and maximally insulin stimulated glucose transport (expressed per ml intracellular water) in the epitrochlearis but not in the soleus. However, the wheel running induced hypertrophy of the soleus and, as a result, both GLUT4 content and the maximal rate of 2DG transport expressed per total soleus muscle mass were significantly increased (Rodnick et al. 1992a). In a subsequent study performed by Slentz et al. (1992) in which rats were trained either by treadmill running or swimming, which, in contrast to wheel running, do not result in hypertrophy of the soleus, both GLUT4 concentration and maximally insulin-stimulated 2DG transport were significantly increased in the trained soleus muscles. Ploug et al. (1990) measured 3MG uptake in perfused hindquarters of rats kept sedentary or trained by swimming. They found that maximally insulin-stimulated 3MG uptake was increased in the red portion of the gastrocenius but not in the soleus, while contraction-stimulated 3MG uptake was increased in the soleus but not in the red gastrocnemius. Brozinick et al. (1993), working with obese Zucker rats, also found, using the perfused hindquarter preparation, that contraction-stimulated 2DG uptake was increased in trained soleus muscles while insulin stimulated 2DG uptake was not increased. These findings are in conflict with those of an earlier study by James et al. (1985) in which treadmill training enhanced the effects of insulin on the uptake of a bolus injection of 2DG by the soleus during a euglycemic clamp. They also disagree with the findings of Slentz et al. (1992) in the split soleus in vitro.

Although it is now well established that the increase in muscle GLUT4 induced by exercise training is associated with an increase in maximal capacity for glucose transport, a number of studies have shown no increase in maximally insulin-stimulated sugar uptake by perfused hindlimb muscles of trained rats when measured 48 h after the last exercise bout (Ivy et al. 1983; Idström et al. 1986; Etgen et al. 1993a). This finding is still unexplained, but may be related to the decrease in insulin responsiveness associated with "glucose toxicity" and glycogen supercompensation. Glycogen supercompensation is the increase in muscle glycogen concentration to above the usual fed level that occurs in response to carbohydrate feeding after glycogen depleting exercise (Bergström and Hultman 1966). When glucose floods into muscle and muscle glycogen supercompensation occurs, there is a decrease in insulin-stimulated glucose transport activity (Fell et al. 1982; Richter et al. 1988b). The mechanism responsible for this decrease in insulin action has not been established, but it could be mediated by "glucose toxicity" through accumulation of glucosamine (Marshall 1989; Marshall et al. 1991a), a phenomenon reviewed in the section on insulin resistance.

With one exception, the studies of the effects of exercise on muscle GLUT4 concentrations have involved prolonged exercise training, usually consisting of 6- to 12-week-long programs of running or swimming. Long training periods were used in these studies because of the general belief that skeletal muscle adapts to exercise relatively slowly. However, to provide a survival advantage, an adaptation must occur sufficiently rapidly to enhance the organism's capability to respond to the environmental stimulus to which it is adapting. Ren et al. (1994), therefore, performed a study to test the hypothesis that the adaptive increase in GLUT4 occurs rapidly in response to an adequate adaptive stimulus. Rats were exercised by means of swimming for two 3-h long periods separated by a 45 min rest period. As shown in Fig. 11, 16 h after the exercise, GLUT4 mRNA was increased twofold and GLUT4 protein was increased 50% in epitrochlearis muscles (Ren et al. 1994). After a second day of the same exercise protocol, GLUT4 protein was increased twofold without a further significant increase in GLUT4 mRNA. Glucose transport activity stimulated maximally with insulin was increased in proportion to the increase in GLUT4 after 1 and 2 days of exercise (Fig. 12). There was also a twofold increase in muscle hexokinase activity in response to two days of exercise. In addition, glucose transport activity stimulated maximally by contractions alone and in combination with insulin was also increased approx. twofold after 2 days of the swimming program (Fig. 13), roughly in proportion to the increase in muscle GLUT4 protein. Treatment with a maximally effective insulin concentration resulted in a twofold greater increase in GLUT4 in the plasma membrane in the 2-day swimmers' muscles than in control muscles. In muscles incubated with glucose plus insulin, the rate of glycogen accumulation was twice as great in muscles of the 2-day swimmers as in muscles from sedentary rats. Availability of muscle glycogen is essential for performance of strenuous, prolonged exercise. Raising muscle glycogen improves the ability to perform prolonged, vigorous exercise, while a low muscle glycogen content results in more rapid fatigue development (Bergström et al. 1967; Karlsson and Saltin 1971). The rapid increase in muscle GLUT4 could provide a survival advantage to an animal suddenly faced with the necessity for a sustained increase in activity (for example, invasion of its territory by predators), by making possible more rapid muscle gly-

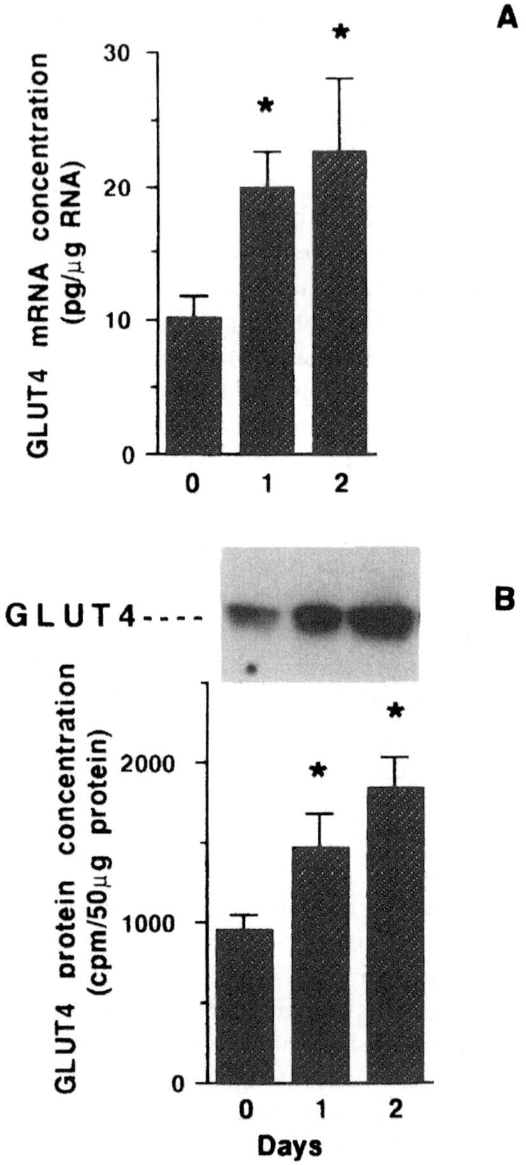

Fig. 11A,B. Effects of 1 and 2 days of swimming on expression of GLUT4 mRNA and GLUT4 protein in epitrochlearis muscles. Rats were either kept sedentary or exercised by means of swimming for 1 or 2 days. Epitrochlearis muscles were dissected out and clamp-frozen 16 h after the swimmers' last exercise bout. **A** GLUT4 mRNA was quantified by RNase protection using a ^{32}P-labeled antisense GLUT4 RNA probe. Each bar represents the average GLUT4 mRNA concentration for epitrochlearis muscles from six rats.

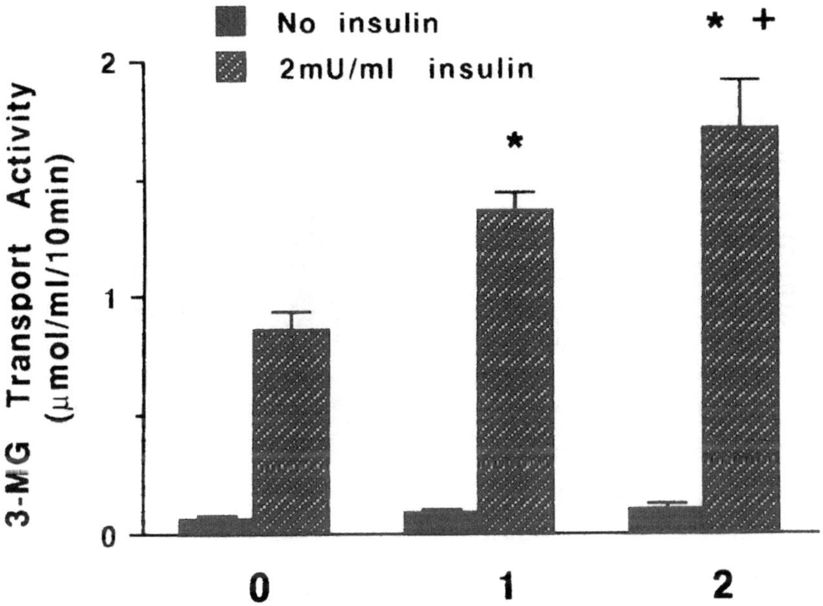

Fig. 12. Stimulation of 3-*O*-[^{3}H]methylglucose (*3MG*) transport by insulin in epitrochlearis muscles from 1- and 2-day swimmers and control rats. Muscles were incubated in oxygenated Krebs-Henseleit bicarbonate buffer containing 8 m*M* glucose and 32 m*M* mannitol with or without 2 mU/ml insulin for 30 min at 35°C. The muscles were then rinsed at 29°C for 10 min in glucose-free medium containing 40 m*M* mannitol and 2 mU/ml insulin. The rate of 3MG transport was then measured at 29°C for 10 min. *Each bar*, mean ± SE for 5–8 muscles. *, $p<0.001$ (1- and 2-day swimmers versus nonexercised controls); +, $p<0.05$ (2-day swimmers versus 1-day swimmers). (From Ren et al. 1994)

cogen replenishment when carbohydrate is eaten between exercise bouts or even during continued lower intensity exercise (Constable et al. 1984).

Fig. 11A,B. ***Continue.*** **B** Muscle homogenates containing 50 μg of protein were subjected to SDS-polyacrylamide gel electrophoresis and then immunoblotted with a polyclonal antibody against GLUT4. Representative autoradiograms of immunoblots of GLUT4 are shown. *Each bar*, mean ± SE for muscles from six rats. *, $p<0.05$ (control versus swimmers). (From Ren et al. 1994)

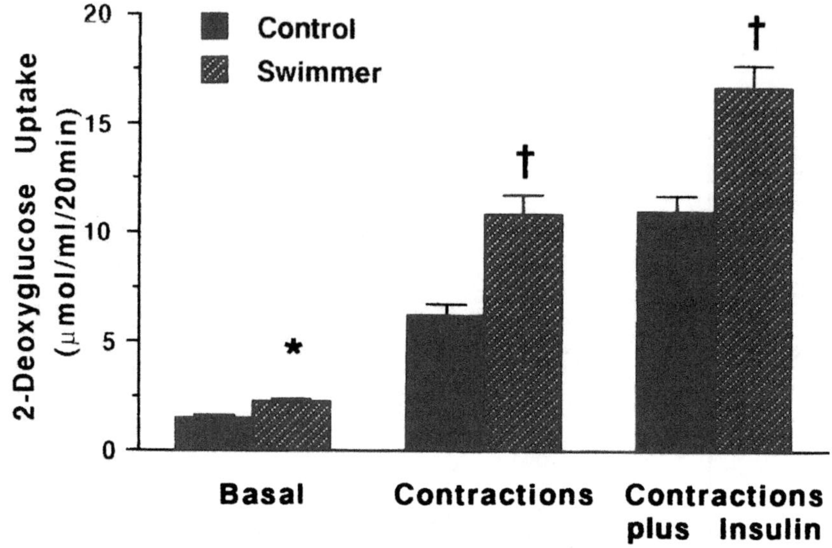

Fig. 13. Stimulation of 2-deoxy-[^{3}H]glucose (2DG) uptake by contractions and contractions plus insulin in epitrochlearis muscles of 2-day swimmers and control rats. Muscles were incubated in medium containing 8 m*M* glucose and 32 m*M* mannitol in the presence or absence of 2 mU/ml insulin at 35°C, stimulated to contract for 10 min, and then incubated for 20 min in the same medium. Next the muscles were rinsed for 10 min in glucose-free medium containing 40 m*M* mannitol and 2 mU/ml insulin, if it was present in the preceding incubation, to remove glucose from the extracellular space. The rate of 2DG uptake was then measured for 20 min. *Each bar*, mean ± SE for 6 or 7 muscles. *, $p<0.05$ (control versus 2-day swimmers); + $p<0.01$ (control versus 2-day swimmers). (From Ren et al. 1994)

6.2.3
Chronic Electrical Stimulation

Muscles that are stimulated continuously via the nerve by means of implanted electrodes undergo a remarkable biochemical and anatomical transformation (Pette et al. 1973; Pette et al. 1975, 1976; Pette and Tyler 1983; Henriksson et al. 1986; Weber and Pette 1990). In studies of this phenomenon, the muscles have generally been stimulated at the firing frequency of nerves that innervate slow-twitch muscles. Some of the biochemical adaptations to chronic low-frequency electrical stimulation are similar to those induced by endurance exercise, including increases in mitochondrial enzymes and hexokinase, and decreases in the enzymes of the glycogenolytic pathway. However, the magnitudes of the changes elicited by chronic stimulation are much greater than those induced by exercise. In addition, chronic electrical stimulation causes an extensive conversion of fast- to slow-twitch fibers and a decrease in muscle fiber size. For a detailed review see (Pette and Vrbová 1992). The effect of chronic electrical stimulation on muscle GLUT4 content has been investigated in two laboratories (Hofmann and Pette 1994; Kong et al. 1994). Hofmann and Pette (1994) found that no change in total GLUT4 protein occurred during the first three days of stimulation. However, by the fifth day, GLUT4 protein concentration had increased more than twofold, without further change over an additional 9 day period. The increase in GLUT4 protein was preceded by an increase in GLUT4 mRNA. Interestingly, the increase in hexokinase activity preceded the increase in GLUT4, being evident after only a few hours of stimulation. Kong et al. (1994) also examined the effects of chronic electrical stimulation on muscle hexokinase and GLUT4 levels. Although their experimental protocol was apparently the same as that of Hoffman and Pette (1994), the findings were somewhat different in that they found a 2.5-fold increase in GLUT4 after only one day of stimulation, and a 14-fold increase after 21 days of stimulation; they also found that GLUT4 protein and hexokinase activity changed roughly in parallel. These differences are likely due to a species difference, as Kong et al. (1994) studied rabbit muscle, while Hofmann and Pette (1994) studied rat muscle.

6.2.4 Hyperthyroidism

Thyrotoxicosis is another inducer of an increase in muscle GLUT4 concentration (Casla et al. 1990; Weinstein et al. 1994). In a study using the perfused rat hindquarter preparation Casla et al. (1990) found that basal, but not maximally insulin stimulated, glucose transport was increased in muscles of thyrotoxic animals. This somewhat surprising finding is, to our knowledge, the only instance in which an increase in muscle GLUT4 was not associated with an increase in the responsiveness of glucose transport to insulin. In contrast, Weinstein et al. (1994), who measured maximally insulin-stimulated 2DG transport in split soleus and extensor digitorum longus muscles from thyrotoxic and control muscles in vitro, found that insulin responsiveness was increased in proportion to the increase in GLUT4.

6.2.5
Muscle Creatine Depletion

The creatine analog β-guanidinopropionic acid (β-GPA) competitively inhibits creatine uptake in skeletal muscle (Fitch et al. 1974). Rats fed β-GPA accumulate β-GPA-P instead of phosphocreatine (PC) in their skeletal muscles, and PC falls to a very low level (Fitch et al. 1974; Fitch et al. 1975). β-GPA-P is a poor substrate for creatine kinase and, as the result of a shift in the equilibrium of the reaction catalyzed by creatine kinase, ATP concentration also decreases. This alteration in the concentrations of the high-energy phosphates results in changes in the enzyme pattern in skeletal muscle. These changes resemble, in a number of respects, those that occur in response to exercise, with increases in the levels of the mitochondrial enzymes and hexokinase, and decreases in the enzymes of the glycogenolytic pathway (Shoubridge et al. 1985). To further evaluate the possibility that GLUT4, hexokinase, and key mitochondrial enzymes may be regulated in concert under some conditions, Ren et al. (1993c) examined the effect of feeding β-GPA on the expression of GLUT4 in rat skeletal muscle. GLUT4 protein concentration, cytochrome c concentration, citrate synthase activity and hexokinase activity were all increased 40%–50% in muscles of rats β-GPA for 6 weeks. 3MG transport maximally stimulated by insulin was increased in parallel with GLUT4 concentration in epitrochlearis muscles of β-GPA fed rats, providing additional evidence that the GLUT4 content of a muscle determines its maximal glucose transport activity when the signaling pathways for activation of glucose transport are intact.

7
Insights from Studies on Transgenic Mice

Both GLUT1 and GLUT4 have been overexpressed in skeletal muscle in mice. These studies have been reviewed in detail by Mueckler (1995). In the studies involving GLUT1, the muscle-specific rat myosin light chain 2 promoter was used to drive transcription of the human GLUT1 cDNA transgene (Marshall et al. 1993). There was a large overexpression of GLUT1 in skeletal muscles; the exact fold increase could not be determined because the concentration of GLUT1 in muscles of normal mice is too low to quantify accurately (Ren et al. 1993a; Marshall et al. 1993). Basal glucose transport activity measured using 3MG or 2DG was increased approx. sevenfold in epitrochlearis and extensor digitorum longus muscles from the GLUT1 transgenic mice. There was a massive accumulation of glycogen in the muscles of the GLUT1 transgenic mice to values tenfold higher than those normally found in fed nontransgenic animals (Ren et al. 1993a). This enormous increase in muscle glycogen occurred even though the active I form of glycogen synthase was only 50% as high as in control muscles, and phosphorylase activity was unchanged. There was also no change in the concentration of glucose-6-P.

These data clearly show that glucose transport is rate limiting for glycogen synthesis in the muscle of wild type mice and that the sevenfold increase in basal glucose transport in the muscles of the GLUT1 transgenic mice was responsible for the huge increase in glycogen accumulation. The concentration of free intracellular glucose in muscle of the GLUT1 transgenic mice was similar to that in the interstitial fluid. Normally there is no free glucose detectable in skeletal muscle, so this finding indicates that the sevenfold increase in basal glucose transport activity resulted in sufficiently rapid entry of glucose to saturate hexokinase and cause equilibration of free glucose across the plasma membrane. The enormous accumulation of glycogen in the GLUT1 transgenic muscles provides evidence that transport normally limits the rate of glycogen synthesis in muscle and that the capacity for glycogen synthesis in the basal state greatly exceeds the rate of glucose transport. Intracellular lactate concentration was increased about twofold, suggesting that transport is also normally rate limiting for glycolysis in vivo.

As might be expected, the GLUT1 mice have an enhanced ability to dispose of an oral glucose load and tend to have lower plasma glucose levels than their nontransgenic littermates (Marshall et al. 1993). The GLUT1 transgenic muscles are completely resistant to the action of a maximally effective insulin concentration on glucose transport activity (Gulve et al. 1994). The muscles of the GLUT1 transgenic mice are also resistant to the stimulation of glucose transport activity by hypoxia and contractile activity. This profound resistance to stimulation of glucose transport in the GLUT1 transgenic muscle is still unexplained. One possibility relates to the rapid glucose uptake and massive glycogen accumulation, which could result in the glucose toxicity phenomenon, possibly mediated by accumulation of glucosamine, as reviewed in the section on insulin resistance.

Olson et al. (1993) have used a human minigene construct to overexpress GLUT4 in skeletal muscle, heart and adipose tissue. The GLUT4 transgenic mice have blood glucose levels that are lower than those of normal mice and also have an enhanced ability to dispose of an oral glucose load. Soleus, epitrochlearis and extensor digitorum longus muscles isolated from the GLUT4 transgenic mice showed two- to fourfold increases in GLUT4 protein concentration (Hansen et al. 1995b). Basal 2DG transport measured in these muscles in vitro was increased in the range of 20%–50%, likely due to some mistargeting of GLUT4 to the plasma membrane in unstimulated cells. It is interesting that, although maximally insulin stimulated 2DG transport was increased roughly twofold (Hansen et al. 1995b), the increase was not proportional to the 3.5- to 4-fold increase in GLUT4 expression in either the EDL or the epitrochlearis muscles. A possible explanation for this discrepancy is that other proteins that are normally coregulated with GLUT4 and are involved in the translocation process may also have to increase in order for translocation to occur in proportion to the increase in GLUT4.

The GLUT4 transgenic mice clearly have much lower rates of muscle glucose uptake in vivo than do the GLUT1 transgenic mice. This is evidenced by dramatic increases in muscle free glucose and in muscle glycogen in the GLUT1 mice (Ren et al. 1993a), but not in the GLUT4 mice. The latter show no increase in free intracellular glucose and only about a 20% increase in muscle glycogen level (Hansen et al. 1995b). This difference is, no doubt, due to the sevenfold increase in basal glucose transport rate in the GLUT1 mice as opposed to only a 20%–50% increase in basal glucose transport rate in the GLUT4 mice. Most of the GLUT4 transporters are located intracellularly in the basal state, whereas in the GLUT1 mice nearly all of the GLUT1 transporters are located in the plasma membrane. Taken together, the findings in mice overexpressing the GLUT1 and the GLUT4 glucose transporters provide support for the view that GLUT1 is responsible for most of the basal glucose transport, and that insulin- and contraction-stimulated glucose transport in skeletal muscle are mediated by GLUT4.

8 Skeletal Muscle Insulin Resistance

8.1 Insulin Deficiency, Muscle GLUT4 Content and Insulin Resistance

Except for an early study by LeMarchand-Brustel and Freychet (1979), who reported that soleus muscles of mice made insulin-deficient by means of streptozotocin (STZ) injection had normal insulin responsiveness and increased insulin sensitivity, there has been general agreement that STZ-induced insulin deficiency causes severe insulin resistance of muscle glucose transport due to a decrease in responsiveness (Wallberg-Henriksson and Holloszy 1985; Klip et al. 1990; Kahn et al. 1991; Richardson et al. 1991; Youn et al. 1994b; Napoli et al. 1995). There is also general agreement that insulin deficiency results in a decrease in muscle GLUT4 protein concentration (Garvey et al. 1989; Bourey et al. 1990; Klip et al. 1990; Kahn et al. 1991; Richardson et al. 1991; Youn et al. 1994b; Napoli et al. 1995). However, there is a remarkable degree of disagreement regarding the time courses and magnitudes of these changes. At one extreme, Richardson et al. (1991) reported that maximally insulin-stimulated muscle glucose transport activity was decreased by approx. 80% within 24 h after STZ injection. At the other extreme, Youn et al. (1994b) reported that there was no decrease in insulin-stimulated glucose uptake by rat muscles until the 8th day after STZ injection when an approx. 40% reduction was detected. Napoli et al. (1995) also found no decrease in insulin-stimulated glucose transport until some time between day 5 and day 10 after STZ injection and actually found an increase in transport 24 h after STZ injection; this is in contrast to Richardson et al. (1991), who found almost complete inhibition of insulin-stimulated glucose transport within 24 h. In a study in which 3MG transport was measured in isolated epitrochlearis muscles in vitro, maximally insulin-stimulated glucose transport activity was decreased 63% three days after STZ injection (Wallberg-Henriksson and Holloszy 1985). This was the only study of this phenomenon in which a sufficiently high dose of STZ was used to result in a plasma insulin level that was too low to measure. In this study, the effect of insulin deficiency on the contraction-activated pathway of glucose transport stimulation was also examined; maximally contraction-stimulated transport was decreased 53%, i.e., roughly to the same extent as insulin-stimulated transport, 3 days after STZ injection (Wallberg-Henriksson and Holloszy 1985).

The decrease in GLUT4 protein concentration in striated muscle that occurs with insulin deficiency appears to be due to a decrease in GLUT4 transcription (Garvey et al. 1989; Neufer et al. 1993). In most of the studies on the effect of insulin deficiency on GLUT4, the concentration of this protein was decreased in the range of 40%–60%, 5–10 days after STZ injection, although there was considerable variability. For example, Kahn et al. (1991) found no change in GLUT4 seven days after STZ and a large decrease after 14 days; Youn et al. (1994b) found no change in GLUT4 in the soleus, and 20%–30% decreases in the epitrochlearis and extensor digitorum longus 14 days after STZ injection; while Richardson et al. (1991) detected a significant decrease in muscle GLUT4 by the fourth day, and an approx. 50% decline by the seventh day after STZ, with the largest decline occurring in the soleus. The differences in the magnitudes of the responses between studies might be due to variability in the degree of insulin deficiency; however, the differences in time course and muscle involvement are puzzling.

In the time course studies, development of insulin resistance began earlier than the decline in GLUT4 concentration; also, the magnitude of the insulin resistance that developed was generally greater than could be explained by the reduction in GLUT4 concentration (Richardson et al. 1991; Kahn et al. 1991). This greater effect on transport than on GLUT4 does not appear to be due to a defect in the insulin signaling pathway, as Folli et al. (1993) found that insulin-stimulated IRS-1 phosphorylation was increased 50%, and IRS-1-associated PI 3-kinase activity was increased twofold compared to normal controls. Instead, the defect appears to be due to a decrease in the proportion of GLUT4 incorporated into the plasma membrane (Klip et al. 1990; Napoli et al. 1995). These findings seem in keeping with the evidence that insulin-stimulated and contraction-stimulated glucose transport are similarly impaired (Wallberg-Henriksson and Holloszy 1985). Noninsulin dependent diabetes is generally not associated with a decrease in GLUT4 in skeletal muscle (Pedersen et al. 1990; Handberg et al. 1990; Andersen et al. 1993), and it seems likely therefore that the decrease in muscle GLUT4 is due to insulin deficiency per se, rather than to the associated hyperglycemia. On the other hand, it seems likely that the component of the decrease in responsiveness of glucose transport to insulin and contractions that can not be explained by the decrease in GLUT4 is mediated by the effect of the hyperglycemia associated with the insulin deficiency (Dimitrakoudis et al. 1992; Napoli et al. 1995).

8.2 Denervation/Immobilization

Denervation of skeletal muscle rapidly induces insulin resistance of glucose transport. Profound reductions in insulin-stimulated glucose uptake have been observed as early as 24 h following denervation (Buse and Buse 1959; Burant et al. 1984; Turinsky 1987; Henriksen et al. 1991; Coderre et al. 1992), although several groups have reported that a defect in sugar transport is not detectable until 2–3 days after denervation (Forsayeth and Gould 1982; Smith and Lawrence 1984; Davis and Karl 1988). Stimulation of muscle glycogen synthesis (Burant et al. 1984; Smith and Lawrence 1984; Davis and Karl 1988) and activation of glycogen synthase (Burant et al. 1984; Smith and Lawrence 1984) by insulin are also impaired within 24 h of denervation. Both the sensitivity and responsiveness of glucose transport and glycogen synthesis to activation by insulin appear to be reduced by denervation (Forsayeth and Gould 1982; Burant et al. 1984).

The time of onset and the severity of insulin resistance following devervation may be dependent on muscle fiber type. Studies in the denervated rat hindlimb preparation by Turinsky (1987) and Megeney et al. (1993, 1994) suggest that, in general, reductions in insulin stimulated glucose transport are greater and are detected earlier in muscles containing a high proportion of type I and type IIa compared to muscles consisting primarily of type IIb fibers. Several different findings indicate that the impairment in insulin action following muscle denervation is mediated by a postreceptor defect. Rat skeletal muscle insulin binding (Buse and Buse 1959; Forsayeth and Gould 1982; Burant et al. 1984) and activation of the insulin receptor tyrosine kinase by insulin (Burant et al. 1986) are not reduced by denervation. A postreceptor defect is also supported by the observation that rat soleus muscle glucose transport stimulated by vanadate, an insulin-mimetic agent that acts distally to the insulin-binding step, is markedly reduced 1–3 days following denervation (Henriksen et al. 1991). In addition, Henriksen et al. (1991) demonstrated that stimulation of soleus muscle 2DG uptake by muscle contractions is decreased approx. 30% at 1 and 3 days following denervation. Together, these findings suggest that the defect associated with denervation is not specific to the insulin signaling pathway but may involve a step common to the insulin- and contraction/hypoxia-stimulated pathways.

Since denervation affects both insulin- and contraction-stimulated glucose transport, an obvious potential explanation for the defect is a decrease in muscle GLUT4 expression. Several studies have shown 40%–80% decreases in skeletal muscle GLUT4 mRNA (Block et al. 1991; Coderre et al. 1992) and protein (Henriksen et al. 1991; Block et al. 1991; Coderre et al. 1992; Megeney et al. 1993) 2–3 days following hindlimb denervation; however, the effect of denervation on the resistance of glucose transport to activation by insulin or contractions is significant by 24 h (Burant et al. 1984; Henriksen et al. 1991), a time when significant changes in transporter message or protein levels have not been observed (Block et al. 1991; Henriksen et al. 1991; Coderre et al. 1992). Thus, these studies suggest that the early decrease in transport activation is independent of changes in GLUT4 expression, although longer periods of denervation are clearly associated with significant reductions in GLUT4 mRNA and protein. In addition to changes in GLUT4, denervation results in increases in GLUT1 mRNA and protein that are apparent between 1 (Coderre et al. 1992) and 3 days (Block et al. 1991) following denervation. This increase in GLUT1 expression may partially explain the increase in basal glucose transport in denervated muscle reported by some investigators (Smith and Lawrence 1984; Turinsky 1987; Davis and Karl 1988; Coderre et al. 1992).

Defects in muscle glucose transport regulation can also be induced by immobilization. Decreases in insulin-stimulated 2DG uptake, glucose oxidation, and glycogen synthesis in the mouse soleus have been reported within 24 h of hindlimb immobilization (Seider et al. 1982; Nicholson et al. 1984). Ploug et al. (1995) have recently reported that insulin binding, insulin receptor tyrosine kinase activity and muscle GLUT1 and GLUT4 content were not changed by 48 h of immobilization. There was also no decrease in 3MG transport into muscles of the perfused hindquarter when stimulated maximally with insulin and contractions in combination (Ploug et al. 1995). There did, however, appear to be modest defects in insulin-stimulated or contraction-stimulated glucose uptake in some muscles, but not in others (Vissing et al. 1988; Ploug et al. 1995).

8.3 Glucocorticoid Excess

Patients treated with glucocorticoid hormones or suffering from Cushing's disease commonly exhibit hepatic and peripheral insulin resistance (Amatruda et al. 1985). In humans, chronic glucocorticoid excess results in a decrease in maximally insulin-stimulated glucose utilization (Pagano et al. 1983; Nosadini et al. 1983). A decrease in insulin stimulation of glucose uptake has also been demonstrated in adipocytes and skeletal muscle isolated from rats treated with glucocorticoids (Riddick et al. 1962). The time of onset of the insulin resistance is fairly rapid, with marked reductions in insulin-stimulated glucose transport activity in rat skeletal muscle apparent within 2–5 days of the initiation of glucocorticoid administration. Riddick et al. (1962) observed that maximally insulin-stimulated 2DG and xylose uptakes in the isolated rat diaphragm were reduced approx. 45% following 4 days of hydrocortisone treatment (4 daily intramuscular injections of 5 mg/100 g BW). Similarly, administration of dexamethasone in the drinking water at a dose of 0.8 mg/day for 2 days resulted in a 38% decrease in basal and a 40%–60% decrease in maximally insulin-stimulated glucose transport activity in the isolated rat soleus muscle compared to pair-fed controls (Weinstein et al. 1995). Reductions in insulin responsiveness (approx. 50% decrease) were also observed in rats treated with 0.1 mg/day or 1.0 mg/day dexamethasone for 1 week (Haber and Weinstein 1992).The decrease in responsiveness was studied using both 3MG and 2DG to measure transport activity with similar results. In addition, the decreased response of glucose transport to insulin is not associated with a decrease in GLUT4 protein content, as GLUT4 levels were either unchanged or slightly increased in muscles of rats treated with dexamethasone for 2 days (Weinstein et al. 1995) or 1 week (Haber and Weinstein 1992).

There is some controversy regarding the effects of glucocorticoids on skeletal muscle insulin receptor content, insulin binding and insulin-stimulated receptor tyrosine kinase activity. Block and Buse (1989) reported no change in the number, insulin binding affinity, or kinase activity of solubilized skeletal muscle insulin receptors isolated following cortisone treatment (100 mg/kg BW/day for 5 days via subcutaneous injection), suggesting a glucocorticoid-induced postreceptor defect. However, in a study by Giorgino et al. (1993) utilizing the same cortisol administration protocol, immunoblots of skeletal muscle showed that total insulin receptor protein was increased 36%, but the total level of receptor tyrosine phosphorylation following in vivo insulin stimulation was decreased by 31%. In addition, muscle IRS-1 content was decreased approx. 50% by cortisone treatment, but the total level of IRS-1 tyrosine phosphorylation was not changed.

These finding suggest that glucocorticoid excess affects insulin action in skeletal muscle at both receptor and postreceptor levels. Saad et al. (1993) used similar methods for assessing insulin activation of the insulin receptor, IRS-1 and PI 3-kinase in skeletal muscle following 5 days of dexamethasone treatment (1 mg/kg BW for 5 days). These investigators found no significant change in muscle insulin receptor level or phosphorylation after in vivo insulin stimulation in dexamethasone-treated animals. Muscle levels of IRS-1 and PI 3-kinase were decreased by 60% and 35%, respectively. Although IRS-1 phosphorylation was not changed significantly, insulin-stimulated IRS-1-associated PI 3-kinase activity was decreased by 41% in the muscle of dexamethasone-treated rats. Thus, these findings suggest that glucocorticoid excess affects early insulin signaling events distal to the insulin receptor. In an interesting recent study by Weinstein et al. (1995), 2 days of treatment with dexamethasone resulted in similar decreases (approx. 60%) in maximally insulin-stimulated and hypoxia-stimulated glucose transport activity, suggesting that at least over this time period, the rate limiting step that is affected by corticosteroids is beyond the insulin signaling pathway.

8.4 Hyperglycemia/Glucose Toxicity

There is increasing evidence that hyperglycemia can cause insulin resistance. Studies of human type I diabetics (Yki-Jarvinen et al. 1987) or insulin-deficient rats (Rossetti et al. 1987) have shown that insulin-stimulated whole body glucose disposal is impaired by hyperglycemia. That skeletal muscle insulin resistance contributes to this hyperglycemia-induced reduction in whole body glucose disposal is suggested by studies examining the effects of glucose perfusion in the isolated rat hindquarter preparation. Richter et al. (1988a, b) found that insulin-stimulated glucose uptake in the perfused rat hindquarter was decreased approx. 65%–80% following a 5–7 h perfusion with 12 m*M* glucose. Reductions in 3MG uptake and increased concentrations of free glucose and G-6-P indicate that both transport and disposal of glucose were impaired in skeletal muscle of the glucose-perfused animals (Richter et al. 1988a, b). Impairments in insulin-stimulated whole body glucose disposal (Davidson et al. 1994) and skeletal muscle glucose uptake (assessed using the 2DG bolus technique; Hager et al. 1991; Davidson et al. 1994) have also been observed following 48–72 h of in vivo glucose infusion.

The relative contributions of glucose and insulin to the development of insulin resistance in the perfused hindquarter preparation were studied by Richter et al. (1988a). These investigators found that perfusion with glucose alone was almost as effective in reducing insulin-stimulated glucose uptake as perfusion with glucose plus 20 mU/ml insulin, indicating that the inhibition was mediated primarily by the hyperglycemia per se (Richter et al. 1988a). The severity of the resistance in this preparation is related to the concentration of glucose in the perfusate; exposure to increasing concentrations of glucose (0, 5, or 25 m*M* glucose) during a 2-h perfusion results in progressive reductions in maximally insulin-stimulated glucose uptake (Hansen et al. 1992). This decrease was observed regardless of the concentration of insulin present during the 2-h infusion, although the inhibitory effects of glucose were enhanced by insulin (Hansen et al. 1992).

The mechanism(s) responsible for the inhibitory effects of hyperglycemia on insulin-stimulated glucose transport in muscle are unclear. The inhibition appears to be mediated by a postreceptor defect, as muscle insulin receptor number and stimulation of receptor tyrosine kinase activity were not different between glucose- and saline-perfused rats (Hansen et al. 1992). Total GLUT4 levels are unchanged following 48 h in vivo glucose infusion (Davidson et al. 1994).

In studies where muscle glycogen concentration was raised in vivo by a 72-h glucose infusion (Hager et al. 1991) or in situ by glucose perfusion of the isolated hindquarter (Richter et al. 1988b; Hansen et al. 1992), the rates of insulin-stimulated glucose uptake and transport appear to be negatively correlated with the muscle glycogen concentration. This inverse relationship has also been demonstrated in rat muscle following glycogen-depleting exercise, where rates of insulin-stimulated glucose uptake in the perfused rat hindquarter (Fell et al. 1982) and 3MG transport in the isolated epitrochlearis muscle (Cartee et al. 1989) are lower in muscles in which glycogen is "supercompensated" compared to muscles in which glycogen is kept low by carbohydrate deprivation. Hespel and Richter (1990) demonstrated a negative relationship between the rate of 3MG transport into rat skeletal muscle during in situ muscle contractions and the precontraction glycogen concentration. In addition, glucose transport in skeletal muscle isolated from transgenic mice overexpressing GLUT1 is completely resistant to stimulation by insulin or by contractions; glycogen levels in quadriceps muscles of the GLUT1 overexpressing mice were 100–150 μmol/g wet wt compared to 15–20 μmol/g wet wt in muscle from non-transgenic littermates (Ren et al. 1993a; Gulve et al. 1994)

One possible mechanism by which glycogen levels could influence stimulation of glucose transport is the binding of GLUT4 vesicles to the muscle glycogen particles. In a preliminary study Coderre et al. (1994) reported that approx. 30% of the GLUT4 in rat skeletal muscle coprecipitated with glycogen, and that transporters could be released from the glycogen particles by amylase digestion. It is conceivable that as muscle glycogen concentration increases, more GLUT4 vesicles are bound to glycogen, resulting in a decrease in the effective number of GLUT4 vesicles available for translocation. Even if GLUT4 binding to glycogen does participate in the regulation of transport activity, the effects of hyperglycemia on skeletal muscle glucose transport cannot be explained entirely by such a mechanism. For example, in the 7 h glucose-perfused hindquarter, insulin-stimulated 3MG uptake decreased progressively during the course of the experiment even though muscle glycogen accumulation reached a plateau after 3–5 h of perfusion (Richter et al. 1988b).

An alternative explanation for the "toxic" effects of hyperglycemia on the activation of glucose transport may be related to increased synthesis of glucosamine. Marshall et al. (1991a) proposed that the downregulation of glucose transport activity observed in adipocytes following prolonged incubation with high insulin and glucose involved the metabolism of hexose-phosphates via the glucosamine pathway. This pathway is initiated by the conversion of fructose-6-phosphate to glucosamine-6-phosphate by the enzyme glutamine:fructose-6-phosphate amidotransferase (GFAT). The amount of glucosamine formed represents a very small fraction of the total glucose metabolized by adipocytes (2%–3%; Marshall et al. 1991a). However, these investigators postulated that if the rate of glucose flux into the cell increased beyond the capacity of the major glucose-utilizing pathways, then early metabolites of glucose metabolism, such as fructose-6-phosphate, accumulate. This results in an increased flux through the hexosamine biosynthetic pathway, with subsequent inhibition of insulin-stimulated glucose transporter translocation and glucose transport into the cell. In this way glucosamines could function as part of a "glucose sensing," negative-feedback system for regulation of glucose transport activity reviewed in Marshall et al. (1991b). Strong evidence for the participation of hexosamines in the loss of insulin-stimulatable glucose transport activity in adipocytes was provided by experiments in which inhibition of GFAT activity prevented the development of insulin resistance in cells cultured in the presence of glucose and insulin (Marshall et al. 1991a). In addition, prolonged incubation with glucosamine, in the absence of glucose, induced insulin resistance in adipocytes (Marshall et al. 1991a). At the time this is written, the contribution of the hexosamine pathway to various states of skeletal muscle insulin resistance is unclear. The potential for involvement of glucosamines in mediating skeletal muscle insulin resistance is suggested by the finding that a 2-h incubation of isolated rat diaphragms with glucosamine reduces insulin-stimulated 2DG transport (Robinson et al. 1993). Similarly, Rossetti et al. (1995) recently published a study in which in vivo glucosamine infusion resulted in reductions in insulin-stimulated whole body glucose disposal in normoglycemic rats as assessed by the euglycemic clamp technique. Interestingly, although the glucosamine infusion decreased insulin-stimulated glucose disposal to rates that were similar to those measured in rats made chronically hyperglycemic by pancreatectomy, the effects of the two stimuli in combination were not additive, suggesting that hyperglycemia and elevated glucosamines were inhibiting insulin action through a common mechanism.

8.5 Obesity

Insulin resistance of skeletal muscle glucose uptake is a characteristic of obesity in both humans and laboratory animals (Kemmer et al. 1979; Kraegen et al. 1986; Dohm et al. 1988). Although the presence of muscle insulin resistance in the obese state is widely recognized, the mechanisms of this reduction in insulin action are not well understood.

8.5.1 Genetic Models of Obesity

The genetically obese Zucker rat (fa/fa) has been widely used as a model to study insulin resistance. This animal model exhibits severe obesity, hyperinsulinaemia, hyperglycemia, impaired glucose tolerance and insulin resistance. A defect in the stimulation of skeletal muscle glucose uptake by insulin has been demonstrated by several groups (Davidson and Karjala 1978; Kemmer et al. 1979; Sherman et al. 1988; Brozinick et al. 1992, 1994a; Dolan et al. 1993). This resistance of muscle glucose uptake to insulin is mediated, at least in part, by a defect in GLUT4 translocation (King et al. 1992, 1993; Brozinick et al. 1994a). The smaller increase in plasma membrane GLUT4 in response to insulin cannot be explained by a decrease in the total available pool of transporters, as GLUT4 protein expression is not different between obese and lean Zucker rats (Friedman et al. 1990;

Banks et al. 1992; Brozinick et al. 1992; Kahn and Pedersen 1993). In addition, the defect in glucose transport appears to be specific to the insulin-stimulated pathway, as several groups have shown that skeletal muscle glucose uptake (Brozinick et al. 1992, 1994a; Dolan et al. 1993) and GLUT4 translocation (King et al. 1993; Brozinick et al. 1994a) in response to exercise or in situ muscle contractions are normal. Insulin binding is decreased (about 25%–35%; Crettaz et al. 1980) and insulin receptor tyrosine kinase activity is reduced (Slieker et al. 1990) in skeletal muscle from obese Zucker rats compared to lean controls. To the best of our knowledge, no other information regarding defects in the insulin signaling pathway in these animals is available at the time this is written.

Several studies have shown that exercise training increases insulin-stimulated (Ivy et al. 1986; Cortez et al. 1991; Banks et al. 1992; Brozinick et al. 1993) and contraction-stimulated (Brozinick et al. 1993) glucose uptake measured in the perfused hindquarter of the obese Zucker rat, although stimulation of transport in the soleus by insulin is apparently not improved by treadmill training in these animals (Crettaz et al. 1979; Brozinick et al. 1993; Torgan et al. 1995). The increase in muscle glucose uptake with training is mediated, at least in part, by an increase in total GLUT4 protein content (Banks et al. 1992; Brozinick et al. 1993; Torgan et al. 1995).

Another animal model of insulin resistance is the ob/ob mouse. This autosomal recessive mutation is characterized by obesity, hyperglycemia, hyperinsulinemia and peripheral insulin resistance (Mordes and Rossini 1985). As with the obese Zucker rat, the ob/ob mouse exhibits reduced skeletal muscle insulin binding (Le Marchand-Brustel and Freychet 1978; Le Marchand-Brustel et al. 1978a; Saad et al. 1992), apparently due to a decrease in receptor number. Muscle IRS-1 protein levels are not different, but insulin-stimulated IRS-1 tyrosine phosphorylation is 50% lower in ob/ob mice compared to lean controls. Insulin-stimulated 2DG uptake measured in the isolated soleus muscle is reduced (Le Marchand-Brustel and Freychet 1978; Le Marchand-Brustel et al. 1978b).

8.5.2
Experimental Models of Obesity

A high level of fat in the Western diet is a contributing factor in the development of insulin resistance and obesity. The feeding of a high fat diet to laboratory rats has been used as a model of the effects of dietary fat in humans, and there is evidence of reduced insulin-mediated glucose metabolism in insulin sensitive tissues isolated from animals fed high-fat diets. In one early study of the effects of 6 wks of fat feeding (72% of calories as fat) it was found that skeletal muscle insulin-stimulated glucose uptake, assessed in vivo (following intraperitoneal injection of insulin plus [^{14}C]glucose) or in vitro (incubated hemidaphragm), was reduced by a high fat diet (Susini and Lavau 1978). Euglycemic hyperinsulinemic clamps performed in rats fed a diet high in fat (60% calories as fat) for 3 weeks showed that fat feeding markedly reduced whole body glucose disposal compared to chow fed controls (Kraegen et al. 1986; Storlien et al. 1986). Over this duration on the high fat diet, no significant differences in body weight, plasma insulin or plasma glucose were detected. The relative contributions of individual tissues to the reductions in insulin-stimulated glucose uptake were assessed following a bolus injection of 2DG; reductions in insulin-stimulated 2DG accumulation in skeletal muscle of the fat-fed animals were observed in both studies (Kraegen et al. 1986; Storlien et al. 1986), suggesting that muscle glucose transport is affected by fat feeding. A defect at the level of transport has been confirmed in several studies assessing 2DG transport in vitro. Reductions in maximally insulin-stimulated 2DG transport (30%–35%) have been demonstrated in the isolated rat soleus muscle following 10–14 days of high fat feeding (60%–67% of calories; Grundleger and Thenen 1982; Maegawa et al. 1986). In both studies, the effect appeared to be on insulin responsiveness, with no significant change in sensitivity of glucose transport to insulin.

The cellular mechanisms of insulin resistance induced by feeding a high-fat diet are not well defined. Fat feeding has been reported to reduce (Grundleger and Thenen 1982; Maegawa et al. 1986) or have no effect on (Okamoto et al. 1992) insulin binding assessed in isolated skeletal muscles. This decrease in binding is apparently due to a decrease in receptor number, not a change in receptor affinity (Grundleger and Thenen 1982). The effects of fat feeding on insulin receptor tyrosine kinase activity are not clear. Okamoto et al. (1992) reported that insulin-stimulated tyrosine kinase activity, measured using wheat germ agglutinin-purified skeletal muscle receptors, was decreased 30% following 4 weeks of high fat feeding. However, in the same study, tyrosine phosphorylation of the insulin receptor and the in vivo substrate pp190 following insulin stimulation was not decreased in skeletal muscle of the fat fed animals. There is also some disagreement in the literature regarding the effect of fat feeding on muscle GLUT4 expression. One group has reported a decrease in muscle GLUT4 mRNA (47%) and protein (34%) in rats fed a high fat diet (80% of total calories) for 7 weeks (Kahn and Pedersen 1993). However, others have found no effect of 3–4 weeks of high fat feeding on muscle GLUT4 protein content (Okamoto et al. 1992; Kusunoki et al. 1993; Rosholt et al. 1994). The reasons for the discrepancy are unclear, although it may be related to the different durations on the high fat diet. Although it is well accepted that fat feeding results in reduced insulin-stimulated glucose transport in muscle, the effects of a high fat diet on glucose transport activated by the exercise/hypoxia pathway are not as clear. Kusunoki et al. (1993) reported that the effect of exercise on muscle glucose uptake (assessed using the bolus 2DG injection technique) was not altered by 3 weeks of fat feeding. However, Rosholt et al. (1994) found that glucose transport in sarcolemmal vesicles prepared from hindlimb muscles of fat fed rats following a bout of exercise or stimulated with insulin was reduced, suggesting that both pathways are affected.

Obesity has also been induced in mice by administration of gold thioglucose (GTG; Le Marchand et al. 1978b). As the fat fed obese rat, GTG obese mice exhibit a decrease in skeletal muscle insulin binding (Le Marchand-Brustel and Freychet 1978; Le Marchand-Brustel et al. 1978b) and a reduction in insulin stimulated 2DG uptake in isolated soleus muscles (Le Marchand-Brustel and Freychet 1978; Le Marchand-Brustel et al. 1978b). Heydrick et al. (1993) have reported that insulin receptor tyrosine kinase activity and insulin stimulation of PI 3-kinase activity are decreased in muscles of GTG obese mice. Interestingly, these defects can apparently be reversed by 48 h starvation (Heydrick et al. 1993).

8.5.3 Human Obesity

Studies using the euglycemic, hyperinsulinemic clamp technique (DeFronzo et al. 1978; Kolterman et al. 1980; Zuniga-Guajardo et al. 1986) have shown that insulin-stimulated glucose uptake is often impaired in obese individuals. Several studies of skeletal muscle isolated from morbidly obese patients and nonobese controls suggest that defects in muscle glucose transport contribute to the reduced insulin-stimulated glucose disposal observed in the obese state. Dohm et al. (1988) developed a human muscle preparation using samples of the rectus abdominus obtained from morbidly obese and nonobese patients undergoing surgery. Thin strips of this muscle are incubated in vitro for measurement of glucose transport activity. Maximally insulin-stimulated glucose transport activity is significantly impaired in muscles isolated from obese individuals compared to muscles from lean controls (Dohm et al. 1988; Friedman et al. 1992; Azevedo et al. 1995; Goodyear et al. 1995b). Insulin-stimulated transport was impaired to a similar extent in muscle strips from obese NIDDM and obese nondiabetic patients (Dohm et al. 1988); this obesity-associated impairment in muscle glucose transport can be reversed by weight loss (Friedman et al. 1992).

Several studies have demonstrated impaired insulin signaling in skeletal muscle of obese patients. Skeletal muscle insulin receptor number (Caro et al. 1987; Goodyear et al. 1995b) and insulin-stimulated receptor tyrosine kinase activity (Arner et al. 1987; Caro et al. 1987; Goodyear et al. 1995b) are reduced in obese patients compared to nonobese controls. It has been shown that expression of IRS-1 and PI 3-kinase is decreased (54% and 64%, respectively) in the rectus abdominus muscle isolated from obese subjects compared to lean controls (Goodyear et al. 1995b). Insulin-stimulation of tyrosine phosphorylation of IRS-1 was 35% lower, and insulin-stimulated IRS-1 associated PI 3-kinase activity was about 70% lower in muscles of obese subjects (Goodyear et al. 1995b). Altered muscle protein tyrosine phosphatase activity may also contribute to the obesity-associated insulin resistance in human skeletal muscle (McGuire et al. 1991; Kusari et al. 1994). It is not clear whether these defects are responsible for mediating the insulin resistance, or are a consequence of chronic exposure of muscle to the hyperinsulinemia that exists in these obese individuals. There is no evidence that decreased expression of muscle GLUT4 protein is the mechanism responsible for the reductions in insulin-stimulated transport observed in obese patients. Several different groups have found no differences in GLUT4 mRNA (Pedersen et al. 1990; Garvey et al. 1992) or protein (Handberg et al. 1990; Pedersen et al. 1990; Friedman et al. 1992; Garvey et al. 1992) expression in obese nondiabetic, obese NIDDM and lean nondiabetic patients. Little is known about whether the defect in glucose transport activation in human muscle from obese individuals is specific to the insulin pathway, but one group has reported that hypoxia-stimulated 2DG uptake in rectus abdominus strips was not significantly different between obese and lean subjects (Azevedo et al. 1995).

9 Unanswered Questions

Despite the rapid progress in recent years, many of the key questions regarding the regulation of glucose transport in skeletal muscle are still unanswered. While the initial steps in the insulin signaling pathway have been identified, the subsequent steps and the signal that is generated to initiate exocytosis of the GLUT4 containing vesicles remain to be discovered. The proteins in the GLUT4 vesicles that serve as receptors for the signals generated by insulin and exercise, and the mechanism by which the interactions between the signaling molecules and the GLUT4 vesicle proteins activate exocytosis, are also still a mystery. The process responsible for movement of the GLUT4 vesicles to, and fusion with, the cell membrane are likewise still unknown. While the structure of the glucose transporter after it has become a component of the plasma membrane has been elucidated to a considerable extent, the mechanism by which it transports glucose from the interstitial fluid into the cytoplasm remains speculative. Another unanswered question relates to the mechanisms by which exercise induces an increase in muscle insulin sensitivity. This phenomenon is of particular interest because of its potential health benefits. We look forward to further developments in this research area with great interest.

References

Almers W (1990) Exocytosis. Annu Rev Physiol 52:607–624

Amatruda JM, Livingston JN, Lockwood DH (1985) Cellular mechanisms in selected states of insulin resistance: human obesity, glucocorticoid excess, and chronic renal failure. Diabetes Metab Rev 1:293–317

Andersen PH, Lund S, Vestergaard H, Junker S, Kahn B, Pedersen O (1993) Expression of the major insulin regulatable glucose transporter (GLUT4) in skeletal muscle on non insulin dependent diabetic patients and healthy subjects before and after insulin infusion. J Clin Endocrinol Metab 77:27–32

Ariano MA, Armstrong RB, Edgerton VR (1973) Hindlimb muscle fiber populations of five animals. J Histochem Cytochem 21:51–55

Arner P, Pollare T, Lithell H, Livingston JN (1987) Defective insulin receptor tyrosine kinase in human skeletal muscle in obesity and Type 2 (non-insulin-dependent) diabetes mellitus. Diabetologia 30:437–440

Azevedo JR, Carey JO, Pories WJ, Morris PG, Dohm GL (1995) Hypoxia stimulates glucose transport in insulin-resistant human skeletal muscle. Diabetes 44:695–698

Baldini G, Hohman R, Charron MJ, Lodish HF (1991) Insulin and nonhydrolyzable GTP analogs induce translocation of GLUT4 to the plasma membrane in α-toxin-permeabilized rat adipose cells. J Biol Chem 266:4037–4040

Baldwin SA, Baldwin JM, Gorga FR, Lienhard GE (1979) Purification of the cytochalasin B binding component of the human erythrocyte monosaccharide transport system. Biochim Biophys Acta 552:183–188

Banks EA, Brozinick JT Jr, Yaspelkis BB III, Kang HY, Ivy JL (1992) Muscle glucose transport, GLUT-4 content, and degree of exercise training in obese Zucker rats. Am J Physiol 263:E1010–E1015

Barnard RJ, Youngren JF (1992) Regulation of glucose transport in skeletal muscle. FASEB J 6:3238–3244

Baron AD, Steinberg H, Brechtel G, Johnson A (1994) Skeletal muscle blood flow independently modulates insulin-mediated glucose uptake. Am J Physiol 266:E248–E253

Bashan N, Burdett E, Guma A, Sargeant R, Tumiati L, Liu Z, Klip A (1993) Mechanisms of adaptation of glucose transporters to changes in the oxidative chain of muscle and fat cells. Am J Physiol 264:C430–C440

Bell GI, Burant CF, Takeda J, Gould GW (1993) Structure and function of mammalian facilitative sugar transporters. J Biol Chem 268:19161–19164

Bergström J, Hultman E (1966) Muscle glycogen synthesis after exercise: an enhancing factor localized to the muscle cells in man. Nature 210:309–310

Bergström J, Hermansen L, Hultman E, Saltin B (1967) Diet, muscle glycogen and physical performance. Acta Physiol Scand 71:140–150

Bergström J, Hultman E, Roch-Norlund AE (1972) Muscle glycogen synthetase in normal subjects. Scand J Clin Lab Invest 29:231–236

Bihler I, Charles P, Sawh PC (1980) Effects of the calcium ionophore A-23187 on the regulation of sugar transport in muscle. Cell Calcium 1:327–336

Birnbaum MJ (1989) Identification of a novel gene encoding an insulin-responsive glucose transporter protein. Cell 57:305–315

Block NE, Buse MG (1989) Effects of hypercortisolemia and diabetes on skeletal muscle insulin receptor function in vitro and in vivo. Am J Physiol 256:E39–E48

Block NE, Menick DR, Robinson KA, Buse MG (1991) Effect of denervation on the expression of two glucose transporter isoforms in rat hindlimb muscle. J Clin Invest 88:1546–1552

Bonen A, Tan MH, Watson-Wright WM (1984) Effects of exercise on insulin-binding and glucose metabolism in muscle. Can J Physiol Pharmacol 62:1500–1504

Bonen A, Tan MH, Clune P, Kirby RL (1985) Effects of exercise on insulin-binding to human muscle. Am J Physiol 248:E403–E408

Bonen A, Clark MG, Henriksen EJ (1994) Experimental approaches in muscle metabolism: hindlimb perfusion and isolated muscle incubations. Am J Physiol 266:E1–E16

Bornemann A, Ploug T, Schmalbruch H (1992) Subcellular localization of GLUT4 in nonstimulated and insulin-stimulated soleus muscle of rat. Diabetes 41:215–221

Bourey RE, Koranyi L, James DE, Mueckler M, Permutt MA (1990) Effects of altered glucose homeostasis on glucose transporter expression in skeletal muscle of the rat. J Clin Invest 86:542–547

Brozinick JT Jr, Etgen GJ Jr, Yaspelkis BB III, Ivy JL (1992) Contraction-activated glucose uptake is normal in insulin-resistant muscle of the obese Zucker rat. J Appl Physiol 73:382–387

Brozinick JT Jr, Etgen GJ Jr, Yaspelkis BB III, Kang HY, Ivy JL (1993) Effects of exercise training on muscle GLUT-4 protein content and translocation in obese Zucker rats. Am J Physiol 265:E419–E427

Brozinick JT Jr, Etgen GJ Jr, Yaspelkis BB III, Ivy JL (1994a) Glucose uptake and GLUT-4 protein distribution in skeletal muscle of the obese Zucker rat. Am J Physiol 267:R236–R243

Broznick JT Jr, Etgen GJ, Yaspelkis BB III, Ivy JL (1994b) The effects of muscle contraction and insulin on glucose-transporter translocation in rat skeletal muscle. Biochem J 297:539–545

Burant CF, Lemon SK, Treutelaar MK, Buse MG (1984) Insulin resistance of denervated rat muscle: a model for impaired receptor-function coupling. Am J Physiol 247:E657–E666

Burant CF, Treutelaar MK, Buse MG (1986) In vitro and in vivo activation of the insulin receptor kinase in control and denervated skeletal muscle. J Biol Chem 261:8985–8993

Buse MG, Buse J (1959) Glucose uptake and response to insulin of the isolated rat diaphragm: the effect of denervation. Diabetes 8:218–225

Cain CC, Trimble WS, Lienhard GE (1992) Members of the VAMP family of synaptic vesicle proteins are components of glucose transporter-containing vesicles from rat adipocytes. J Biol Chem 267:11681–11684

Calderhead DM, Kitagawa K, Lienhard GE, Gould GW (1990) Translocation of the brain-type glucose transporter largely accounts for insulin stimulation of glucose transport in BC3H-1 myocytes. Biochem J 269:597–601

Carlsen RC, Larson Db, Walsh DA (1985) A fast-twitch oxidative glycolytic skeletal muscle with a robust inward calcium current. Can J Physiol Pharmacol 63:958–965

Caro JF, Sinha MK, Raju SM, Ittoop O, Pories WJ, Flickinger EG, Meelheim D, Dohm GL (1987) Insulin receptor kinase in human skeletal muscle from obese subjects with and without noninsulin dependent diabetes. J Clin Invest 79:1330–1337

Carruthers A (1990) Facilitated diffusion of glucose. Physiol Rev 70:1135–1176

Cartee GD, Holloszy JO (1990) Exercise increases susceptibility of muscle glucose transport to activation by various stimuli. Am J Physiol 258:E390–E393

Cartee GD, Young DA, Sleeper MD, Zierath J, Wallberg-Henriksson H, Holloszy JO (1989) Prolonged increase in insulin-stimulated glucose transport in muscle after exercise. Am J Physiol 256:E494–E499

Cartee GD, Douen AG, Ramlal T, Klip A, Holloszy JO (1991) Stimulation of glucose transport in skeletal muscle by hypoxia. J Appl Physiol 70:1593–1600

Casla A, Rovira A, Wells JA, Dohm GL (1990) Increased glucose transporter (GLUT4) protein expression in hyperthyroidism. Biochem Biophys Res Commun 171:182–188

Charron MJ, Brosius FC III, Alper SL, Lodish HF (1989) A glucose transport protein expressed predominantely in insulin-responsive tissues. Proc Natl Acad Sci USA 86:2535–2539

Charron MJ, Kahn BB (1990) Divergent molecular mechanisms for insulin-resistant glucose transport in muscle and adipose cells vivo. J Biol Chem 265:7994–8000

Cheng K, Creacy S, Larner J (1983) Insulin-like effects of lithium ion on isolated rat adipocytes. II. Specific activation of glycogen synthase. Mol Cell Biochem 56:183–189

Clausen T (1968) The relationship between the transport of glucose and cations across cell membranes in isolated tissues. Biochem Biophys Acta 150:66–72

Clausen T, Elbrink J, Dahl-Hansen AB (1975) The relationship between the transport of glucose and cations across cell membranes in isolated tissues. IX. The role of cellular calcium in the activation of the glucose transport system in rat soleus muscle. Biochem Biophys Acta 375:292–308

Coderre L, Monfar MM, Chen KS, Heydrick SJ, Kurowski TG, Ruderman NB, Pilch PF (1992) Alteration in the expression of GLUT1 and GLUT4 protein and messenger RNA levels in denervated rat muscles. Endocrinology 131:1821–1825

Coderre L, Vallega G, Pilch PF (1994) Association of GLUT4 vesicles with glycogen particles in skeletal muscle. Identification of a contraction-sensitive pool. Diabetes 43 [Suppl 1]:159a

Conlee RK, Hickson RC, Winder WW, Hagberg JM, Holloszy JO (1978) Regulation of glycogen resynthesis in muscle of rats following exercise. Am J Physiol 235:R145–R150

Constable SH, Young JC, Higuchi M, Holloszy JO (1984) Glycogen resynthesis in leg muscles of rats during exercise. Am J Physiol 247:R880–R883

Constable SH, Favier RJ, Cartee GD, Young DA, Holloszy JO (1988) Muscle glucose transport: interactions of vitro contractions, insulin and exercise. J Appl Physiol 64:2329–2332

Cormont M, Tanti J-F, Gremeaux T, Van Obberghen E, Le Marchand-Brustel Y (1991) Subcellular distribution of low molecular weight guanosine triphosphate-binding proteins in adipocytes: colocalization with the glucose transporter Glut 4. Endocrinology 129:3343–3350

Cormont M, Gremeaux T, Tanti J, Van Obberghen E, Le Marchand-Brustel Y (1992) Polymyxin B inhibits insulin-induced glucose transporter and IGF-II receptor translocation in isolated adipocytes. Eur J Biochem 207:185–193

Cormont M, Tanti J, Zahraoui A, Van Obberghen E, Tavitian A, Le Marchand-Brustel Y (1993) Insulin and okadaic acid induce Rab4 redistribution in adipocytes. J Biol Chem 268:19491–19497

Cortez MY, Torgan CE, Brozinick JT Jr, Ivy JL (1991) Insulin resistance of obese Zucker rats exercise trained at two different intensities. Am J Physiol 261:E613–E619

Crettaz M, Horton ES, Wardzala LJ, Horton ED, Jeanrenaud B (1979) Physical training of Zucker rats: lack of alleviation of muscle insulin resistance. Am J Physiol 237:E414–E420

Crettaz M, Prentki M, Zaninetti D, Jeanrenaud B (1980) Insulin resistance in soleus muscle from obese Zucker rats. Involvement of several defective sites. Biochem J 186:525–534

Cushman SW, Wardzala LJ (1980) Potential mechanism of insulin action on glucose transport in the isolated rat adipose cell. J Biol Chem 255:4758–4762

Davidson MB, Karjala RG (1978) Primary insulin antagonism of glucose transport in muscle from the older-obese rat. Metabolism 27:1994–2005

Davidson MB, Bouch C, Venkatesan N, Karjala RG (1994) Impaired glucose transport in skeletal muscle but normal GLUT4 tissue distribution in glucose-infused rats. Am J Physiol 267:E808–E813

Davis TA, Karl IE (1988) Resistance of protein and glucose metabolism to insulin in denervated rat muscle. Biochem J 254:667–675

De Camilli P, Jahn R (1990) Pathways to regulated exocytosis in neurons. Annu Rev Physiol 52:625–645

DeFronzo RA, Soman V, Sherwin RS, Hendler R, Felig P (1978) Insulin binding to monocytes and insulin action in human obesity, starvation, and refeeding. J Clin Invest 62:204–213

Dela F, Ploug T, Handberg A, Petersen LN, Larsen JJ, Mikines KJ, Galbo H (1994) Physical training increases muscle GLUT4 protein and mRNA in patients with NIDDM. Diabetes 43:862–865

Dela F, Handberg A, Mikines KJ, Vinten J, Galbo H (1993) GLUT4 and insulin receptor binding and kinase activity in trained human muscle. J Physiol (Lond) 469:615–624

DeMartino GN, Croall DE (1987) Calcium-dependent proteases: a prevalent proteolytic system of uncertain function. NIPS 2:82–85

Dettbarn C, Palade P (1991) Effects of alkaline pH on sarcoplasmic reticulum Ca^{2+} release and Ca2+ uptake. J Biol Chem 266:8993–9001

Dimitrakoudis D, Ramlal T, Rastogi S, Vranic M, Klip A (1992) Glycaemia regulates the glucose transporter number in the plasma membrane of rat skeletal muscle. Biochem J 284:341–348

Dohm GL, Tapscott EB, Pories WJ, Dabbs DJ, Flickinger EG, Meelheim D, Fushiki T, Atkinson SM, Elton CW, Caro JF (1988) An in vitro human muscle preparation suitable for metabolic studies. J Clin Invest 82:486–494

Dolan PL, Tapscott EB, Dorton PJ, Dohm GL (1993) Contractile activity restores insulin responsiveness in skeletal muscle of obese Zucker rats. Biochem J. 289:423–426

Douen AG, Ramlal T, Klip A, Young DA, Cartee GD, Holloszy JO (1989) Exercise-induced increase in glucose transporters in plasma membranes of rat skeletal muscle. Endocrinology 124:449–454

Douen AG, Ramlal T, Cartee GD, Klip A (1990a) Exercise modulates the insulin-induced translocation of glucose transporters in rat skeletal muscle. FEBS Lett 261:256–260

Douen AG, Ramlal T, Rastogi S, Bilan PJ, Cartee GD, Vranic M, Holloszy JO, Klip A (1990b) Exercise induces recruitment of the "insulin-responsive glucose transporter." J Biol Chem 265:13427–13430
Draznin B, Sussman K, Kao M, Lewis D, Sherman N (1987) The existence of an optimal range of cytosolic free calcium for insulin-stimulated glucose transport in rat adipocytes. J Biol Chem 262:14385–14388
Draznin B, Lewis D, Houlder N, Sherman N, Adamo M, Garvey WT, LeRoith D, Sussman K (1989) Mechanism of insulin resistance induced by sustained levels of cytosolic free calcium in rat adipocytes. Endocrinology 125:2341–2349
Dulin WE, Clark JJ (1961) Studies concerning a possible humoral factor produced by working muscles. Its influence on glucose utilization. Diabetes 10:289–297
Ebashi S (1976) Excitation-contraction coupling. Annu Rev Physiol 38:293–313
Etgen GJ, Bronzinick JT, Kang HY, Ivy JL (1993a) Effects of exercise training on skeletal muscle glucose uptake and transport. Am J Physiol 264:C727–C733
Etgen GJ, Memon AR, Thompson CM Jr, Ivy JL (1993b) Insulin- and contraction-stimulated translocation of GTP-binding proteins and GLUT4 protein in skeletal muscle. J Biol Chem 268:20164–20169
Ezaki O, Higuchi M, Nakatsuka H, Kawanaka K, Itakura H (1992) Exercise training increases glucose transporter content in skeletal muscles more efficiently from aged obese rats than young lean rats. Diabetes 41:920–926
Fell RD, Terblanche SE, Ivy JL, Young JC, Holloszy JO (1982) Effects of muscle glycogen content on glucose uptake by muscle following exercise. J Appl Physiol 52:434–437
Fitch CD, Jellinek M, Mueller EJ (1974) Experimental depletion of creatine and phosphocreatine from skeletal muscle. J Biol Chem 249:1060–1063
Fitch CD, Jellinek M, Fitts RH, Baldwin KM, Holloszy JO (1975) Phosphorylated β-guanidinopropionate as a substitute for phosphocreatine in rat muscle. Am J Physiol 228:1123–1125
Folli F, Saad MJA, Backer JM, Kahn CR (1992) Insulin stimulation of phosphatidylinositol 3-kinase activity and association with insulin receptor substrate 1 in liver and muscle of the intact rat. J Biol Chem 267:22171–22177
Folli F, Saad MJA, Backer JM, Khan CR (1993) Regulation of phosphatidylinositol 3-kinase activity in liver and muscle of animal models of insulin-resistant and insulin-deficient diabetes mellitus. J Clin Invest 92:1787–1794
Forsayeth JR, Gould MK (1982) Inhibition of insulin-stimulated xylose uptake in denervated rat soleus muscle: a post-receptor effect. Diabetologia 23:511–516
Friedman JE, Sherman WM, Reed MJ, Elton CW, Dohm GL (1990) Exercise-training increases glucose transporter protein GLUT4 in skeletal muscle of obese Zucker (fa/fa) rats. FEBS Lett 268:13–16
Friedman JE, Dohm GL, Leggett-Frazier N, Elton CW, Tapscott EB, Pories WP, Caro JF (1992) Restoration of insulin responsiveness in skeletal muscle of morbidly obese patients after weight loss. J Clin Invest 89:701–705
Fukumoto H, Kayno T, Buse JB, Edwards Y, Pilch PF, Bell GI, Seino S (1989) Cloning and characterization of the major insulin-responsive glucose transporter expressed

in human skeletal muscle and other insulin-responsive tissues. J Biol Chem 264:7776–7779

Furler SM, Jenkins AB, Storlien LH, Kraegen EW (1991) In vivo location of the rate-limiting step of hexose uptake in muscle and brain tissue of rats. Am J Physiol 261:E337–E347

Fushiki T, Wells JA, Tapscott EB, Dohm GL (1989) Changes in glucose transporters in muscle in response to exercise. Am J Physiol 256:E580–E587

Gao J, Gulve EA, Holloszy JO (1994a) Contraction-induced increase in muscle insulin sensitivity: requirement for a serum factor. Am J Physiol 266:E186–E192

Gao J, Ren J, Gulve EA, Holloszy JO (1994b) Additive effect of contractions and insulin on GLUT4 translocation into the sarcolemma. J Appl Physiol 77:1597–1601

Garetto LP, Richter EA, Goodman MN, Ruderman NB (1984) Enhanced muscle glucose metabolism after exercise in the rat: the two phases. Am J Physiol 246:E471–E475

Garthwaite SM, Holloszy JO (1982) Increased permeability to sugar following muscle contraction. Inhibitors of protein synthesis prevent reversal of the increase in 3-methylglucose transport rate. J Biol Chem 257:5008–5012

Garvey WT, Huecksteadt TP, Birnbaum MJ (1989) Pretranslational suppression of an insulin-responsive glucose transporter in rats with diabetes mellitus. Science 245:60–63

Garvey WT, Maianu L, Hancock JA, Golichowski AM, Baron A (1992) Gene expression of GLUT4 in skeletal muscle from insulin-resistant patients with obesity, IGT, GDM, and NIDDM. Diabetes 41:465–475

Giorgino F, Almahfouz A, Goodyear LJ, Smith RJ (1993) Glucocorticoid regulation of insulin receptor and substrate IRS-1 tyrosine phosphorylation in rat skeletal muscle in vivo. J Clin Invest 91:2020–2030

Goldstein MS, Mullick V, Huddleston B, Levine R (1953) Action of muscular work on transfer of sugars across cell barriers: comparison with action of insulin. Am J Physiol 173:212–216

Gomperts BD (1990) G_E: a GTP-binding protein mediating exocytosis. Annu Rev Physiol 52:591–606

Goodman MN, Ruderman NB (1979) Insulin sensitivity of rat skeletal muscle: effects of starvation and aging. Am J Physiol 236:E519–E523

Goodyear LJ, Hirshman MF, King PA, Horton ED, Thompson CM, Horton ES (1990a) Skeletal muscle plasma membrane glucose transport and glucose transporters after exercise. J Appl Physiol 68:193–198

Goodyear LJ, King PA, Hirshman MF, Thompson CM, Horton ED, Horton ES (1990b) Contractile activity increases plasma membrane glucose transporters in absence of insulin. Am J Physiol 258:E667–E672

Goodyear LJ, Hirshman MF, Horton ES (1991) Exercise-induced translocation of skeletal muscle glucose transporters. Am J Physiol 261:E795–E799

Goodyear LJ, Hirshman MF, Valyou PM, Horton ES (1992) Glucose transporter number, function, and subcellular distribution in rat skeletal muscle after exercise training. Diabetes 41:1091–1099

Goodyear LJ, Giorgino F, Balon TW, Condorelli G, Smith RJ (1995a) Effects of contractile activity on tyrosine phosphoproteins and PI 3-kinase activity in rat skeletal muscle. Am J Physiol 268:E987–E995

Goodyear LJ, Giorgino F, Sherman LA, Carey J, Smith RJ, Dohm GL (1995b) Insulin receptor phosphorylation, insulin receptor substrate-1 phosphorylation, and phosphatidylinositol 3-kinase activity are decreased in intact skeletal muscle strips from obese subjects. J Clin Invest 95:2195–2204

Greco-Perotto R, Wertheimer E, Jeanrenaud B, Cerasi E, Sasson S (1992) Glucose regulates its transport in L8 myocytes by modulating cellular trafficking of the transporter GLUT-1. Biochem J 286:157–163

Green A (1986) The insulin-like effect of sodium vanadate on adipocyte glucose transport is mediated at a post-insulin receptor level. Biochem J 238:663–669

Grimditch GK, Barnard RJ, Kaplan SA, Sternlicht E (1986) Effect of training on insulin binding to rat skeletal muscle sarcolemmal vesicles. Am J Physiol 250:E570–E575

Grundleger ML, Thenen SW (1982) Decreased insulin binding, glucose transport, and glucose metabolism in soleus muscle of rats fed a high fat diet. Diabetes 31:232–237

Gulve EA, Holloszy JO (1993) Sphingosine inhibits glucose transport activated by insulin or contractile activity in rat skeletal muscle. Biochem Life Sci Adv 12:75–80

Gulve EA, Cartee GD, Zierath JR, Corpus VM, Holloszy JO (1990) Reversal of enhanced muscle glucose transport after exercise: roles of insulin and glucose. Am J Physiol 259:E685–E691

Gulve EA, Ren J, Marshall BA, Gao J, Hansen PA, Holloszy JO, Mueckler M (1994) Glucose transport activity in skeletal muscles from transgenic mice overexpressing GLUT1. J Biol Chem 269:18366–18370

Haber RS, Weinstein SP (1992) Role of glucose transporters in glucocorticoid-induced insulin resistance. GLUT4 isoform in rat skeletal muscle is not decreased by dexamethasone. Diabetes 41:728–735

Hager SR, Jochen AL, Kalkhoff RK (1991) Insulin resistance in normal rats infused with glucose for 72 h. Am J Physiol 260:E353–E362

Handberg A, Vaag A, Damsbo P, Beck Nielsen H, Vinten J (1990) Expression of insulin regulatable glucose transporters in skeletal muscle from type 2 (non-insulin-dependent) diabetic patients. Diabetologia 33:625–627

Hansen BF, Hansen SA, Ploug T, Bak JF, Richter EA (1992) Effects of glucose and insulin on development of impaired insulin action in muscle. Am J Physiol 262:E440–E446

Hansen PA, Gulve EA, Holloszy JO (1994) Suitability of 2-deoxyglucose for in vitro measurement of glucose transport activity in skeletal muscle. J Appl Physiol 76:979–985

Hansen PA, Gulve E, Gao J, Schluter J, Mueckler M, Holloszy JO (1995a) Kinetics of 2-deoxyglucose transport in skeletal muscle: effects of insulin and contractions. Am J Physiol 268:C30–C35

Hansen PA, Gulve EA, Marshall BA, Gao J, Pessin JE, Holloszy JO, Mueckler M (1995b) Skeletal muscle glucose transport and metabolism are enhanced in transgenic mice overexpressing the GLUT4 glucose transporter. J Biol Chem 270:1679–1684

Haugaard ES, Mickel RA, Haugaard N (1974) Actions of lithium ions and insulin on glucose utilization, glycogen synthesis and glycogen synthase in the isolated rat diaphragm. Biochem Pharmacol 23:1675–1685

Helmreich E, Cori CF (1957) Studies of tissue permeability. II. The distribution of pentoses between plasma and muscle. J Biol Chem 224:663–679

Henriksen EJ, Holloszy JO (1991) Effect of diffusion distance on measurement of rat skeletal muscle glucose transport vitro. Acta Physiol Scand 143:381–386

Henriksen EJ, Rodnick KJ, Holloszy JO (1989a) Activation of glucose transport in skeletal muscle by phospholipase C and phorbol ester. Evaluation of the regulatory roles of protein kinase C and calcium. J Biol Chem 264:21536–21543

Henriksen EJ, Sleeper MD, Zierath JR, Holloszy JO (1989b) Polymyxin B inhibits stimulation of glucose transport in muscle by hypoxia or contractions. Am J Physiol 256:E662–E667

Henriksen EJ, Bourey RE, Rodnick KJ, Koranyi L, Permutt MA, Holloszy JO (1990) Glucose transporter protein content and glucose transport capacity in rat skeletal muscles. Am J Physiol 259:E593–E598

Henriksen EJ, Rodnick KJ, Mondon CE, James DE, Holloszy JO (1991) Effect of denervation or unweighting on GLUT4 protein in rat soleus muscle. J Appl Physiol 70:2322–2327

Henriksson J, Chi MM-Y, Hintz CS, Young DA, Kaiser R, Salmons S, Lowry OH (1986) Chronic stimulation of mammalian muscle: changes in enzymes of six metabolic pathways. Am J Physiol 251:C614–C632

Hespel P, Richter EA (1990) Glucose uptake and transport in contracting, perfused rat muscle with different pre-contraction glycogen concentrations. J Physiol (Lond) 427:347–359

Hespel P, Vergauwen L, Vandenberghe K, Richter EA (1995) Important role of insulin and flow in stimulating glucose uptake in contracting skeletal muscle. Diabetes 44:210–215

Heydrick SJ, Jullien D, Gautier N, Tanti J, Giorgetti S, Van Obberghen E, Le Marchand-Brustel Y (1993) Defect in skeletal muscle phosphatidylinositol-3-kinase in obese insulin-resistant mice. J Clin Invest 91:1358–1366

Hidaka H, Sasaki Y, Tanaka T, Endo T, Ohno S, Fujii Y, Nagata T (1981) N-(6-aminohexyl)-5-chloro-1-naphthalenesulfonamide, a calmodulin antagonist, inhibits cell proliferation. Proc Natl Acad Sci USA 78:4354–4357

Hirshman MF, Wallberg-Henriksson H, Wardzala LJ, Horton ED, Horton ES (1988) Acute exercise increases the number of plasma membrane glucose transporters in rat skeletal muscle. FEBS Lett 238:235–239

Hirshman MF, Goodyear LJ, Wardzala LJ, Horton ED, Horton ES (1990) Identification of an intracellular pool of glucose transporters from basal and insulin-stimulated rat skeletal muscle. J Biol Chem 265:987–991

Hofmann S, Pette D (1994) Low-frequency stimulation of rat fast-twitch muscle enhances the expression of hexokinase II and both the translocation and expression of glucose transporter 4 (GLUT-4). Eur J Biochem 219:307–315
Holloszy JO (1967) Biochemical adaptations in muscle. Effects of exercise on mitochondrial O2 uptake and respiratory enzyme activity in skeletal muscle. J Biol Chem 242:2278–2282
Holloszy JO, Booth FW (1976) Biochemical adaptations to endurance exercise in muscle. Annu Rev Physiol 38:273–291
Holloszy JO, Coyle EF (1984) Adaptations of skeletal muscle to endurance exercise and their metabolic consequences. J Appl Physiol 56:831–839
Holloszy JO, Narahara HT (1965) Studies of tissue permeability. X. Changes in permeability to 3-methylglucose associated with contraction of isolated frog muscles. J Biol Chem 240:3493–3500
Holloszy JO, Narahara HT (1967a) Nitrate ions: potentiation of increased permeability to sugar associated with muscle contraction. Science 155:573–575
Holloszy JO, Narahara HT (1967b) Enhanced permeabilty to sugar associated with muscle contraction. Studies of the role of Ca^{++}. J Gen Physiol 50:551–561
Holman GD, Kozka IJ, Clark AE, Flower CJ, Saltis J, Habberfield AD, Simpson IA, Cushman SW (1990) Cell surface labeling of glucose transporter isoform GLUT4 by bis-mannose photolabel. J Biol Chem 265:18172–18179
Horn RS, Walaas O, Walaas E (1973) The influence of sodium, potassium and lithium on the response of glycogen synthetase I to insulin and epinephrine in the isolated rat diaphragm. Biochem Biophys Acta 313:296–309
Horton EG (1983) Increased insulin sensitivity without altered insulin-binding in rat soleus muscle. Excerpta Med Int Congr Ser 577:182
Houmard JA, Shinebarger MH, Dolan PL, Leggett-Frazier N, Bruner RK, McCammon MR, Israel RG, Dohm GL (1993) Exercise training increases GLUT-4 protein concentration in previously sedentary middle-aged men. Am J Physiol 264:E896–E901
Hughes VA, Fiatarone MA, Fielding RA, Kahn BB, Ferrara CM, Shepherd P, Fisher EC, Wolfe RR, Elahi D, Evans WJ (1993) Exercise increases muscle GLUT-4 levels and insulin action in subjects with impaired glucose tolerance. Am J Physiol 264:E855–E862
Idström J-P, Elander A, Soussi B, Schersten T, Bylund-Fellenius A-C (1986) Influence of endurance training on glucose transport and uptake in rat skeletal muscle. Am J Physiol 251:H903–H907
Ingle DJ, Nezamis JE, Rice KL (1950) Work output and blood glucose values in normal and in diabetic rats subjected to the stimulation of muscle. Endocrinology 46:505–509
Ivy JL, Holloszy JO (1981) Persistent increase in glucose uptake by rat skeletal muscle following exercise. Am J Physiol 241:C200–C203
Ivy JL, Young JC, McLane JA, Fell RD, Holloszy JO (1983) Exercise-training and glucose uptake by skeletal muscle in rats. J Appl Physiol 55:1393–1396

Ivy JL, Sherman WM, Cutler CL, Katz AL (1986) Exercise and diet reduce muscle insulin resistance in obese Zucker rat. Am J Physiol 251:E299–E305

Jahn R, Sudhof TC (1994) Synaptic vesicles and exocytosis. Annu Rev Physiol 17:219–246

James DE, Kraegen EW, Chisholm DJ (1985) Effects of exercise-training on vivo insulin action in individual tissues of the rat. J Clin Invest 76:657–666

James DE, Lederman L, Pilch PF (1987) Purification of insulin-dependent exocytic vesicles containing the glucose transporter. J Biol Chem 262:11817–11824

James DE, Brown R, Navarro J, Pilch PF (1988) Insulin-regulatable tissues express a unique insulin-sensitive glucose transport protein. Nature 333:183–185

James DE, Strube M, Mueckler M (1989) Molecular cloning and characterization of an insulin-regulatable glucose transporter. Nature Lond 338:83–87

Jhun BH, Rampal AL, Lui H, Lachaal M, Jung CY (1992) Effects of insulin on steady state kinetics of GLUT4 subcellular distribution in rat adipocytes. J Biol Chem 267:17710–17715

Kahn BB, Pedersen O (1993) Suppression of GLUT4 expression in skeletal muscle of rats that are obese from high fat feeding but not from high carbohydrate feeding or genetic obesity. Endocrinology 132:13–22

Kahn BB, Rossetti L, Lodish HF, Charron MJ (1991) Decreased in vivo glucose uptake but normal expression of GLUT1 and GLUT4 in skeletal muscle of diabetic rats. J Clin Invest 87:2197–2206

Kahn CR (1978) Insulin resistance, insulin insensitivity, and insulin unresponsiveness: a necessary distinction. Metabolism [Suppl] 2:1893–1902

Karlsson J, Saltin B (1971) Diet, muscle glycogen, and endurance performance. J Appl Physiol 31:203–206

Karnieli EK, Zarnowski MJ, Hissin PJ, Simpson IA, Salans LB, Cushman SW (1981) Insulin-stimulated translocation of glucose transport systems in the isolated rat adipose cell. J Biol Chem 256:4772–4777

Kasahara M, Hinkle PC (1977) Reconstitution and purification of the D-glucose transporter from human erythrocytes. J Biol Chem 253:7384–7390

Katz A, Broberg S, Sahlin K, Wahren J (1986) Leg glucose uptake during maximal dynamic exercise in humans. Am J Physiol 251:E65–E70

Kemmer FW, Berger M, Herberg L, Gries FA, Wirdeier A, Becker K (1979) Glucose metabolism in perfused skeletal muscle. Demonstration of insulin resistance in the obese Zucker rat. Biochem J. 178:733–741

Kern M, Wells JA, Stephens JM, Elton CW, Friedman JE, Tapscott EB, Pekala PH, Dohm GL (1990) Insulin responsiveness in skeletal muscle is determined by glucose transporter (GLUT 4) protein level. Biochem J 270:397–440

King PA, Hirshman MF, Horton ED, Horton ES (1989) Glucose transport in skeletal muscle membrane vesicles from control and exercised rats. Am J Physiol 257:C1128–C1134

King PA, Horton ED, Hirshman MF, Horton ES (1992) Insulin resistance in obese Zucker rat (fa/fa) skeletal muscle is associated with a failure of glucose transporter translocation. J Clin Invest 90:1568–1575

King PA, Betts JJ, Horton ED, Horton ES (1993) Exercise, unlike insulin, promotes glucose transporter translocation in obese Zucker rat muscle. Am J Physiol 265:R447–R452

Kipnis DM, Cori CF (1959) Studies of tissue permeability. V. The penetration and phosphorylation of 2-deoxyglucose in the rat diaphragm. J Biol Chem 234:171–177

Klee CB, Newton DL, Ni WC, Haiech J (1986) Regulation of the calcium signal by calmodulin. Ciba Found Symp 122:162–170

Klip A (1987) Hexose transport across skeletal muscle sarcolemma. In: Kidwai AM (ed) Sarcolemmal biochemistry, 2nd edn. CRC, Boca Raton, pp 129–153

Klip A, Paquet MR (1990) Glucose transport and glucose transporters in muscle and their metabolic regulation. Diabetes Care 13:228–243

Klip A, Logan WJ, Li G (1982) Hexose transport in L6 muscle cells. Kinetic properties and the number of [^{3}H]cytochalasin B binding sites. Biochem Biophys Acta 687:265–280

Klip A, Ramlal T, Young DA, Holloszy JO (1987) Insulin-induced translocation of glucose transporters in rat hindlimb muscles. FEBS Lett 224:224–230

Klip A, Ramlal T, Bilan PJ, Cartee GD, Gulve EA, Holloszy JO (1990) Recruitment of GLUT4 glucose transporters by insulin in diabetic rat skeletal muscle. Biochem Biophys Res Commun 172:728–736

Knight DE (1986) Calcium and exocytosis. Ciba Found Symp 122:250–265

Kohn PG, Clausen T (1971) The relationship between the transport of glucose and cations across cell membranes in isolated tissues. VI. The effect of insulin, ouabain and metabolic inhibitors on the transport of 3-O-methylglucose and glucose in rat. Biochim Biophys Acta 225:277–290

Kolterman OG, Insel J, Saekow M, Olefsky JM (1980) Mechanisms of insulin resistance in human obesity. J Clin Invest 65:1272–1284

Kong X, Manchester J, Salmons S, Lawrence JC Jr (1994) Glucose transporters in single skeletal muscle fibers. J Biol Chem 269:12963–12967

Kono T, Robinson FW, Blevins TL, Izaki O (1982) Evidence that translocation of the glucose transport activity is the major mechanism of insulin action on glucose transport in fat cells. J Biol Chem 257:10942–10947

Kraegen EW, James DE, Jenkins AB, Chisholm DJ (1985) Dose-response curves for in vivo insulin sensitivity in individual tissues in rats. Am J Physiol 248:E353–E362

Kraegen EW, James DE, Storlien LH, Burleigh KM, Chisholm DJ (1986) In vivo insulin resistance in individual peripheral tissues of the high fat fed rat: assessment by euglycaemic clamp plus deoxyglucose administration. Diabetologia 29:192–198

Kusari J, Kenner KA, Suh K, Hill DE, Henry RR (1994) Skeletal muscle protein tyrosine phosphatase activity and tyrosine phosphatase 1B protein content are associated with insulin action and resistance. J Clin Invest 93:1156–1162

Kusunoki M, Storlien LH, MacDessi J, Oakes ND, Kennedy C, Chisholm DJ, Kraegen EW (1993) Muscle glucose uptake during and after exercise is normal in insulin-resistant rats. Am J Physiol 264:E167–E172

Lam K, Carpenter CL, Ruderman NB, Friel JC, Kelly KL (1994) The phosphatidylinositol 3-kinase serine kinase phosphorylates IRS-1. J Biol Chem 269:20648–20652

Laurie SM, Cain CC, Lienhard GE, Castle JD (1993) The glucose transporter GluT4 and secretory carrier membrane proteins (SCAMPs) colocalize in rat adipocytes and partially segregate during insulin stimulation. J Biol Chem 268:19110–19117

Lawrence JC (1992) Signal transduction and protein phosphorylation in the regulation of cellular metabolism by insulin. Annu Rev Physiol 54:177–193

Le Marchand-Brustel Y, Freychet P (1978) Studies of insulin insensitivity in soleus muscles of obese mice. Metabolism 27:1982–1993

Le Marchand-Brustel Y, Freychet P (1979) Effect of fasting and streptozotocin diabetes on insulin binding and action in the isolated mouse soleus muscle. J Clin Invest 64:1505–1515

Le Marchand Y, Freychet P, Jeanrenaud B (1978a) Longitudinal study on the establishment of insulin resistance in hypothalamic obese mice. Endocrinology 102:74–85

Le Marchand-Brustel Y, Jeanrenaud B, Freychet P (1978b) Insulin binding and effects in isolated soleus muscle of lean and obese mice. Am J Physiol 234:E348–E358

Lee AD, Gulve EA, Chen M, Schluter J, Holloszy JO (1995a) Effects of Ca^{2+} ionophore ionomycin on insulin-stimulated and basal glucose transport in muscle. Am J Physiol 268:R997–R1002

Lee AD, Hansen PA, Holloszy JO (1995b) Wortmannin inhibits insulin-stimulated but not contraction-stimulated glucose transport activity in skeletal muscle. FEBS Lett 361:51–54

Lee J, Pilch PF (1994) The insulin receptor: structure, function, and signaling. Am J Physiol 266:C319–C334

Lienhard GE, Slot JW, James DE, Mueckler MM (1992) How cells absorb glucose. Sci Am 266:86–91

Liu C, Hermann TE (1978) Characterization of ionomycin as a calcium ionophore. J Biol Chem 253:5892–5894

Lund S, Holman GD, Schmitz O, Pedersen O (1993) Glut 4 content in the plasma membrane of rat skeletal muscle: comparative studies of the subcellular fractionation method and the exofacial photolabelling technique using ATP-BMPA. FEBS Lett 330:312–318

Lupien JR, Hirshman MF, Horton ES (1990) Effects of norepinephrine infusion on in vivo insulin sensitivity and responsiveness. Am J Physiol 259:E210–E215

Maegawa H, Kobayashi M, Ishibashi O, Takata Y, Shigeta Y (1986) Effect of diet change on insulin action: difference between muscles and adipocytes. Am J Physiol 251:E616–E623

Marette A, Richardson JM, Ramlal T, Balon TW, Vranic M, Pessin JE, Klip A (1992) Abundance, localization, and insulin-induced translocation of glucose transporters in red and white muscle. Am J Physiol 263:C443–C452

Marshall BA, Ren J, Johnson DW, Gibbs EM, Lillquist JS, Soeller WC, Holloszy JO, Mueckler M (1993) Germline manipulation of glucose homeostasis via alteration of glucose transporter levels in skeletal muscle. J Biol Chem 268:18442–18445

Marshall S (1989) Kinetics of insulin action on protein synthesis in isolated adipocytes. Ability of glucose to selectively desensitize he glucose transport system without altering insulin stimulation of protein synthesis. J Biol Chem 264:2029–2036

Marshall S, Bacote V, Traxinger RR (1991a) Discovery of a metabolic pathway mediating glucose-induced desensitization of the glucose transport system. J Biol Chem 266:4706–4712

Marshall S, Garvey WT, Traxinger RR (1991b) New insights into the metabolic regulation of insulin action and insulin resistance: role of glucose and amino acids. FASEB J 5:3031–3036

McGuire MC, Fields RM, Nyomba BL, Raz I, Bogardus C, Tonks NK, Sommercorn J (1991) Abnormal regulation of protein tyrosine phosphatase activities in skeletal muscle of insulin-resistant humans. Diabetes 40:939–942

Megeney LA, Neufer PD, Dohm GL, Tan MH, Blewett CA, Elder GCB, Bonen A (1993) Effects of muscle activity and fiber composition on glucose transport and GLUT-4. Am J Physiol 264:E583–E593

Megeney LA, Prasad MA, Tan MH, Bonen A (1994) Expression of the insulin-regulatable transporter GLUT-4 in muscle is influenced by neurogenic factors. Am J Physiol 266:E813–E816

Mellgren RL (1987) Calcium-dependent proteases: an enzyme system active at cellular membranes? FASEB J 1:110–115

Mitsumoto Y, Klip A (1992) Developmental regulation of the subcellular distribution and glycosylation of GLUT1 and GLUT4 glucose transporters during myogenesis of L6 muscle cells. J Biol Chem 267:4957–4962

Molé PA, Oscai LB, Holloszy JO (1971) Adaptation of muscle to exercise. Increase in levels of palmityl CoA synthetase, and in the capacity to oxidize fatty acids. J Clin Invest 50:2323–2330

Mooseker MS, Coleman TR, Conzelman KA (1986) Calcium and the regulation of cytoskeletal assembly, structure and contractibility. Ciba Found Symp 122:232–245

Mordes JP, Rossini AA. (1985) Animal models of diabetes mellitus. In: Marble A, Krall LP, Bradley RF, Christlieb HR, Soeldner JS (eds) Joslin's diabetes mellitus. Lea and Febiger, Philadelphia, pp 110–137

Morgan HE, Whitfield CF (1973) Regulation of sugar transport in eukaryotic cells. Curr Top Membr Transp 4:255–303

Morgan HE, Henderson MJ, Regen DM, Park CR (1961) Regulation of glucose uptake in muscle. I. The effects of insulin and anoxia on glucose transport and phosphorylation in the isolated, perfused heart of normal rats. J Biol Chem 236:253–261

Morgan HE, Regen DM, Park CR (1964) Identification of a mobile carrier-mediated sugar transport system in muscle. J Biol Chem 239:369–374

Mueckler M (1994) Facilitative glucose transporters. FEBS Eur J Biochem 219:713–725

Mueckler M, Caruso C, Baldwin SA, Panico M, Blench M, Morris HR, Allard WJ, Lienhard GE, Lodish HF (1985) Sequence and structure of a human glucose transporter. Science 299:941–945

Mueckler MM (1995) Glucose transport and glucose homeostasis: new insights from transgenic mice. NIPS 10:22–29

Napoli R, Hirshman MF, Horton ES (1995) Mechanisms and time course of impaired skeletal muscle glucose transport activity in streptozocin diabetic rats. J Clin Invest 96:427–437

Narahara HT, Ozand P, Cori CF (1960) Studies of tissue permeability. VII. The effect of insulin on glucose penetration and phosphorylation in frog muscle. J Biol Chem 235:3370–3378

Narahara HT, Özand P (1963) Studies of tissue permeability. IX. The effect of insulin on the penetration of 3-methylglucose-H^3 in frog muscle. J Biol Chem 238:40–49

Nesher R, Karl IE, Kipnis DM (1980) Epitrochlearis muscle. II. Metabolic effects of contraction and catecholamines. Am J Physiol 239:E461–E467

Nesher R, Karl IE, Kipnis DM (1985) Dissociation of effects of insulin and contraction on glucose transport in rat epitrochlearis muscle. Am J Physiol 249:C226–C232

Neufer PD, Carey JO, Dohm GL (1993) Transcriptional regulation of the gene for glucose transporter GLUT4 in skeletal muscle. J Biol Chem 268:13824–13829

Nicholson WF, Watson PA, Booth FW (1984) Levels of blood-borne factors and cytosol glucocorticoid receptors during the initiation of muscle atrophy in rodent hindlimbs. Pfluegers Arch 401:321–323

Nolte LA, Gulve EA, Holloszy JO (1994) Epinephrine-induced in vivo muscle glycogen depletion enhances insulin sensitivity of glucose transport. J Appl Physiol 76:2054–2058

Nosadini R, Del Prato S, Tiengo A, Valerio A, Muggeo M, Opocher G, Mantero F, Duner E, Marescotti C, Mollo F, Belloni F (1983) Insulin resistance in Cushing's syndrome. Clin Endocrinol Metab 57:529–536

Okada T, Kawano Y, Sakakibara T, Hazeki O, Ui M (1994) Essential role of phosphatidylinositol 3-kinase in insulin-induced glucose transport and antilipolysis in rat adipocytes. J Biol Chem 269:3568–3573

Okamoto M, Kono S, Inoue G, Hayashi T, Kosaki A, Maeda I, Kubota M, Kuzuya H, Imura H (1992) Effects of a high-fat diet on insulin receptor kinase and the glucose transporter in rats. J Nutr Biochem 3:241–250

Olson AL, Liu ML, Moye RWS, Buse JB, Bell GI, Pessin JE (1993) Hormonal/metabolic regulation of the human GLUT4/muscle-fat facilitative glucose transporter gene in transgenic mice. J Biol Chem 268:9839–9846

Özand P, Narahara HT, Cori CF (1962) Studies of tissue permeability. VIII. The effect of anaerobiosis on glucose uptake in frog sartorius muscle. J Biol Chem 237:3037–3043

Pagano G, Cavallo-Perin P, Cassader M, Bruno A, Ozzello A, Masciola P, Dall'omo AM, Imbimbo B (1983) An in vivo and in vitro study of the mechanism of prednisone-induced insulin resistance in healthy subjects. J Clin Invest 72:1814–1820

Palade P (1987) Drug induced Ca^{2+} release from isolated sarcoplasmic reticulum II. Releases involving a Ca^{2+} induced Ca^{2+} release channel. J Biol Chem 262:6142–6148

Park CR, Reinwein D, Henderson MJ, Cadenas E, Morgan HE (1959) The action of insulin on the transport of glucose through the cell membrane. Am J Med 26:674–684

Pedersen O, Bak JF, Andersen PH, Lund S, Moller DE, Flier JS, Kahn BB (1990) Evidence against altered expression of GLUT1 or GLUT4 in skeletal muscle of patients with obesity or NIDDM. Diabetes 39:865–870

Pette D, Staron RS (1990) Cellular and molecular diversities of mammalian skeletal muscle fibers. Rev Physiol Biochem Pharmacol 116:1–76

Pette D, Tyler KR (1983) Response of succinate dehydrogenase activity in fibers of rabbit tibials anterior muscle to chronic nerve stimulation. J Physiol (Lond) 338:1–9

Pette D, Vrbová G (1992) Adaptation of mammalian skeletal muscle fibers to chronic electrical stimulation. Rev Physiol Biochem Pharmacol 120:115–202

Pette D, Smith ME, Staudte HW, Vrbova G (1973) Effects of long term electrical stimulation on some contractile and metabolic characteristics of fast rabbit muscles. Pfluegers Arch 338:257–272

Pette D, Ramirez BU, Muller W, Simon R, Exner GU (1975) Influence of intermittent long-term stimulation of contractile, histochemical and metabolic properties of fibre populations in fast and slow rabbit muscles. Eur J Physiol 361:1–7

Pette D, Muller W, Leisner E, Vrbova G (1976) Time dependent effects on contractile properties, fiber population, myosin light chains and enzymes of energy metabolism in intermittently and continuously stimulated fast twitch muscles of the rabbit. Pfluegers Arch 364:103–112

Philipson KD, Frank JS, Nishimoto AY (1983) Effects of phospholipase C on the Na^+-Ca^{2+} exchange and Ca^{2+} permeability of cardiac sarcolemmal vesicles. J Biol Chem 258:5905–5910

Ploug T, Galbo H, Richter EA (1984) Increased muscle glucose uptake during contractions: no need for insulin. Am J Physiol 247:E726–E731

Ploug T, Galbo H, Vinten J, Jørgensen M, Richter EA (1987) Kinetics of glucose transport in rat muscle: effects of insulin and contractions. Am J Physiol 253:E12–E20

Ploug T, Stallknecht BM, Pedersen O, Kahn BB, Ohkuwa T, Vinten J, Galbo H (1990) Effect of endurance-training on glucose transport capacity and glucose transporter expression in rat skeletal muscle. Am J Physiol 259:E778–E786

Ploug T, Galbo H, Ohkuwa T, Tranum-Jensen J, Vinten J (1992) Kinetics of glucose transport in rat skeletal muscle membrane vesicles: effects of insulin and contractions. Am J Physiol 262:E700–E711

Ploug T, Ohkuwa T, Handberg A, Vissing J, Galbo H (1995) Effect of immobilization on glucose transport and glucose transporter expression in rat skeletal muscle. Am J Physiol 268:E980–E986

Post RL, Morgan HE, Park CR (1961) Regulation of glucose uptake in muscle. III. The interaction of membrane transport and phosphorylation in the control of glucose uptake. J Biol Chem 236:269–272

Ralston E, Beushausen S, Ploug T (1994) Expression of the synaptic vesicle proteins VAMPs/synaptobrevins 1 and 2 in non-neural tissues. J Biol Chem 269:15403–15406

Randle PJ, Smith GH (1958) Regulation of glucose uptake by muscle. I. The effects of insulin anaerobiosis and cell poisons on the uptake of glucose and release of potassium by isolated rat diaphragm. Biochem J 70:490–508

Rasmussen H, Barrett PQ (1984) Calcium messenger system: an integrated view. Physiol Rev 64:938–984

Ren J, Marshall BA, Gulve EA, Gao J, Johnson DW, Holloszy JO, Mueckler M (1993a) Evidence from transgenic mice that glucose transport is rate-limiting for glycogen deposition and glycolysis in skeletal muscle. J Biol Chem 268:16113–16115

Ren J, Youn JH, Gulve EA, Henriksen EJ, Holloszy JO (1993b) Effects of alkaline pH on the stimulation of glucose transport in rat skeletal muscle. Biochim Biophys Acta 1145:199–204

Ren J, Semenkovich CF, Gulve EA, Gao J, Holloszy JO (1994) Exercise induces rapid increases in GLUT4 expression, glucose transport capacity, and insulin-stimulated glycogen storage in muscle. J Biol Chem 269:14396–14401

Ren JM, Semenkovich C, Holloszy JO (1993c) Adaptation of muscle to creatine depletion: effect on GLUT-4 glucose transporter expression. Am J Physiol 264:C146–C150

Rennie MJ, Idstrom JP, Mann GE, Schersten T, Bylund-Fellenius AC (1983) A paired-tracer dilution method for characterizing membrane transport in the perfused rat hindlimb. Biochem J 214:737–743

Reusch JE-B, Begum N, Sussman KE, Draznin B (1991) Regulation of GLUT-4 phosphorylation by intracellular calcium in adipocytes. Endocrinology 129:3269–3273

Richardson J, Balon TW, Treadway JL, Pessin JE (1991) Differential regulation of glucose transporter activity and expression in red and white skeletal muscle. J Biol Chem 266:12690–12694

Richter EA, Garetto LP, Goodman MN, Ruderman NB (1982) Muscle glucose metabolism following exercise in the rat. Increased sensitivity to insulin. J Clin Invest 69:785–793

Richter EA, Garetto LP, Goodman MN, Ruderman NB (1984) Enhanced muscle glucose metabolism after exercise: modulation by local factors. Am J Physiol 246:E476–E482

Richter EA, Ploug T, Galbo H (1985) Increased muscle glucose uptake after exercise. No need for insulin during exercise. Diabetes 34:1041–1048

Richter EA, Hansen BF, Hansen SA (1988a) Glucose-induced insulin-resistance of skeletal muscle glucose transport and uptake. Biochem J 252:733–737

Richter EA, Hansen SA, Hansen BF (1988b) Mechanisms limiting glycogen storage in muscle during prolonged insulin stimulation. Am J Physiol 255:E621–E628

Riddick FA, Reisler DM, Kipnis DM (1962) The sugar transport system in striated muscle. Effect of growth hormone, hydrocortisone and alloxan diabetes. Diabetes 2:171–178

Robinson KA, Sens DA, Buse MG (1993) Pre-exposure to glucosamine induces insulin resistance of glucose transport and glycogen synthesis in isolated rat skeletal muscles. Study of mechanisms in muscle and in rat-1 fibroblasts overexpressing the human insulin receptor. Diabetes 42:1333–1346

Robinson LJ, Pang S, Harris DS, Heuser J, James DE (1992) Translocation of the glucose transporter (GLUT4) to the cell surface in permeabilized 3T3-L1 adipocytes: effects of ATP, insulin, and GTPgamma and localization of GLUT4 to clathrin lattices. J Cell Biol 117:1181–1196

Rodnick KJ, Holloszy JO, Mondon CE, James DE (1990) Effects of exercise-training on insulin-regulatable glucose-transporter protein levels in rat skeletal muscle. Diabetes 39:1425–1429

Rodnick KJ, Henriksen EJ, James DE, Holloszy JO (1992a) Exercise-training, glucose transporters and glucose transport in rat skeletal muscles. Am J Physiol 262:C9–C14

Rodnick KJ, Slot JW, Studelska DR, Hanpeter DE, Robinson LJ, Geuze HJ, James DE (1992b) Immunocytochemical and biochemical studies of GLUT4 in rat skeletal muscle. J Biol Chem 267:6278–6285

Rosholt MN, King PA, Horton ES (1994) High-fat diet reduces glucose transporter responses to both insulin and exercise. Am J Physiol 266:R95–R101

Rossetti L (1989) Normalization of insulin sensitivity with lithium in diabetic rats. Diabetes 38:648–652

Rossetti L, Smith D, Shulman GI, Papachristou D, DeFronzo RA (1987) Correction of hyperglycemia with phlorizin normalizes tissue sensitivity to insulin in diabetic rats. J Clin Invest 79:1510–1515

Rossetti L, Hawkins M, Chen W, Gindi J, Barzilai N (1995) In vivo glucosamine infusion induces insulin resistance in normoglycemic but not in hyperglycemic conscious rats. J Clin Invest 96:132–140

Rothenberg PL, Lane WS, Karasik A, Backer JM, White MF, Kahn CR (1991) Purification and partial sequence analysis of pp 185, the major cellular substrate of the insulin receptor tyrosine kinase. J Biol Chem 266:8302–8311

Ruderman NB, Houghton CRS, Hems R (1971) Evaluation of the isolated perfused rat hindquarter for the study of muscle metabolism. Biochem J 124:639–651

Saad MJA, Araki E, Miralpeix M, Rothenberg PL, White MF, Kahn CR (1992) Regulation of insulin receptor substrate-1 in liver and muscle of animal models of insulin resistance. J Clin Invest 90:1839–1849.

Saad MJA, Folli F, Kahn JA, Kahn CR (1993) Modulation of insulin receptor, insulin receptor substrate-1, and phosphatidylinositol 3-kinase in liver and muscle of dexamethasone-treated rats. J Clin Invest 92:2065–2072

Sarabia V, Lam L, Burdett E, Leiter LA, Klip A (1992) Glucose transport in human skeletal muscle cells in culture. Stimulation by insulin and metformin. J Clin Invest 90:1386–1395

Schatzman RC, Raynor RL, Kuo JF (1983) N-(6-aminohexyl)-5-chloro-1-naphthalene-sulfonamide (W-7), a calmodulin antagonist, also inhibits phospholipid-sensitive calcium-dependent protein kinase. Biochim Biophys Acta 755:144–147

Schultz TA, Lewis SB, Westbie DK, Wallin JD, Gerich JE (1977) Glucose delivery: a modulator of glucose uptake in contracting skeletal muscle. Am J Physiol 233:E514–E518

Seider MJ, Nicholson WF, Booth FW (1982) Insulin resistance for glucose metabolism in disused soleus muscle of mice. Am J Physiol 242:E12–E18

Sherman WM, Katz AL, Cutler CL, Withers RT, Ivy JL (1988) Glucose transport: locus of muscle insulin resistance in obese Zucker rats. Am J Physiol 255:E374–E382

Shoubridge EA, Challis AJ, Hayes DJ, Radda GK (1985) Biochemical adaptation in the skeletal muscle of rats depleted of creatine with the substrate analogue β-guanidinopropionic acid. Biochem J 232:125–131

Simpson IA, Cushman SW (1986) Hormonal regulation of mammalian glucose transport. Ann Rev Biochem 55:1059–1089

Slentz CA, Gulve EA, Rodnick KJ, Henriksen EJ, Youn JH, Holloszy JO (1992) Glucose transporters and maximal transport are increased in endurance-trained rat soleus. J Appl Physiol 73:486–492

Slieker LJ, Roberts EF, Shaw WN, Johnson WT (1990) Effect of streptozocin-induced diabetes on insulin-receptor tyrosine kinase activity in obese Zucker rats. Diabetes 39:619–625

Slot JW, Geuze HJ, Gigengack S, James DE, Lienhard GE (1991a) Translocation of the glucose transporter GLUT4 in cardiac myocytes of the rat. PNAS 88:7815–7819

Slot JW, Geuze HJ, Gigengack S, Lienhard GE, James DE (1991b) Immuno-localization of the insulin regulatable glucose transporter in brown adipose tissue of the rat. J Cell Biol 113:123–135

Smith RL, Lawrence JC (1984) Insulin action in denervated rat hemidiaphragms. J Biol Chem 259:2201–2207

Sorenson SS, Christensen F, Clausen T (1980) The relationship between the transport of glucose and cations across cell membranes in isolated tissues. X. Effect of glucose transport stimuli on the efflux of isotopically-labelled calcium and 3-o-methylglucose from soleus muscles and epididymal fat pads of the rat. Biochim Biophys Acta 602:433–445

Standaert ML, Farese RV, Cooper DR, Pollet RJ (1988) Insulin-induced glycerolipid mediators and the stimulation of D-glucose transport and insulin binding in isolated rat adipocytes. J Biol Chem 263:8696–8705

Steenbergen C, Murphy E, Levy L, London RE (1987) Elevation in cytosolic-free calcium concentration early in myocardial ischemia in perfused rat heart. Circ Res 60:700–707

Storlien LH, James DE, Burleigh KM, Chisholm DJ, Kraegen EW (1986) Fat-feeding causes widespread in vivo insulin resistance, decreased energy expenditure, and obesity in rats. Am J Physiol 251:E576–E583

Sudhof, TC, Jahn R (1991) Proteins of synaptic vesicles involved in exocytosis and membrane recycling. Neuron 6:665–677

Sun X-J, Rothenberg P, Kahn CR, Backer JM, Araki E, Wilden P, Cahill DA, Goldstein BJ, White MF (1991) Structure of the insulin receptor substrate IRS-1 defines a unique signal transduction protein. Nature 352:73–77

Sun X-J, Miralpeix M, Myers JR, Glasheen EM, Backer JM, Kahn CR, White MF (1992) Expression and function of IRS-1 in insulin signal transmission. J Biol Chem 267:22662–22672

Susini C, Lavau M (1978) In-vitro and in-vivo responsiveness of muscle and adipose tissue to insulin in rats rendered obese by a high-fat diet. Diabetes 27:114–120

Suzuki K, Kono T (1980) Evidence that insulin causes translocation of glucose transport activity to the plasma membrane from an intracellular storage site. Proc Natl Acad Sci USA 77:2542–2545

Tabata I, Schluter J, Gulve EA, Holloszy JO (1994) Lithium increases susceptibility of muscle glucose transport to stimulation by various agents. Diabetes 43:903–907

Tanaka T, Ohmura T, Yamakado T, Hidaka H (1982) Two types of calcium-dependent protein phosphorylations modulated by calmodulin antagonists. Naphthalenesulfonamide derivatives. Mol Pharmacol 22:408–412

Thies RS, Molina JM, Ciaraldi TP, Freidenberg GR, Olefsky JM (1990) Insulin receptor autophosphorylation and endogenous substrate phosphorylation in human adipocytes from control, obese and NIDDM subjects. Diabetes 39:250–258

Thoidis G, Kotliar N, Pilch PF (1993) Immunological analysis of GLUT4-enriched vesicles. J Biol Chem 269:11691–11696

Torgan CE, Etgen GJ, Kang HY, Ivy JL (1995) Fiber type-specific effects of clenbuterol and exercise training on insulin-resistant muscle. J Appl Physiol 79:163–167

Treadway JL, James DE, Burcel E, Ruderman NB (1989) Effect of exercise on insulin receptor-binding and kinase activity in skeletal muscle. Am J Physiol 256:E138–E144

Turinsky J (1987) Dynamics of insulin resistance in denervated slow and fast muscles vivo. Am J Physiol 252:R531–R537

Valant P, Erlij D (1983) K^+-stimulated sugar uptake in skeletal muscle: role of cytoplasmic Ca^{2+}. Am J Physiol 245:C125–C132

Vissing J, Ohkuwa T, Ploug T, Galbo H (1988) Effect of prior immobilization on muscular glucose clearance in resting and running rats. Am J Physiol 255:E456–E462

Walker PM, Idström J-P, Schersten T, Bylund-Fellenius A-C (1982) Glucose uptake in relation to metabolic state in perfused rat hindlimb at rest and during exercise. Eur J Appl Physiol Occup Physiol 48:163–176

Walker PS, Ramlal T, Donovan JA, Doering TP, Sandra A, Klip A, Pessin JE (1989) Insulin and glucose-dependent regulation of the glucose transport system in the rat L6 skeletal muscle cell line. J Biol Chem 264:6587–6595

Wallberg-Henriksson H, Holloszy JO (1984) Contractile activity increases glucose uptake by muscle in severely diabetic rats. J Appl Physiol 57:1045–1049

Wallberg-Henriksson H, Holloszy JO (1985) Activation of glucose transport in diabetic muscle: responses to contraction and insulin. Am J Physiol 249:C233–C237

Wallberg-Henriksson H, Constable SH, Young DA, Holloszy JO (1988) Glucose transport into rat skeletal muscle: interaction between exercise and insulin. J Appl Physiol 65:909–913

Wardzala LJ, Jeanrenaud B (1981) Potential mechanism of insulin action on glucose transport in the isolated rat diaphragm. J Biol Chem 256:7090–7093

Wardzala LJ, Cushman SW, Salans LB (1978) Mechanism of insulin action on glucose transport in the isolated rat adipose cell. Enhancement of the number of functional transport systems. J Biol Chem 253:8002–8005

Weber FE, Pette D (1990) Changes in free- and bound-forms and total amount of hexokinase isozyme II of rat muscle in response to contractile activity. Eur J Biochem 191:85–90

Webster BA, Vigna SR, Paquette T (1986) Acute exercise, epinephrine, and diabetes enhance insulin-binding to skeletal muscle. Am J Physiol 250:E187–E197

Weinstein SP, O'Boyle E, Haber RS (1994) Thyroid hormone increases basal and insulin-stimulated glucose transport in skeletal muscle. The role of GLUT4 glucose transporter expression. Diabetes 43:1185–1189

Weinstein SP, Paquin T, Pritsker A, Haber RS (1995) Glucocorticoid-induced insulin resistance: dexamethasone inhibits the activation of glucose transport in rat skeletal muscle by both insulin- and non-insulin-related stimuli. Diabetes 44:441–445

Westfall MW, Sayeed MM (1990) Effect of Ca^{2+}-channel agonists and antagonists on skeletal muscle sugar transport. Am J Physiol 258:R462–R468

White MF, Kahn CR (1994) The insulin signaling system. J Biol Chem 269:1–4

White MF, Maron R, Kahn CR (1985) Insulin rapidly stimulates tyrosine phosphorylation of a Mr-185,000 protein in intact cells. Nature 318:183–186

Wilde MW, Slonczewski JL, Carson M, Zigmond SH (1986) Glycogen phosphorylase: a noninvasive indicator of cytoplasmic calcium. Methods Enzymol 141:18–25

Wilson CM, Cushman SW (1994) Insulin stimulation of glucose transport activity in rat skeletal muscle: increase in cell surface GLUT4 as assessed by photolabelling. Biochem J 299:755–759

Yang J, Holman GD (1993) Comparison of GLUT4 and GLUT1 subcellular trafficking in basal and insulin-stimulated 3T3-L1 cells. J Biol Chem 268:4600–4603

Yeh J-I, Gulve EA, Rameh L, Birnbaum MJ (1995) The effects of wortmannin on rat skeletal muscle. J Biol Chem 270:2107–2111

Yki-Jarvinen H, Helve E, Koivisto VA (1987) Hyperlgycemia decreases glucose uptake in type I diabetes. Diabetes 36:892–896

Youn JH, Gulve EA, Holloszy JO (1991) Calcium stimulates glucose transport in skeletal muscle by a pathway independent of contraction. Am J Physiol 260:C555–C561

Youn JH, Gulve EA, Henriksen EJ, Holloszy JO (1994a) Interactions between effects of W-7, insulin and hypoxia on glucose transport in muscle. Am J Physiol 267:R888–R894

Youn JH, Kim JK, Buchanan TA (1994b) Time courses of changes in hepatic and skeletal muscle insulin action and GLUT4 protein in skeletal muscle after STZ injection. Diabetes 43:564–571

Young DA, Uhl JJ, Cartee GD, Holloszy JO (1986) Activation of glucose transport in muscle by prolonged exposure to insulin: effects of glucose and insulin concentration. J Biol Chem 261:16049–16053

Young DA, Wallberg-Henriksson H, Sleeper M, Holloszy JO (1987) Reversal of the exercise-induced increase in muscle permeability to glucose. Am J Physiol 253:E331–E335

Young JC, Garthwaite SM, Bryan JE, Cartier L-J, Holloszy JO (1983) Carbohydrate-feeding speeds reversal of enhanced glucose uptake in muscle after exercise. Am J Physiol 245:R684–R688

Ziel FH, Venkatesan N, Davidson MB (1988) Glucose transport is rate limiting for skeletal muscle glucose metabolism in normal and STZ-induced diabetic rats. Diabetes 37:885–890

Zorzano A, Balon TW, Garetto LP, Goodman MN, Ruderman NB (1985) Muscle -aminoisobutyric acid transporter after exercise: enhanced stimulation by insulin. Am J Physiol 249:E546–E552

Zorzano A, Balon TW, Goodman MN, Ruderman NB (1986a) Glycogen depletion and increased insulin sensitivity and responsiveness in muscle after exercise. Am J Physiol 251:E664–E669

Zorzano A, Balon TW, Goodman MN, Ruderman NB (1986b) Additive effects of prior exercise and insulin on glucose and AIB uptake by muscle. Am J Physiol 251:E21–E26

Zuniga-Guajardo S, Jimenez J, Zinman B (1986) Effects of massive obesity on insulin sensitivity and insulin clearance and the metabolic response to insulin as assessed by the euglycemic clamp technique. Metabolism 35:278–282

Editor-in-charge: Professor D. Pette

Molecular Pathophysiology of Voltage-Gated Ion Channels

F. Lehmann-Horn and R. Rüdel

Department of Physiology, University of Ulm, 89069 Ulm, Germany

Contents

1
Introduction

Voltage-gated ion channels, ion-conducting proteins equipped with a membrane-spanning pore, gates and voltage sensors, are present in the surface and inner membranes of all excitable and of most nonexcitable cells.

The membranes of nerve and muscle cells, the paradigms of excitable cells, are equipped with sodium, potassium, and calcium channels which, by virtue of their voltage sensitivity, allow them to generate action potentials. Conducted along the cell surface, these "spikes" are the bits of information conveyed from cell to cell in the nervous system; in muscle cells, they induce contraction. In vertebrate skeletal muscle, additional voltage-gated chloride channels are responsible for 75%–80% of the membrane conductance at rest, thus contributing to a fast repolarization phase of the action potentials.

Nonexcitable cells, such as pancreatic islet cells or lymphocytes, also control their specific functions of secretion and immune activity, respectively, by voltage-gated ion channels. In contrast to excitable cells, however, control of function is exerted by small or slow alterations of the resting potential.

The excitation process is one of the basic principles brought to perfection by Nature in the development of the animal kingdom. Since the essential domains of the various ion channels as its mediators were highly conserved for more than 600 million years, these molecules were long imagined to be so sophisticated that mutations in the corresponding genes would not be compatible with life. This idea had to be abandoned when Fontaine et al. (1990) discovered that hyperkalemic periodic paralysis, a hereditary human muscle disorder characterized by episodes of weakness of the skeletal musculature, is linked to *SCN4A*, the gene encoding the α subunit of the adult skeletal muscle sodium channel. Soon afterwards, two further hereditary muscle diseases, paramyotonia congenita (Ebers et al. 1991; Koch et al. 1991; Ptáček et al. 1991b) and potassium-aggravated myotonia (Lerche et al. 1993; Ptáček et al. 1994b; Ricker et al. 1994), were found to be linked to the same gene. Hyperkalemic periodic paralysis is also known in the horse and is caused by a mutation in the corresponding equine gene (Rudolph et al. 1992b). Moreover, two other hereditary muscle diseases characterized by defective excitation are linked to other ion channel genes. Firstly, myotonia congenita, already described more than 100 years ago (Thomsen 1876) is caused by various mutations in *CLCN1*, the gene coding for the major skeletal muscle chloride channel (Koch et al. 1992) and several mouse models of the disease, in particular the myotonic *adr* mouse, are caused by mutations in the corresponding murine gene (Steinmeyer et al. 1991a). Secondly, hypokalemic periodic paralysis, the classical form of human hereditary periodic paralysis, is linked to *CACNL1A3*, the gene encoding the α_1 subunit of the dihydropyridine receptor, the voltage-gated skeletal muscle L-type calcium channel (Fontaine et al. 1994).

In addition to these hereditary "channelopathies" that all involve skeletal muscle channel genes, there are also diseases of the heart and the central and peripheral nervous system, which are based on mutations of genes coding for voltage-gated ion channels. Long QT syndrome type 3, an inherited cardiac arrhythmia in man, is linked to *SCN5A*, the gene encoding the α subunit of the heart muscle sodium channel (Jiang et al. 1994), and a "motor endplate disease" mouse mutant is linked to *Scn8a*, a sodium channel gene expressed in brain and spinal cord (Burgess et al. 1995). Furthermore episodic ataxia type 1 (EA-1), a paroxysmal cerebellar disease associated with myokymia, is caused by mutations in the gene encoding a voltage-gated potassium channel (Browne et al. 1994, 1995).

This review concentrates on naturally occurring mutations affecting voltage-gated ion channels and their pathological consequences. Ligand-gated ion channels and receptors (e.g., the acetylcholine receptor) or transporters such as the one involved in cystic fibrosis, are not considered. Reviews of the structure and function of voltage-gated ion channels in general have appeared for sodium (Catterall 1992; Keynes 1994), potassium (Pongs 1992; Chandy and Gutman 1995), calcium (Hofmann et al. 1994; Melzer et al. 1995) and chloride (Pusch and Jentsch 1994) channels. Reviews on muscle ion channel disorders have also been published (Barchi 1995; Hoffman et al. 1995; Hudson et al. 1995). Extensive descriptions of the clinical features of the muscle channelopathies, including therapeutic considerations, are contained in Chap. 49 (Rüdel et al. 1994) and Chap. 50 (Lehmann-Horn et al. 1994) of the 2nd edition of the textbook *Myology*.

2 Sodium Channelopathies

2.1 The Multigene Family Encoding Voltage-Gated Sodium Channels

The human genome contains a number of almost identical genes coding for slightly different voltage-gated sodium channels. The different channel genes are specific for expression in the various tissues and this specialization may help to restrict the consequence of a mutation to a single type of cell or tissue. The various isoforms of voltage-gated sodium channels are heteromeric proteins consisting of a large, heavily glycosylated α subunit and one or two small β subunits. In the brain, a noncovalently associated β_1 subunit (36 kDa) and a disulfide-linked β_2 subunit (33 kDa) copurify with the α subunit, whereas in skeletal and heart muscle, only the β_1 subunit has been identified.

Eight different genes are known to encode α subunits, most of them being expressed in brain, peripheral nerve and muscle. A gene cluster occurs on human chromosome 2q21-24.1 and contains several genes, *SCN1A* (2q24, Malo et al. 1994a), *SCN2A* (2q23-24.1, Ahmed et al. 1992), *SCN3A* (2q24, Malo et al. 1994b) for brain channels (type I, II, III) and *SCN6A* (2q21-23, George et al. 1992a, 1994a) for heart muscle and myometrium. *SCN7A* has not yet been localized in the human genome, but the product of the homologous murine gene has been detected in mouse glia cells (Potts et al. 1993). *SCN8A* maps to chromosome 12q13 and is a candidate gene for inherited neurodegenerative disease as a deletion in the mouse homolog causes a motor endplate disease (Burgess et al. 1995).

Three genes encode three known isoforms of sodium channels in skeletal muscle which differ in their sensitivity to various toxins. Expression depends on the innervation of the muscle fibers. *SCN4A*, the gene encoding the tetrodotoxin (TTX)-sensitive sodium channel α subunit (hSkM1) of adult skeletal muscle, is located on human chromosome 17q23 (Fontaine et al. 1990; George et al. 1992b). The gene contains 24 exons distributed over about 30 kb. As with many genes, the genomic structure becomes more condensed towards the 3' end, with the last 30% of the coding sequence appearing in a single exon (George et al. 1992b; McClatchey et al. 1992a; Wang et al. 1992). Intron-exon boundaries are known; primer sets consisting of intron sequences for amplification of all 24 exons by use of PCR are available (George et al. 1993b). The cDNA encodes 1836 amino acids in a single open reading frame. *SCN4A* is only expressed in skeletal muscle and its product is the only sodium channel α subunit detectable in the fully differentiated and innervated muscle. Its function is modified by the β_1 subunit (Makita et al. 1994a).

Two other genes encode skeletal muscle α subunits under certain circumstances. *SCN5A*, located on human chromosome 3p21 (George et al. 1995), is expressed in fetal skeletal muscle (SkM2) and in adult cardiac muscle (hH1; Gellens et al. 1992; Rogart et al. 1989) and its gene product is characterized by low TTX sensitivity. *SCN6A*, located on chromosome 2q21-23 (George et al. 1994a), is expressed in fetal skeletal muscle, in cardiac muscle, and myometrium (Gautron et al. 1992).

A single gene on human chromosome 19q13.1, *SCN1B*, encodes the β_1 subunit expressed in brain, heart, and skeletal (β_1=β) muscle (McClatchey et al. 1993; Makita et al. 1994a,b). Molecular information on the β_2 subunit is now available (Isom et al. 1995a).

2.2 Structure and Function of hSkM1, the α Subunit of the Adult Human Skeletal Muscle Sodium Channel

The subunit hSkM1 is a large glycoprotein of approximately 260 kDa, 25%–30% of its mass being carbohydrate (in contrast, the MW of the α subunit expressed in brain, nerve, and in muscle other than adult skeletal is up to 280 kDa). The sequence shows four regions of internal homology, each encompassing 225–327 amino acids (Fig. 1). These so-called repeats (I–IV) consist of six hydrophobic segments (S1–S6), putative transmembrane helices located within highly conserved regions in each repeat. Each S4 helix is unique in containing a repeating motif with a positively charged lysine or arginine at every third position. The motif occurs four times in I-S4, five times in II-S4 and III-S4, and eight times in IV-S4. These 22 positive charges are identical in most sodium channels of this gene family and make S4 a natural candidate for the gating voltage sensor (Stühmer et al. 1989). The charges seem to move outward in response to depolarization (Durell and Guy 1992; Yang and Horn 1995), thus playing an essential part in voltage-dependent activation of the channel. Also, most of the hydrophobic residues of the transmembrane segments are conserved.

The carboxy and amino terminals and the linkers joining the repeats are located intracellularly. Their sequences are quite divergent among species as well as channel isoforms, with the exception of the sequence

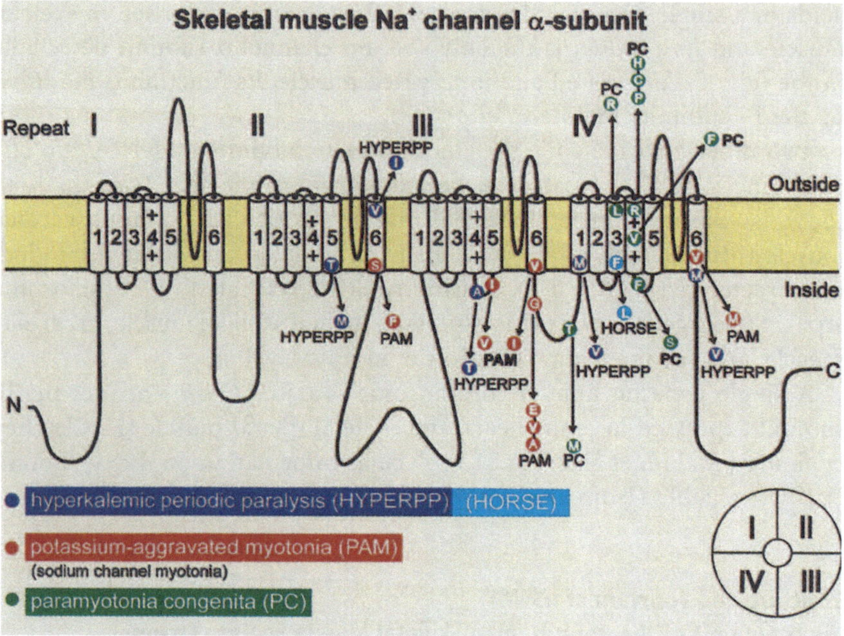

Fig. 1. Substitutions predicted in the α subunit of the voltage-gated sodium channel of skeletal muscle (transmembrane segments indicated by cylinders). *HYPERPP*, Hyperkalemic periodic paralysis; *PAM*, potassium-aggravated myotonia; *PC*, paramyotonia congenita; *HORSE*, position of the mutation causing hyperkalemic periodic paralysis in quarter horses. Conventional one-letter abbreviations are used for wild-type amino acids and substitutions (*arrows*). Mutations discovered by many laboratories in 104 families. V → I is listed in Table 1 as paramyotonia congenita

of the linker between repeats III-S6 and IV-S1 to be discussed below. Also most of the short intra- and extracellular loops connecting the transmembrane segments in each repeat are poorly conserved except for the linkers between segments S5 and S6 of each repeat. These latter linkers consist of an extracellular part and the so-called SS1-SS2 or P-region that dips into the membrane. The highly conserved intramembrane P-regions of the four repeats are thought to form the lining of the channel pore, as in potassium channels (Backx et al. 1992; Heinemann et al. 1992).

A special part of the protein to which a certain function has been assigned, is the above-mentioned intracellular linker connecting repeat III-S6 with repeat IV-S1. Most likely, this mobile chain acts as the inactivation gate of the channel in a way that has been compared with a tethered ball that can plug an opening (Armstrong and Bezanilla 1977). The intracellular orifice of the pore or its surrounding protein parts may act as the acceptor of the ball. In the resting state, the ball is away from the pore and, subsequent to activation, it swings into the mouth to block the ion pathway (similar to N-type inactivation in potassium channels).

The high conservation of the transmembrane segments and of parts of the linker between III-S6 and IV-S1 suggests that the exact amino acid sequence of these parts is essential for proper channel function.

2.3
Structure-Function Relations of the α Subunit Discovered by Artificial Site-Directed Mutagenesis and Functional Expression

The first experimental studies of the structure-function relations of the channel protein were performed by producing base exchanges in the cDNA at sites of special interest. Functional studies of heterologously expressed mutant channels, performed with a patch-clamp, led to the following results:

(a) The TTX-binding site is in the 5' region of the S5-S6 interlinker of repeat I. This was shown by creation of chimeras between the TTX-sensitive SkM1 and the TTX-insensitive SkM2 (Backx et al. 1992; Chen et al. 1992; Satin et al. 1992). An aromatic amino acid (Tyr or Phe) in the C-terminus of the S5-S6 loop produced the sensitive phenotype, a cysteine the insensitive phenotype.

(b) The pore has a typical conductance and is highly selective for sodium although it is structurally similar to the pores of the voltage-gated potassium and calcium channels (Noda et al. 1989; Yellen et al. 1991). Mutations in the 3' portions of the S5-S6 interlinkers of each repeat reduced the channel conductance suggesting that portions of all four interlinkers contribute to the lining of the pore (Terlau et al. 1991). The substitution of glutamates for lysine and alanine of the P-region converted the channel to be selective for calcium (Heinemann et al. 1992).

(c) The hypothesis of a voltage sensor function of the S4 helices characterized by their high charge density was confirmed since reduction of the number of positive charges decreased the voltage dependence of channel activation (Stühmer et al. 1989). However, replacement of certain neutral amino acids altered activation in a similar way suggesting a complex interaction during activation by charge movements rather than a simple dependence on the net charge of a dipole.

(d) The ball-and-chain model (Armstrong and Bezanilla 1977), originally developed to explain the abolition of sodium channel inactivation following enzymatic removal or functional blockage of the cytosolic gate (Armstrong et al. 1973; Vassilev et al. 1988, 1989) and later verified for the N-terminal of the fast inactivating potassium channel (Hoshi et al. 1990; Patton et al. 1993), had to be modified on account of the mutagenesis experiments. Deletions in the N-terminus of the linker between repeats III and IV, i.e., the inactivation gate, removed inactivation whereas the positively charged particles, the supposed "ball," were not required for proper inactivation (Stühmer et al. 1989; Moorman et al. 1990). However, replacement of three hydrophobic amino acids (Ile-Phe-Met by Gln-Gln-Gln) in the same part of the linker completely eliminated inactivation (West et al. 1992). The single substitution (Phe-1489-Gln in the rat brain type IIA sodium channel, corresponding to phenylalanine 1311 of hSkM1) alone abolished fast inactivation (West et al. 1992). Hydrophobic parts of the intracellular orifice of the pore or its surrounding protein parts may act as acceptor of the hydrophobic motif. In the closed resting state, the lid is away from the pore, and following activation the lid moves into the mouth to inactivate the channel.

2.4 Structure and Function of the β Subunits and Their Effects on the α Subunit

The β_1 subunit is a small glycoprotein of 38 kDa, containing 218 amino acids in the rat. Sequence analysis suggests a single transmembrane segment with an extracellular N-terminus containing glycosylation sites (Isom et al. 1992). The human and rat cDNAs exhibit 96% homology (McClatchey et al. 1993). The β and α subunits interact in a 1:1 stoichiometry by a noncovalent binding. Perhaps β binds to only one of the four repeats (Kraner et al. 1985). The structure of β_2 which is covalently attached to the α subunit (Hartshorne et al. 1982) by disulfide bonds has just been clarified (Isom et al. 1995a).

Injection of the cRNA encoding hSkM1 into *Xenopus* oocytes results in the expression of functional sodium channels (Trimmer et al. 1989). However, the inactivation of the macroscopic sodium current conducted by these channels is slower than that of native channels or channels produced by coinjection of total mRNA or the β subunit cRNA (Krafte et al. 1988, 1990; Isom et al. 1992; Cannon et al. 1993b). This slowed inactivation is caused by channel re-openings occuring as single events or as occasional bursts (ten to hundreds of repetitive openings). It occurs much more frequently with the isolated α subunit. The occurrence of various gating modes in a single population of channels seems to reflect the existence of at least two conformations of the α subunit. The physiologically preferred gating mode (no re-openings) can be stabilized by coexpression with the β_1 subunit (Isom et al. 1994, 1995b; Ji et al. 1994; Schreibmayer et al. 1994) which accelerates inactivation in *Xenopus* oocytes by an extracellular interaction (Chen and Cannon 1995). In addition, the β_1 subunit increases functional expression of the α subunit resulting in a larger current amplitude. It shifts the steady-state activation and inactivation curves of rat and human sodium channels expressed towards more negative potentials (about 5–10 mV), and accelerates recovery from inactivation (Wallner et al. 1993; Makita et al. 1994a), whereas it had no effect on the gating of the human heart sodium channel (Makita et al. 1994a). Coexpression of β_1 in mammalian cells without endagenous β_1 production has similar effects except for the almost missing acceleration of the already fast inactivation of the channel (Isom et al. 1995b). Different protein glycosylation is discussed to be responsible for the slow inactivation of the isolated α subunit in oocytes.

2.5
Human Diseases Due to Mutations in SCN4A

As soon as *SCN4A* was cloned, a genetic approach was available to test the hypothesis of primary sodium channel defects in some hereditary myotonias and periodic paralyses. Abnormal inactivation of sodium currents had been the major result of electrophysiological studies performed on excised muscle specimens of patients having paramyotonia or periodic paralysis (Lehmann-Horn et al. 1987a,b). Linkage of hyperkalemic periodic paralysis to *SCN4A* (Fontaine et al. 1990) provided further evidence for the existence of a sodium channel disease in man. Three groups then showed independently that paramyotonia congenita is also linked to *SCN4A* (Ebers et al. 1991; Koch et al. 1991; Ptáček et al. 1991b; see Rüdel et al. 1993). Finally, molecular biology revealed that the myotonia in various families that were previously diagnosed as having "abnormal forms" of myotonia congenita (which is now known to be a chloride channel disease, see below) is in fact the consequence of mutations in *SCN4A*. Several clinical variants of such "sodium channel myotonias" (myotonia fluctuans, Ricker et al. 1990, 1994; Lennox et al. 1992), myotonia permanens (Lerche et al. 1993), acetazolamide-responsive myotonia (Ptáček et al. 1994b) have been described that are clinically different from classical hyperkalemic periodic paralysis or paramyotonia congenita. All types of sodium channel myotonia differ distinctly from the chloride channel myotonias, in that ingestion of potassium by the patient enhances the myotonia (without leading to weakness, as would be characteristic for hyperkalemic periodic paralysis). Therefore, the sodium channel myotonias are grouped together as "potassium-aggravated myotonias."

All sodium channel diseases have in common that they are transmitted with dominant mode of inheritance and that the basic pathology may include both hyper- and hypoexcitability in the same patient at different times. Fortunately for the patients, the corresponding symptoms, muscle stiffness (myotonia) and muscle weakness (paralysis), are not present all the time. Rather, they are elicited by typical stimuli. A typical trigger for an episode of weakness in hyperkalemic periodic paralysis would be rest after a heavy work load; stiffness and weakness in paramyotonia congenita is triggered by muscle exercise during exposure of the muscles to cold, and ingestion of potassium-rich food may induce muscle stiffness in patients having myotonia fluctuans. Even more fortunately for the patients, all these symptoms disappear spontaneously within an hour or so. Nevertheless, the episodes hamper the patient's life considerably, although they may be prevented to a certain extent by proper behavior and symptomatic treatment with drugs (for review see Lehmann-Horn et al. 1994).

2.5.1 Hyperkalemic Periodic Paralysis

The disease was first described by Tyler et al. (1951) and was extensively investigated by Gamstorp (1956) who clearly differentiated it from "paroxysmal familial paralysis" and named it "adynamia episodica hereditaria." Clinically, the most striking difference of the two diseases is that, during the paralytic episode, serum potassium decreases in the former and increases in the latter. To stress this distinction, the names hypokalemic periodic paralysis and hyperkalemic periodic paralysis are now preferred for these two diseases. It should be mentioned that although the respective mutations concern different channels, the symptoms associated with these two diseases may be so similar that, before the era of molecular biology, differential diagnosis of patients was sometimes very difficult.

Hyperkalemic periodic paralysis has three distinct variants. It can occur (a) without myotonia, (b) with clinical or electromyographic myotonia, or (c) with paramyotonia. In some patients, a chronic progressive myopathy may develop (Bradley et al. 1989) which seems to be caused by a single mutation (Ptáček et al. 1991a; Lehmann-Horn et al. 1993). A few families have been reported in which a change of the serum potassium was barely detectable during the episodes of paralysis (Poskanzer and Kerr 1961). Such normokalemic periodic paralysis should probably be regarded as a special condition of hyperkalemic periodic paralysis (Lehmann-Horn et al. 1993).

The diagnosis is based on the presence of typical attacks of weakness or paralysis, the positive family history, and the myotonic or paramyotonic phenomena, if present. Except for some older patients with progressive myopathy, the muscles are well-developed.

An elegant provocative method, often used in experiments designed to study the pathophysiology of paralytic attacks, consists of exercise on a bicycle ergometer for 30 min so that the pulse increases to 120–160 bpm, followed by absolute rest in bed (Ricker et al. 1989). The serum potassium level rises during exercise and then declines to almost the pre-exercise value, as in healthy individuals. At 10–20 min after the onset of the bed rest, a second hyperkalemic period occurs in the patients but not in normal subjects. It is during this period that the patients become paralyzed.

2.5.2 Paramyotonia Congenita

The hallmarks of this disease, as first described by Eulenburg (1886) and later confirmed in many families by Becker (1970) are: (a) paradoxical myotonia, defined as myotonia that appears during exercise and increases with continued exercise, (b) severe worsening of the exercise-induced myotonia by cold, (c) a predilection of the myotonia for the face, neck, and distal upper extremity muscles, and (d) weakness after prolonged exercise and exposure to cold.

Percussion myotonia is not pronounced, but many patients exhibit the lid lag phenomenon and in the course of repeated strong contractions of the orbicularis oculi the eyelids open more and more slowly. In a number of families the symptoms are clearly different from those found in most cases of paramyotonia: (a) Some patients experience myotonic stiffness during work even under warm conditions. Such patients require long-term medication. (b) In some kinships cold induces stiffness but no weakness (Koch et al. 1995). (c) Still other patients are immediately paralyzed by cold. (d) In some kinships, the patients have not only paramyotonic symptoms, but also temperature-independent paralytic attacks, resembling those in hyperkalemic periodic paralysis. The attacks usually begin early in the day and can last for several hours. Oral intake of potassium can induce such attacks in these patients distinguished by "paramyotonic hyperkalemic periodic paralysis."

These special forms led to the suggestion that hyperkalemic periodic paralysis and paramyotonia congenita are two facets of the same disease (DeSilva et al. 1990). On the other hand, the symptoms in the "pure" forms are so different that it seems reasonable to keep both nosological entities, although it is now known that the two diseases are allelic.

Permanent weakness and muscle atrophy are not signs of paramyotonia congenita. The electromyogram always shows myotonic discharges in all muscles, even at a normal muscle temperature. The serum creatine kinase is often elevated, sometimes to 5–10 times above normal.

2.5.3
Potassium-Aggravated Myotonia

For many families with dominant myotonia thought to have a subtype of Thomsen's disease, a muscle chloride channel disease, molecular genetics revealed mutations in SCN4A. In contrast to Thomsen's disease, these patients developed severe stiffness following oral ingestion of potassium. The spectrum of the degree of myotonia is large, ranging from the mild myotonia fluctuans to the very severe myotonia permanens.

Myotonia Fluctuans and Acetazolamide-Responsive Myotonia. The clinical signs of myotonia fluctuans resemble those of myotonia congenita (see below) with the peculiarity that the stiffness tends to fluctuate from day to day. The patients never experience muscle weakness and are not very sensitive to cold as regards muscle stiffness. Their muscle stiffness is provoked by exercise: usually it occurs during rest about half an hour after the exercise and lasts approximately another hour. Ingestion of potassium aggravates myotonia but does not induce weakness as in hyperkalemic periodic paralysis. Also other depolarizing agents such as suxamethonium can induce or aggravate myotonia so that severe ventilation problems may occur during general anesthesia. The incidence of such events is high in myotonia fluctuans families (Ricker et al. 1994; Iaizzo and Lehmann-Horn 1995; Vita et al. 1995), and there seems to be no other biological reason for this than the frequent absence of clinical myotonia in these patients. Thus, the anesthesiologists are unaware of the condition. Therefore it is worth mentioning that even during the spells of absence of clinical myotonia, latent myotonia can be consistently recorded by the use of electromyography. In acetazolamide-responsive myotonia (Trudell et al. 1987), also described as atypical myotonia congenita (Ptáček et al. 1992b), the muscle stiffness also fluctuates and, in addition, muscle pain is induced by exercise. Both stiffness and pain are alleviated by acetazolamide.

Myotonia Permanens. The definition of this disease is the consequence of genotyping a patient earlier thought to have a "myogenic" type of Schwartz-Jampel syndrome (Spaans et al. 1990); in fact he was carrying a mutation in *SCN4A* (Lerche et al. 1993; Mitrovic et al. 1995). Continuous myotonic activity is detectable in the electromyogram of these patients causing persistent severe myotonia. Muscle hypertrophy, particularly in the neck and shoulder, is very marked. During attacks of severe muscle stiffness the patients suffer from impaired ventilation due to stiffness of the thoracic muscles. In particular, children can suffer from acute hypoventilation and this may lead to cyanosis and unconsciousness, so that such episodes were in the past mistaken for an epileptic seizure. In spite of the misdiagnosis, anti-epileptic medication, for example, administration of carbamazepine, was successful in these cases because of its antimyotonic effects. The patients could probably not survive without persisting treatment. The aggravation of myotonia by potassium was proven in vitro (Lerche et al. 1993). Since the myotonia is so severe in these patients, potassium must never be administered as a diagnostic tool.

A further indication of the severity of this disease is that the few patients known to date were sporadic cases having a de novo mutation, i.e., their proven biologic parents did not carry the mutation; none of these patients has offspring.

2.6 Quarter Horse Hyperkalemic Periodic Paralysis Due to a Mutation in Scn4a

An equivalent condition to human hyperkalemic periodic paralysis in man has become known in the quarter horse, a very common breed of race horses in the United States (Cox 1985; Pickar et al. 1991). The symptoms are similar to those described above for the human disease, but the condition seems to be sometimes more serious than in man as some affected horses have died from attacks as a result of shock or ischemia. The disease, the most frequent hereditary disorder of horses, is linked to the equine homologue *Scn4a* (Rudolph et al. 1992a,b).

Since the discovery of this homology, affected horses have been used for the study of the cellular and physiological factors dictating the onset and severity of attacks and the relationship between exercise, systemic potassium, catecholamines, and other factors influencing muscle metabolism. Thus, the hyperkalemic horse has been used to show the first correlation of levels of mutant mRNA relative to normal mRNA as a likely determinant of clinical severity in dominantly inherited disease (Zhou et al. 1994).

2.7 Pathophysiology of Sodium Channel Diseases

Electrophysiology on excised muscle specimens from such patients revealed a noninactivating component of the sodium current as a specific abnormality (Lehmann-Horn et al. 1983, 1987a,b, 1991; Cannon et al. 1991). This indicated that both key symptoms, stiffness and weakness, are caused by the same mechanism, namely a long-lasting depolarization of the muscle fiber membranes. When the depolarization is mild, say 5–10 mV, this may fulfil exactly the condition for the voltage-dependent sodium channels to open again after an action potential, i.e., for repetitive firing which is the basis for the involuntary muscle activity that the patient experiences as muscle stiffness. This hyperexcitable state can be computer-simulated (Cannon et al. 1993a) and mimicked by anemone toxin (Cannon and Corey 1993). When the depolarization is strong, say 20–30 mV, the majority of the intact sodium channels adopt the state of inactivation, i.e., the muscle fibers become inexcitable, which is the basis of the muscle weakness. When all fibers of a muscle are inexcitable, this results in complete paralysis (for unknown reason, the diaphragm is spared, and fortunately the heart muscle in which a different gene is expressed).

2.8
Naturally Occurring Mutations in SCN4A

With the knowledge of the special function of some domains of the channel protein, it is interesting to see where the various point mutations are situated and how they might produce the different symptoms.

Nineteen disease-causing amino acid changes have been identified to date (Fig. 1, Table 1), and five additional silent amino acid substitutions were found in healthy controls. Five of the mutations result in hyperkalemic periodic paralysis, eight in paramyotonia congenita, and six in potassium-aggravated myotonias. At first glance, no obvious correlations can be made between the location of amino acid change and the clinical phenotype. However, the distribution of the various substitutions along the protein molecule might display certain systematic features. The number of substitutions increases as one proceeds from the N to the C terminus: none of the substitutions is situated in repeat I, and only 3 substitutions are located in repeat II, all of them in transmembrane segments. Three other substitutions are situated in repeat III, two of them in an intracellular interlinker and one in transmembrane segment 6. The majority of the substitutions are situated in the linker connecting repeats III and IV (n=5) and in repeat IV (n=9).

None of the substitutions concerns an extracellular loop, whereas 7 substitutions concern intracellular loops. Of the 12 substitutions in the transmembrane segments, 6 are situated near the intracellular membrane face, 5 are near the extracellular face, and only one substitution is located in the middle of a transmembrane segment (in IV-S4).

The most common mutation causes hyperkalemic periodic paralysis and predicts Thr-704-Met in segment II-S5. The mutation may cause the myotonic or the nonmyotonic form of the disease (Ptáček et al. 1991a; Feero et al. 1993; Wang et al. 1993; Plassart et al. 1994). In either case progressive myopathy is found in older and sometimes even in younger patients. The second most common mutation, also causing HyperPP and predicting Met-1592-Val in IV-S6, is always associated with myotonia and does not lead to permanent weakness (Rojas et al. 1991; Feero et al. 1993; Wang et al. 1993). The rare third and fourth mutations, predicting Ala-

Table 1. Mutations of *SCN4A*, the gene encoding the α subunit of the human skeletal muscle sodium channel

Genotype	Channel domain	Substitution	Exon no.	Phenotype	First report
Hyperkalemic periodic paralysis					
C2188T	IIS5i	Thr-704-Met	13	Permanent weakness (non)-myotonic, most frequent	Ptáček et al. 1991a
G2341A	IIS6	Val-781-Ile	13	Cardiomyopathy?	Baquero et al. 1995
G3466A	(IIIS4/5)i	Ala-1156-Thr	19	Reduced penetrance	McClatchey et al.1992b
A4078G	IVS1	Met-1360-Val	23	Reduced penetrance	Lehmann-Horn et al.1993
A4774G	IVS6i	Met-1592-Val	24	Myotonic, frequent	Rojas et al. 1991
Paramyotonia congenita					
G3877A	IIIS6i	Val-1293-Ile	21		Koch et al. 1995
C3938T	(III/IV)i	Thr-1313-Met	22	Frequent	McClatchey et al.1992c
T4298G	IVS3	Leu-1433-Arg	24		Ptáček et al. 1993
C4342T	IVS4	Arg-1448-Cys	24	Potential atrophy	Ptáček et al. 1992a
G4343A	IVS4	Arg-1448-His	24		Ptáček et al. 1992a
G4343C	IVS4	Arg-1448-Pro	24	Potential atrophy	Lerche et al. 1996b
G4372T	IVS4	Val-1458-Phe	24		Heine et al. 1996
T4418C	IVS4/5i	Phe-1473-Ser	24		Heine et al. 1996
Potassium-aggravated myotonias					
C2411T	IIS6i	Ser-804-Phe	14	Overlap Myotonia fluctuans	McClatchey et al.1992b Ricker et al. 1994
A3478G	(III/IV)i	Ile-1160-Val	19	Acetazolamide-responsive	Ptáček et al. 1994b
G3917A	(III/IV)i	Gly-1306-Glu	22	Myotonia permanens	Lerche et al. 1993
G3917C	(III/IV)i	Gly-1306-Ala	22	Myotonia fluctuans	Lerche et al. 1993
G3917T	(III/IV)i	Gly-1306-Val	22	Overlap Myotonia	McClatchey et al. 1992c Lerche et al. 1993
G4765A	IVS6i	Val-1589-Met	24	Myotonia	Heine et al. 1993

1156-Thr in the linker between III-S4 and III-S5 (McClatchey et al. 1992b) and Met-1360-Val in IV-S1 (Lehmann-Horn et al. 1993), respectively, are characterized by incomplete clinical penetrance in females, although "unaffected" family members show electrical myotonia in the electromyogram indicating that penetrance is really 100% (Lehmann-Horn et al. 1993). The fifth mutation, predicting Val-783-Ile, is a sporadic case with hyperkalemic periodic paralysis and cardiac dysrhythmia (Baquero et al. 1995). One family that was convincingly diagnosed as having hyperkalemic periodic paralysis, was not linked to *SCN4A* (Wang et al. 1993). Genetic heterogeneity is the most probable explanation.

Three of the 8 point mutations leading to paramyotonia congenita involve Arg-1448 in segment IV-S4, replacing it by histidine, cysteine or proline (Ptáček et al. 1992a; Wang et al. 1993; Meyer-Kleine et al. 1994a; Wang J et al. 1995; Lerche et al. 1996b). The more the substitution differs from the normal arginine the more severe the clinical symptoms: a patient having the proline change is the most severely afflicted paramyotonia patient observed to date, with fixed ankle joint contractures (probably secondary to myotonia), and induction of severe myotonia and weakness by even tepid temperatures (Wang J et al. 1995; Lerche et al. 1996b).

A frequent paramyotonia-causing mutation results in Thr-1313-Met (McClatchey et al. 1992c; Ptáček et al. 1993; Plassart et al. 1994; Tahmoush et al. 1994) and affects the cytoplasmic loop between repeats III and IV, i.e., the inactivation gate. Other mutations predict Leu-1433-Arg in IV-S3 (Ptáček et al. 1993), Val-1293-Ile in III-S6 (Koch et al. 1995; Heine et al. 1996), Val-1458-Phe in IV-S4 (Heine et al. 1996), and Phe-1473-Ser in the intracellular loop connecting S4 and S5 of repeat IV (Heine et al. 1996, Mitrovic et al. 1996).

Two further mutations, predicting Gly-1306-Glu (Lerche et al. 1993) and Gly-1306-Ala (Lerche et al. 1993; Ricker et al. 1994; Mitrovic et al. 1995; Vita et al. 1995), cause myotonia permanens and myotonia fluctuans, respectively. A third, Gly-1306-Val (McClatchey et al. 1992c; Lerche et al. 1993; Plassart et al. 1994), causes a moderate degree of myotonia. The mutations substitute one of a pair of glycines (Gly-1306/Gly-1307) in the III-S6 to IV-S1 linker and thus affect the putative inactivation gate at the same position as the Gly-1306-Val substitution causing paramyotonia congenita. This pair of unbranched amino acids is believed to act as a kind of "hinge" of the inactivation gate and, therefore, should be essential for proper channel inactivation (West et al. 1992). As discussed above for the three Arg-1448 substitutions, the more the substituted amino acid differs from the normal glycine, the greater the membrane hyperexcitability, and the more severe the clinical symptoms: a change from glycine to glutamic acid, an amino acid with a long and charged side-chain, causes myotonia permanens, the most severe form of all known nondystrophic myotonic diseases; valine, an amino acid with a side-chain of intermediate size, causes moderate myotonia; whereas alanine, with a short side-chain, results in a benign, often "subclinical" form of myotonia fluctuans.

The other two mutations lead to moderate myotonia, Val-1589-Met in IV-S6 (Heine et al. 1993; Mitrovic et al. 1994), or myotonia fluctuans, Ser-804-Phe in II-S6 (Ricker et al. 1994). A family having mutation Ser-804-Phe had earlier been described as having "features of paramyotonia congenita and myotonia congenita" (McClatchey et al. 1992b).

Finally, a mutation predicting Ile-1160-Val in the linker of segments 4 and 5 of repeat III was found to cause acetazolamide-responsive myotonia (Ptáček et al. 1994b).

2.9 Frequency of Natural Mutations in SCN4A

Mutations of human genes may arise either as a consequence of endogenous error-prone processes, such as DNA replication and repair, or as a result of exposure to exogenous factors, for example, chemical mutagens or ionizing/UV irradiation. A systematic investigation of spontaneous point mutations causing human genetic disease disclosed that the dinucleotide CpG (p denotes cytosin 5' to guanin 3' binding) is a "hot spot" for mutations in the human genome (Cooper and Krawczak 1990). This has been explained to be the result of the hypermutability of methylated CpG, deamination of 5-methylcytosine (5mC) to thymidine in this doublet giving rise to C $\rightarrow$ T or G $\rightarrow$ A substitutions depending upon in which strand the 5mC is mutated (Duncan and Miller 1980). Thymidine being a "normal" nucleoside is thought to be less readily detectable and removable by cellular repair mechanisms. Evaluation of 139 point mutations causing human genetic diseases other than sodium channelopathies, and consistent with methylation-mediated deamination of 5mC, yielded 21 CG $\rightarrow$ TG and 23 CG $\rightarrow$ CA mutations, i.e., a total of 44 or 31.7% (Cooper and Krawczak 1990).

Of the 19 mutations detected in *SCN4A*, 8 (40%) contained a mutated base in the CpG dinucleotide. Thus, the frequency of CpG mutations in *SCN4A* corresponds well with that found in general for disease-causing point mutations (Heine 1995). With the exception of the "Ravensberg" families (Arg-1448-His, Meyer-Kleine et al. 1994a), a founder effect for the mutations could be excluded (Wang et al. 1993). Considering the number of independently originating *SCN4A* mutations in the 94 nonrelated families entering this review, 60 concerned the CpG dinucleotide. Thus, the percentage of mutations in a CpG nucleotide in *SCN4A* is even higher, i.e., around 64% of the total number.

Second in frequency amongst the human disease-causing point mutations (Cooper and Krawczak 1990) were those affecting a GG dinucleotide. Interestingly, mutations of this type are responsible for the three substitutions for the amino acid Gly-1306-Ala/Val/Glu in the supposed inactivation gate (Heine 1995).

2.10 Properties of Mutant Sodium Channels in Heterologous Expression Systems

When wild-type *SCN4A* is transfected into mammalian cells, for example, human embryonic kidney (HEK293) cells, inactivation is normal (Ukomadu et al. 1992).

Patch-clamp recordings of sodium currents from HEK293 cells transfected with mutant channels revealed that the mutations leading to the clinically different phenotypes of hyperkalemic periodic paralysis, paramyotonia congenita or one of the potassium-aggravated myotonias, yielded similar results: The major defect of all investigated mutants was a more or less impaired inactivation (Cannon and Strittmatter 1993; Cummins et al. 1993; Chahine et al. 1994; Mitrovic et al. 1994, 1995; Yang et al. 1994). The macroscopic changes of inactivation were evident as (a) a slowing of the decay of the current transients, elicited by depolarizing voltage pulses, (b) an increased steady-state current following the current, (c) shifts of the steady-state inactivation and/or activation curves increasing the window current, and (d) uncoupling of inactivation from activation. Single-channel recordings revealed that in particular point (b) was owing to a shift in the equilibrium between two physiological gating modes known for inactivation. A qualitatively similar but quantitative smaller late activity has been also seen with membrane patches from native normal skeletal and heart muscle as well as from brain cells of various species (Patlak and Ortiz 1986; Nilius 1988; Moorman et al. 1990; Zhou et al. 1991; Saint et al. 1992; Ju et al. 1994).

Careful investigation revealed that the changes of inactivation are not quite the same with different mutants. For example, slowing of the current decay was most pronounced with substitutions for Arg-1448 causing paramyotonia (Chahine et al. 1994; Lerche et al. 1996b), whereas a large persistent sodium current was found for the mutants Met-1592-Val (HyperPP, Cannon and Strittmatter 1993) and Val-1589-Met (PAM, Mitrovic et al. 1994).

There is reason to believe that it is hinged-lid inactivation of the channels that is affected by the thirteen amino acid substitutions on the cytoplasmic surface of the channel. Particularly, the mutations affecting Gly-1306 may directly affect the hinge of the inactivation gate (West et al. 1992; Lerche et al. 1993; Mitrovic et al. 1995). In contrast, the mutations within the voltage sensor IV/S4 (Arg-1448-His/Cys/Pro) uncouple activation and inactivation which results in a relatively voltage-independent time course of inactivation (Chahine et al. 1994; Lerche et al. 1996b). Thus, the disease – causing mutations in the human muscle sodium channel have uncovered functions of the channel protein which had previously not been recognized in studies of in-vitro mutagenesis.

Under the experimental conditions of voltage-clamp of heterologous cells, neither extracellular potassium nor low temperature had a marked and reproducible effect on any of the mutant channels investigated. Therefore, these triggering factors probably exert their effects indirectly, for example, they physiologically cause slight and sustained membrane depolarization followed by an increase in the persistent sodium current.

2.11
Long QT3 Syndrome, a Disease Linked to SCN5A Encoding the α Subunit of the Human Cardiac Muscle Sodium Channel

In the congenital disorder of long QT syndrome (LQT), the cardiac action potential is prolonged as clinically evidenced by a prolonged QT interval of the electrocardiogram. The condition is transmitted as an autosomal dominant trait. Patients are predisposed to syncope and sudden death owing to ventricular arrhythmias.

Linkage studies have demonstrated genetic heterogeneity for the disease by the identification of so far three different loci: LQT1 on chromosome 11p15.5, LQT2 on 7q35-36, and LQT3 on 3p21-24. Whilst no candidate gene is known for LQT1, mutations in *HERG* have been claimed to cause LQT2 (Curran et al. 1995). The *HERG* gene product is believed to be a potassium channel conducting the rapidly activating delayed rectifier current (IKr) in cardiac muscle (Sanguinetti et al. 1995).

LQT3 was shown to cosegregate with polymorphisms within *SCN5A*, the gene encoding the α subunit of the major cardiac sodium channel (Jiang et al. 1994). Analysis of the DNA of affected members of two unrelated LQT3 families revealed an in-frame deletion in *SCN5A* which results in the omission of three highly conserved amino acids (KPQ) in the cytoplasmic linker between repeats III and IV, the inactivation gate (Wang Q et al. 1995).

Expression of the mutated gene in *Xenopus* oocytes demonstrated abnormal sodium channel gating and, as a result, a small persistent inward current (Bennett et al. 1995). Such a persistent sodium current may also flow during the plateau phase of the action potential of cardiac myocytes. The plateau is normally maintained by a delicate balance between inward and outward currents, and repolarization takes place when the outward currents prevail over the inward currents. The sustained inward current generated by the mutant sodium channel prolongs the action potential and thus lengthens the QT interval. Thus in LQT, the first recognized form of a myocardial ion channel disease, the defect in channel function seems to be very similar to defects described in the skeletal muscle sodium channel diseases.

2.12
A Mouse Mutant Having Motor Endplate Disease, Linked to Scn8a

Motor endplate disease (MED), transmitted as a recessive trait, is characterized by progressive weakness and wasting of the skeletal muscles. It begins around day 9 after birth at the hind limbs and ends fatally 10–15 days later. The cerebellar Purkinje cells are degenerated and show a reduction in sodium-dependent spontaneous action potentials. The motor neurons have an increased temperature sensitivity of action potentials as well as a reduced conduction velocity and prolonged refractory period (Angaut-Petit et al. 1982). The alterations in the motor nerves probably account for the failures in neuromuscular transmission. A progressive abnormality of the skeletal muscle innervation is marked sprouting. As the muscle weakness becomes more severe, the terminal arborizations become increasingly more complex, in particular more elongated than normal, and may be in contact with several muscle fibers (Duchen 1970). The light microscopic changes are in many ways similar to those seen in muscle after local injection of botulinum toxin.

A gene encoding a sodium channel α subunit was isolated from the flanking region of a transgene-induced allele of the *med* gene and was called *Scn8a*. It is expressed in brain and spinal cord but not in skeletal or heart muscle. *Scn8a* is most closely related to a brain cDNA from the pufferfish *Fugu rubripes*, with 83% overall sequence identity and several shared insertions and deletions that are not present in other mammalian cDNAs. The divergence of *Scn8a* from the other brain sodium channels thus occurred prior to the separation of fish and mammals and predates the origin of the gene cluster on chromosome 2. A deletion at the transgene insertion site spanning 5–10 kb genomic DNA causes loss of expression (Burgess et al. 1995). In MED mice, i.e., animals that are homozygous for the deletion, the absence of the *Scn8a* gene product could account for the above-mentioned symptoms although a multiplicity of other sodium channel isoforms might be expressed.

2.13 Paralysis in Drosophila Due to para Locus Sodium Channel Mutations

Para encodes a functionally predominant class of voltage-gated sodium channels in neurons of the fruit fly *Drosophila melanogaster*. It contains a minimum of 26 exons distributed over more than 60 kb genomic DNA. The transcript undergoes alternative splicing to produce several distinct channel subtypes. The proteins are very similar to those of vertebrate sodium channels (Loughney et al. 1989).

Several insertions of transposable elements within introns of several *Drosophila* genes are known to interfere with transcription or processing of the RNA (Levis et al. 1984). These mutations often allow some degree of expression because a small fraction of the transcripts from the mutated gene is correctly transcribed and processed to produce functional mRNAs. A similar mechanism seems to operate for the *para* insertional mutations, so that they result in a low sodium channel expression rate rather than in the production of altered channels. Thus, in certain *para* mutants, the number of sodium channels may be greatly reduced in neurons whose sodium channels are expressed by the *para* gene. Excitation of these neurons may physiologically still be possible, but may be abortive under adversary circumstances. Indeed, some *para* mutants become paralyzed at elevated temperature, probably because faster channel kinetics cause conduction block (Loughney et al. 1989).

3 Potassium Channelopathies

3.1 The Multigene Family Encoding Voltage-Gated Potassium Channels

A mutant of the fruit fly *Drosophila melanogaster* episodically presents, and consistently responds to ether anesthesia, with jerking leg movements and was therefore named "Shaker". Electrophysiological investigation of the muscle fibers of this mutant revealed potassium currents that inactivated much faster than those of wild type *Drosophila*. This led to the correct assumption that the Shaker mutation is located in a gene coding for a potassium channel (*Shaker* gene). Cloning and functional expression of this gene was the starting point for the identification of a large gene family encoding voltage-gated potassium channels that have homologs in vertebrates. The human homolog of the *Shaker* gene was termed *Kv1.1* or *KCNA1* according to the nomenclature of mammalian potassium channel genes (Gutman and Chandy 1993). Typically enough, the first human channelopathy detected involving a voltage-gated potassium channel is caused by a mutation in *Kv1.1* (Browne et al. 1994).

In vertebrates, the family consists of six subfamilies, with *Kv1.1* to *Kv1.7* (*KCNA1* to *KCNA7*) belonging to the *Shaker* gene-related subfamily 1. Correspondingly, subfamily 2 (*Kv2* or *KCNB*) is related to *Shab*, subfamily 3 (*Kv3* or *KCNC*) to *Shaw*, and subfamily 4 to *Shal*, genes that are all related to potassium channel genes in *Drosophila*. Further related mammalian subfamilies encode either voltage-gated (*Kv5/KCND*, *Kv6/KCNG*) or ligand-gated potassium channels. The voltage-gated potassium channels inactivate at different rates and to a varying extent (fast N-type and slow C-type inactivation). They are found in almost all eukaryotic cells of the animal and plant kingdoms (Rudy 1988) and are not only involved in the excitability of cells, such as nerve or muscle cells, but also of "inexcitable" cells such as lymphocytes and pancreatic islet cells (Grissmer et al. 1990).

3.2
Structure of the Drosophila-Related Mammalian Potassium Channels

The functional channels are tetrameric complexes of four usually identical domains. Each potassium channel domain is encoded by a gene, whereas in the sodium channel, the whole tetramer is encoded by one gene. Corresponding to the segments S1–S6 in each of the repeats I–IV of the sodium channel, each potassium channel monomer contains six membrane-spanning α-helices, and the segments S5 and S6 and the linkers between them make up the lining of the conducting pore (Pongs 1992; Salkoff et al. 1992). Homo- and heteromultimeric potassium channels may occur in nature, with heteromers belonging to the same subfamily. Most homotetramers, for example, all *Shaker*-related channels except Kv1.4, inactivate slowly. Heterotetramers show fast inactivation if at least one of the four domains, for example, Kv1.4, carries the inactivation ball within its N-terminus (MacKinnon et al. 1993).

3.3
Episodic Ataxia with Myokymia (EA-1), a Kv1.1 Channelopathy, and EA-2, a Possible Potassium Channelopathy

This autosomal dominant human disease is characterized by episodic failure of excitation of cerebellar neurons and sustained hyperexcitability of the second motoneurons. The latter feature leads to repetitive twitching of small muscle fibers, for example, those around the eyes or in the hands. These so-called myokymias respond to anticonvulsants such as carbamazepine. The onset of motion and exercise may provoke attacks of atactic gait that last for seconds to minutes. Acetazolamide is able to reduce the number and severity of attacks in some families (Gancher and Nutt 1986; Brunt and van Weerden 1990). Litt et al. (1994) showed linkage of the disease to chromosome 12p13, and six point mutations have been identified in Kv1.1 for six families: Val-176-Phe, Arg-240-Ser, Phe-250-Ile, Val-410-Ala, Glu-325-Asp, Phe-184-Cys (Browne et al. 1994, 1995; Fig. 2).

A clinically similar disease, episodic ataxia-type 2 (EA-2), has recently been linked to a locus that possibly encodes another potassium channel (Vahedi et al. 1995). This form of episodic ataxia is generally associated with interictal nystagmus. Attacks are provoked by emotional stress and exercise but not by starting movements. Attacks last for several hours to more than a day. Some patients develop progressive ataxia and dysarthria

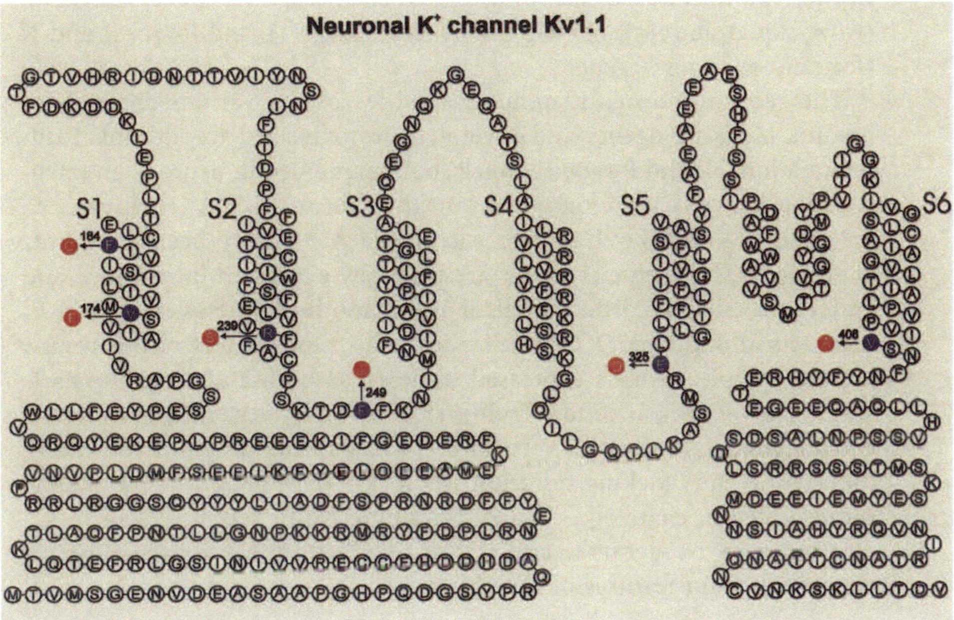

Fig. 2. Substitutions predicted in potassium channel Kv1.1 causing episodic ataxia with myokymia

and may display cerebellar vermian atrophy on magnetic resonance imaging. Acetazolamide is usually very effective in reducing the frequency of attacks.

4
Calcium Channelopathies

4.1
Types of Voltage-Gated Calcium Channels

Initially, voltage-gated calcium channels were distinguished on the basis of their inactivation properties and named T-type when conducting transient current or L-type when conducting long lasting current. A later classification noted their presence in certain tissues, naming them B for brain, N for neuronal, and P for cerebellar Purkinje cells. Then, another classification was introduced according to their electrical threshold of activation (Nowycky et al. 1985), namely low (T-type), high (L-and P-types) and N (for neither L nor T-type).

The recognition of different sensitivities of the various channels to organic blocking agents and several neurotoxins led to additional differentiation. N- and P-type channels, both expressed in neurons and neuroendocrine cells, are blocked by omega-conotoxin GVIA (Hirning et al. 1988) and the funnel web spider toxin omega-Aga-IVA, respectively (Mintz et al. 1992). The L-type channels are generally expressed in neuronal and endocrine cells, in cardiac, skeletal and smooth muscle as well as in fibroblasts and kidney. Of particular interest in the context of this review are the L-type channels expressed in heart, smooth and skeletal muscle which are very sensitive to dihydropyridines (DHP, e.g., nifedipine), phenylalkalamines (PAA, e.g., verapamil), and benzothiazepines (BTZ, e.g., diltiazem). The blocking criterion led to the designation of the muscle L-type calcium channel as "dihydropyridine(DHP) receptor" which is a misnomer as it suggests ligand-activation when in fact voltage-gating is one of the main features of the channel.

4.2 Multigene Family, Structure, and Function of Voltage-Gated Calcium Channels

Calcium channels are oligomeric proteins consisting of maximally five subunits: α_1, α_2, β, γ, and δ (the γ subunit seems only present in skeletal muscle L-type channels).

4.2.1 The α_1 Subunit

The α_1 subunit shows slight variation depending on the tissue in which it is expressed. Structurally, it very much resembles the sodium channel α subunit with its four repeats (I-IV), each with six hydrophobic segments (S1-S6) thought to span the T-tubular membrane. It always contains the channel pore and several voltage sensors. The muscular L-type channel α_1 subunit also contains the receptors for dihydropyridines, phenylalkalamines and benzothiazepines.

Six different α_1 subunit genes (*CACNL1A1-6*) have been identified in vertebrates. *CACNL1A3*, the gene encoding S, the α_1 subunit of skeletal muscle, has been localized to human chromosome 1q31-32 (Drouet et al. 1993; Gregg et al. 1993) and its human cDNA sequenced (Hogan et al. 1994). The mRNA was originally cloned from rabbit muscle (Tanabe et al. 1987). The gene product occurs in two isoforms, a rare 212-kDa protein containing the complete amino acid sequence, and a frequent (approx. 95%) 190-kDa isoform truncated by posttranslational proteolysis at aa 1690 (DeJongh et al. 1991). The truncated form was shown to be functional (Beam et al. 1992). Additional forms seem to exist, at least in postnatal muscle (Malouf et al. 1992). S-specific mRNA sequences have been also detected in other tissues including brain and kidney (see Hofmann et al. 1994).

CACNL1A1 is located on human chromosome 12p13 (Lacerda et al. 1991; Powers et al. 1991; Schultz et al. 1993) and encodes two different splice products, i.e., the α_1 subunits for cardiac (Ca) and smooth (Cb) muscle. This gene is also expressed in most excitable and many nonexcitable cells such as fibroblasts. The L-type channel D encoded by *CACNL1A2* on chromosome 3p21.3–21.2 (Chin et al. 1991; Seino et al. 1992b) seems to be specific for the neuroendocrine system (brain: Williams et al. 1992; pancreatic islet: Seino et al. 1992a) as well as a splice variant of D or a channel encoded by *CACNL1A4*. Two other genes, *CACNL1A5* on chromosome 9q34 and *CACNL1A6* on 1q25-31 seem also to code for L-type α_1 subunits of so far unidentified distribution (Diriong, Genomics, 1995 GDB-No.). Brain cDNAs have been sequenced and functionally expressed for various P, N, and T type channels (A, B, and E, respectively; for review see Hofmann et al. 1994) but their human genes have not yet been localized.

4.2.2
The α_2/δ Subunit

Although the α_1 subunit is able to form a calcium channel by itself, the fact that several other proteins consistently copurify with it has led to the idea that other modifying subunits also exist. The skeletal muscle α_1 subunit copurifies with a protein of approximate molecular mass 175 kDa (α_2/δ) that characteristically shifts to 150 kDa (α_2) upon reduction, with appearance of three peptides of 25, 22, and 17 kDa (δ; Catterall 1991). The α_2/δ protein is produced by a single gene with its 5'-end encoding α_2 and its 3'-end encoding δ. The δ subunit spans the membrane and anchors the extracellularly located α_2 protein in the plasma membrane via disulfide bonds (Jay et al. 1991). Sequences identical or closely related to the skeletal muscle α_2/δ subunit have been found in a variety of tissues ($\alpha_1/\delta b$), including cardiac and smooth muscle, and the nervous system (Ellis et al. 1988). The α_2/δ subunit from rat brain predicts an identical δ protein and a splice variant of the processed α_2 protein. The corresponding human gene has been localized to chromosome 7q21-22 (Iles et al. 1994; Powers et al. 1994).

4.2.3
The β Subunit

The skeletal β subunit (β_1) is a membrane protein, located intracellulary, consisting of 524 amino acids (Ruth et al. 1989). Its deduced amino acid sequence contains stretches of heptad repeat structure characteristic of cytoskeletal proteins. The corresponding human gene, *CACNLB1*, has been localized to chromosome 17q21-22 (Iles et al. 1993a). Differential splicing of the primary transcript of β_1 results in at least three isoforms: β_1a through β_1c (Ruth et al. 1989; Pragnell et al. 1991; Powers et al. 1992; Williams et al. 1992); β_1a is expressed in skeletal muscle whereas the two other isoforms are expressed in brain, heart, and spleen (Powers et al. 1992). The protein does not appear to be significantly glycosylated and is hydrophilic. Thus, it probably represents an extrinsic membrane protein and may normally be associated with the cytoplasmic aspect of the α_1 subunit. When skeletal muscle α_1 subunits were stably transfected in heterologous cells, coexpression of the skeletal muscle β subunits dramatically increased the number of DHP-binding sites but curiously had no effect on the current density (Lacerda et al. 1991; Varadi et al. 1991). However, the β subunit increased the rate of calcium current activation (Lacerda et al. 1991; Varadi et al. 1991). Similar effects have been noted for the α_1 subunit from cardiac muscle. When this protein was expressed in oocytes, rate of activation and amplitude of the calcium current were substantially increased by the coexpression of either the skeletal muscle or cardiac muscle forms of the β subunit (Perez-Reyes et al. 1992; Wei et al. 1991). Transcripts of two other genes (β_1 and β_2) encoding proteins different from the skeletal muscle β subunit have been isolated from a cardiac cDNA library (Hullin et al. 1992). A fourth β subunit, β_4, has been cloned from rat brain (Castellano et al. 1993).

4.2.4
The γ Subunit

This subunit seems to be specific for skeletal muscle. It consists of 222 aa and contains four putative transmembrane domains (Bosse et al. 1990; Jay et al. 1990). The human gene, *CACNLG*, has been localized to chromosome 17q24 (Iles et al. 1993b; Powers et al. 1993). When the γ subunit was coexpressed with cardiac α_1-subunits in *Xenopus* oocytes, rate of inactivation and amplitude of the calcium currents were increased, and the voltage range of inactivation was shifted in the hyperpolarizing direction (Singer et al. 1991; Lerche et al. 1996a).

4.3
The Skeletal Muscle L-Type Calcium Channel

The high density of L-type channels in the transverse tubules of adult skeletal muscle was noticed as an unusually high concentration of DHP binding sites which made the isolation and characterization of the "DHP receptor" possible. It is for this reason that most of our knowledge on the structure and function of voltage-gated calcium channels stems from work on this channel. Linkage of human or animal diseases with genes encoding voltage-gated calcium channel genes concerns either the α_1 subunit or the α_2/δ subunit of the skeletal muscle L-type channel.

In skeletal muscle, this channel is essential for excitation-contraction coupling which, in contrast to heart and smooth muscle, does not require calcium influx from the extracellular space. The channel is involved in voltage-dependent calcium release from the sarcoplasmic reticulum, mediating contraction (for review see Melzer et al. 1995). It is characterized by a high activation threshold and slow inactivation. Its exceptionally slow activation is unique for voltage-gated ion channels.

The currents conducted by this channel have been studied mainly in frog and rat muscle both by the use of the microelectrode voltage clamp on intact fibers and the vaseline-gap clamp on cut fibers.

The channel of intact adult muscle has not yet been characterized on the single-channel level because of its location in the depth of the transverse tubular system. However, its isolation and enrichment in vesicular membranes provided the basis for the reconstitution of functional channels in artificial lipid bilayers. Fluctuations of the current conducted by single channels were usually investigated with high concentrations of permeant divalent cations on both sides of the bilayer. Two separate conductances (9 and 20 pS when measured with symmetrical 90 mM $BaCl_2$) were found in such an isolated channel preparation (Pelzer et al. 1989). For both conductances the open probability depended on the membrane potential. The large conductance was sensitive to dihydropyridines and phenylalkylamines as well as to cAMP-dependent phosphorylation, and was therefore attributed to the L-type channel. The small conductance was not sensitive to these agents and may perhaps be related to the fast calcium current in frog muscle (Cota and Stefani 1986).

4.4
Calcium Channels in Developing Skeletal Muscle

Skeletal muscle fibers are formed either during embryogenesis by fusion of myoblasts or during regeneration by fusion of proliferated satellite cells. The early stages of this process can be followed in tissue culture by studying oligo- or multinucleated myotubes. Myotubes (or myoballs derived from them) exhibit both T-type and DHP-sensitive L-type currents (Cognard et al. 1986). Cultured human muscle cells conduct an additional high voltage-activated calcium current that resembles the N channel current as far as voltage-dependence and kinetics are concerned, but lacks N channel pharmacology (Rivet et al. 1992; Lehmann-Horn et al. 1995b; Sipos et al. 1995). In rat and mouse muscle fibers, the T-type current seems to prevail early during postnatal development, to give way later to L-type current (Beam and Knudson 1988a,b; Gonoi and Hasegawa 1988). In contrast Cognard and colleagues reported that L-type channels start first and then the T-type channels are transiently expressed (Cognard et al. 1992, 1993a,b). Experiments on fibers that had the connection of the transverse tubules to the surface membrane disrupted by glycerol treatment showed that at the later stages of development (e.g., 34 days of culture) the L-type channels are exclusively located in the T system whereas the T-type channels reside in the surface membrane (Romey et al. 1989). Both types are found in the surface membrane as early as day 9 of culture. The spatial separation apparently goes along with the appearance of well-defined triads (Romey et al. 1989).

4.5 Heterologous Expression of the Skeletal Muscle L-Type Calcium Channel

The mRNA for the α_1 subunit of the cardiac L-type calcium channel produced functional channels when introduced into *Xenopus* oocytes (Mikami et al. 1989). On the other hand, the attempt to express cDNA of the (rabbit) skeletal muscle α_1 subunit in the oocyte system did not result in functional channels. Functional expression of this cDNA was possible, however, in myotubes derived from mice with muscular dysgenesis (Tanabe et al. 1988; see below). Moreover, when this cDNA was introduced into the nuclei of the dysgenic myotubes, not only the typical slow calcium current could be recorded, but also EC coupling was restored (Tanabe et al. 1988). About 10% of the cells surviving the injection procedure contracted either spontaneously or upon electrical stimulation. Contracting cells showed slow calcium currents when investigated with the whole-cell technique. These currents were not necessary for contraction since EC coupling remained functional after the calcium current was blocked by 0.5 mM cadmium ions. The restoration of the slow calcium current and skeletal muscle type EC coupling indicates that a protein derived from the same gene is a main constituent of both the calcium channel and the voltage-dependent mechanism that controls internal calcium release.

4.6 Hypokalemic Periodic Paralysis, a Disease Linked to CACNL1A3, the Gene Encoding the Skeletal Muscle L-Type Channel α Subunit

Familial hypokalemic periodic paralysis is the most common form of the periodic paralyses in man. The condition is transmitted as an autosomal dominant trait. Its main symptom is the episodic occurrence of attacks of flaccid weakness. The attacks are associated with a decrease of the potassium concentration in the serum. They can be provoked by excessive intake of carbohydrates, strenuous exercise or by mental stress. Interestingly they do not occur during the physical stress but during a following rest, sometimes several hours later. The attacks vary in frequency, duration and severity. In severe attacks, the vital capacity may be reduced and death can occur from ventilatory failure or cardiac arrhythmia due to hypokalemia. However, usually patients show a normal life span (reviewed by Lehmann-Horn et al. 1994). Independent of the occurrence of attacks, patients develop a late onset, progressive myopathy (Links et al. 1994a). The Burmese cat may be considered as a naturally occurring animal model for the disease (Blaxter et al. 1986; Mason 1988).

The hypokalemia is assumed to be caused by stimulation of the sodium-potassium pump by insulin which is one physiological mechanism by which potassium ions are transported from the extracellular space into the intracellular compartment (Zierler and Andres 1957; Clausen and Kohn 1977; Flatman and Clausen 1979; Grob et al. 1957). Other hormones influencing the transmembrane movement of potassium are thought to contribute and thus to affect the frequency and severity of the clinical symptoms. This hypothesis is based on the observations that the onset of the disease is usually during puberty, penetrance in women is incomplete, attacks become predominant during pregnancy, and adrenaline is a specific triggering agent.

4.6.1
Mutations in CACNL1A3

A systematic genome-wide search in members of three families demonstrated that the disease is linked to chromosome 1q31-32 and cosegregates with the gene encoding the L-type calcium channel (DHP-receptor) α_1 subunit (Fontaine et al. 1994) which is located in this region (Drouet et al. 1993; Gregg et al. 1993). Sequencing of cDNA derived from muscle biopsies of patients revealed so far three mutations (Fig. 3). Two of these are analogous predicting arginine to histidine substitutions within the highly conserved S4 regions of repeats II and IV (Arg-528-His and Arg-1239-His, respectively), the third predicts an arginine to glycine substitution in IVS4 (Arg-1239-Gly; Jurkat-Rott et al. 1994; Ptáček et al. 1994a; Grosson et al. 1996). The substitutions have corresponding counterparts in the α subunit of the sodium channel and those cause paramyotonia congenita by uncoupling activation from inactivation (Chahine et al. 1994). The majority of families carry either the Arg-528-His or the Arg-1239-His substitution (Elbaz et al. 1995).

Expression of cDNA of *CACNL1A3* results in functional channels only when (a) the cell system has a sarcoplasmic reticulum and triads necessary for EC-coupling and contraction and (b) the other four subunits of the pentameric DHP receptor are coexpressed. Thus, for the study of the dysfunction of the mutant *CACLN1A3* gene product, myotubes cultured from muscle specimens of patients are the preparation of choice although they

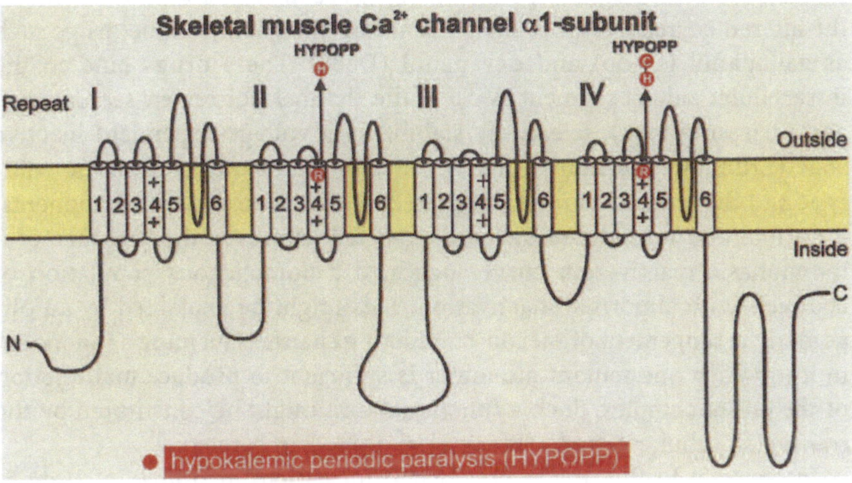

Fig. 3. Substitutions predicted in the α subunit of the L-type calcium channel of skeletal muscle causing hypokalemic periodic paralysis (*HYPOPP*)

contain also normal channels. In such myotubes three voltage-dependent calcium currents can be separated (Rivet et al. 1992; Lehmann-Horn et al. 1995b; Sipos et al. 1995): (a) a rapidly activating and inactivating current with a low-voltage threshold at about –50 mV (T-type current), seen in the majority of cells, (b) a slowly activating and inactivating current with a high voltage threshold of about –20 mV, and sensitive to dihydropyridine, and (c) in one third of the cells, a second type of fast inward current with kinetics similar to the T-type current but with its maximum about 30 mV more positive (3rd-type according to Rivet et al. 1992).

In myotubes from patients, both arginine-to-histidine exchanges seemed to influence channel inactivation but did not alter activation (Lehmann-Horn et al. 1995b; Sipos et al. 1995). The density of the 3rd-type current was markedly increased. As this current is also seen in normal myotubes, it is unlikely that it is conducted by mutant L-type channels. Rather, the authors suggested that calcium channel expression may be altered as a consequence of the mutation. A delay of the downregulation of the gene coding for the channel conducting the 3rd-type calcium current could account for the results.

The L-type calcium current of myotubes with the II-S4 mutation (Arg-528-His) has a normal voltage dependence, however, its inactivation curve is shifted to more negative potentials (Sipos et al. 1995) and thus resembles the altered currents seen in the presence of phenylalkylamine drugs such as gallopamil (D600) and devapamil (D888). These drugs bind to the intracellular side of segment IV-S6 of the channel (for review see Catterall and Striessnig 1992), selectively stabilizing a voltage-dependent inactive state (Erdmann and Lüttgau 1989). Since in the myotubes both the wild-type and the mutant channel is expressed, two L-type current components, a normal and a modified one, are expected. However, the presence of a monophasic inactivation curve suggested a homogenous population of channels with abnormal inactivation. This might be explained by an oligomeric arrangement of the ion-conducting channels forming a functional unit in which one mutant monomer is sufficient to produce malfunction of the entire complex. Such a functional unit might be constituted by the transverse tubular tetrads (Block et al. 1988; Lamb 1992).

In contrast to the results obtained from patient myotubes, a study of the functional consequences of the corresponding Arg-to-His exchange in the cardiac α_1 subunit, which can be easily expressed in a heterologous cell system, did not show this shift (Lerche et al. 1996a). Since the tetradic organization of the DHP receptors is most likely lost in the cells used, the authors suggest that this arrangement is essential to create the specific functional change on inactivation observed in human myotubes.

How inactivation of the L-type calcium current is related to hypokalemia-induced attacks of muscle weakness which characterize familial hypokalemic periodic paralysis can only be speculated upon. Since electrical muscle activity is reduced or even absent during attacks (Engel et al. 1965; Links et al. 1994b), a failure of excitation is more probable than a failure of excitation-contraction coupling. Nevertheless, the hypokalemia-induced membrane depolarization observed in excised muscle fibers (Rüdel et al. 1984) might also reduce calcium release by inactivating sarcolemmal and t-tubular sodium channels, and repolarization of the membrane by activation of ATP-sensitive potassium channels restores force (Grafe et al. 1990).

4.7 The Muscular Dysgenesis (mdg) Mouse

A mouse mutant was discovered with homozygous animals dying at birth because of nonfunctional respiratory muscles. The disorder is inherited as an autosomal recessive trait and was shown to be caused by a deletion of a single nucleotide in the gene encoding the L-type calcium channel (DHP receptor) α_1 subunit (Chaudhari 1992). In the muscles of homozygous animals, the altered mRNA is present only at low levels (Tanabe et al. 1988) and α_1 subunits are immunologically not detectable (Knudson et al. 1989). Action potentials can be elicited in these muscles but they do not trigger contractions. Mice heterozygous for the deletion are not distinguishable from normal mice.

Myotubes can be cultured from the muscle of newborns and studied in primary culture (Beam et al. 1986; Beam and Knudson 1988a,b; Adams and Beam 1989). The DHP-sensitive L-type current and excitation-contraction coupling are both absent (Beam et al. 1986). However, L-type currents were present after rabbit cDNA encoding the α_1 subunit had been injected into nuclei of these myotubes (Tanabe et al. 1988). Ever since this detection, this expression system was successfully used for the characterization of the structure-function relationship of the L-type calcium channel α_1 subunit.

4.8 Malignant Hyperthermia

Susceptibility to malignant hyperthermia (MH) is a subclinical myopathy that is transmitted as an autosomal dominant trait. Life-threatening symptoms may occur when a susceptible individual is exposed to triggering agents such as volatile inhalation anesthetics and depolarizing muscle relaxants. MH crises are characterized by a dysregulation of free myoplasmic calcium that cause (a) increased skeletal muscle metabolism producing hypercapnia, hypoxemia, tachycardia, acidosis and elevated core temperature; and (b) skeletal muscle symptoms such as masseter spasms, generalized muscle rigidity, increased serum creatine kinase (CK), myoglobinuria and hyperkalemia (Denborough and Lovell 1960).

In some of the affected families, the susceptibility to malignant hyperthermia is linked to the gene encoding the skeletal muscle ryanodine receptor (RyR1; MacLennan et al. 1990; McCarthy et al. 1990), a calcium channel situated in the triadic membrane parts of the sarcoplasmic reticulum. This channel is not considered a voltage-gated channel per se although voltage sensitivity is generated in conjunction with the L-type calcium channels situated in the opposing triadic membrane parts of the transverse tubular system to which they are connected by the triadic "feet" structures (Francini-Armstrong 1970; for review see Francini-Armstrong and Jorgensen 1994). Several point mutations in the RyR1 gene have been detected to cause this susceptibility (MH type 1). One of the point mutations (predicting Arg-614-Cys, Gillard et al. 1991) has a counterpart in the pig genome (predicting Arg-615-Cys) which causes the porcine stress syndrome, an animal homologue of MH (Mitchell and Heffron 1982). How the mutations cause the intracellular calcium dysregulation is still a matter of debate (for review see MacLennan and Phillips 1992).

In other families with typical MH susceptibility, linkage of the condition to the RyR1 gene was excluded (Deufel et al. 1992). In one such family, linkage to chromosome 7q and the gene encoding the L-type calcium channel α_2/δ subunit was found (Iles et al. 1994). The lod score for the linkage is close to 3 and therefore establishes this gene as candidate gene for MH type 2. This gene is not yet completely sequenced.

5 Chloride Channelopathies

5.1 Voltage-Gated Chloride Channels in Skeletal Muscle

In nervous tissue, the chloride conductance is so small that it is usually neglected in considerations concerning axonal excitability. In contrast, the chloride conductance of resting adult skeletal muscle is rather large (133 $\mu S/cm^2$ in human external intercostal muscle fibers at the membrane potential of −80 mV, as compared to a potassium conductance of 42 $\mu S/cm^2$, Kwiecinski et al. 1984). Nevertheless, electrophysiological identification and characterization of the channels responsible for this high conductance turned out to be very difficult. Three different types of chloride channel were found in single-channel recordings of human myotubes (Fahlke et al. 1992), and chloride channels of intermediate (Burton et al. 1988) or large (Költgen et al. 1991) conductance were occasionally detected in membrane blebs from adult human muscle fibers, but none of these seemed to be responsible for the high conductance. After many laboratories had failed to find the major chloride channel with the patch-clamp or to reconstitute it using techniques that had turned out successful with potassium channels, the use of the vaseline-gap technique on adult rat skeletal muscle fibers finally led to the desired determination of the properties of homologously expressed major muscle chloride channel

(Fahlke and Rüdel 1995). By that time the channel had also been cloned, functionally expressed and characterized in the heterologous *Xenopus* oocyte system (Steinmeyer et al. 1991b). It had been named ClC-1, and it was shown that it is responsible for the high chloride conductance of the muscle fiber membrane. Recent immunohistochemical analysis with Clc-1 antibodies show that ClC-1 is localized at least in the sarcolemma (Gurnett et al. 1995).

5.2
Muscle Diseases in Man Linked to CLCN1, the Gene Encoding ClC-1

As mentioned, mutations in *CLCN1* lead to several forms of myotonia, diseases characterized by a temporary muscle stiffness that can affect every skeletal muscle of the body. The first description of the classical picture of myotonia congenita was given by Asmus Julius Thomsen (1876) who had the disease himself. He already clearly described the myotonic stiffness and the nonprogressive character of the disease, and correctly stated that the mode of inheritance was dominant.

In the 1950s, Becker (1957) claimed that in many families diagnosed as having myotonia congenita, the mode of inheritance was recessive, as already discussed by Thomasen (1948). In the recessive type, the symptoms were also not congenital but seemed to start at the end of the first decade of life. Becker also found that in the recessive type the myotonia was more generalized than in Thomsen's disease. Therefore he named this type "recessive generalized myotonia" and later it was considered a nosological entity that received its own entry in the standard listing of hereditary diseases (McKusick 1992). This "Becker form" of myotonia is also characterized by a transient weakness, particularly in the arm and hand muscles, that patients experience when they want to make use of their muscles after a period of rest.

It is now clear that both the dominant and the recessive form are caused by mutations in *CLCN1* (Koch et al. 1992). The intensive search for mutations that followed this discovery showed that the dominant form is very rare, as less than 10 different families have been identified at the molecular level up to date. The recessive form is much more common, and the estimation by Becker (1977) of a frequency between 1:23 000 and 1:50 000 might still hold. Males seem to predominate at a rate of 3:1 when the Becker-type propositi are counted. However, family studies disclose that women are affected at the same frequency though to a much lesser degree.

In 1966, DeJong (1966) reported a very benign form of myotonia congenita with dominant mode of inheritance and called it myotonia levior. The variant is characterized by mild symptoms, late onset of myotonia and absence of muscle hypertrophy. The mutation in *CLCN1* that is responsible for this benign form was recently described (Lehmann-Horn et al. 1995a).

5.3 Pathophysiology of Chloride Channel Myotonia

The myotonic reaction can be described in electrophysiological terms as resulting from a lowered electrical threshold and an increased tendency to react to a prolonged current pulse with repetitive activity. The pharmacologist Shirley Bryant (1969) was the first to relate these abnormalities to a congenital absence or reduction of the chloride conductance of the muscle fiber. The "low chloride conductance theory of myotonia," first developed for an animal model, the myotonic goat, and later confirmed by in vitro studies on excised muscles from patients with both dominant and recessive myotonia (Lipicky et al. 1971; Lipicky 1979; Rüdel et al. 1988; Franke et al. 1991), explains the hyperexcitability as follows (Adrian and Bryant 1974). During physiological activity potassium leaves the interior of the muscle fibers and accumulates in the lumen of the transverse tubules. In a normal fiber with a high chloride conductance, tubular potassium accumulation causes depolarization of about 0.1 mV per action potential, not enough to be of physiological importance. In myotonic fibers with reduced chloride conductance, the same potassium accumulation causes about ten times as much depolarization. In other words, while in normal fibers the chloride conductance tends to clamp the membrane potential to the chloride equilibrium potential near –80 mV (Bryant 1976), the lack of this shunt in myotonic fibers leads to after-depolarization and subsequent repetitive activity.

Variation of the parameters for a mathematical membrane model that included the T system showed that the mere lowering of the chloride conductance did not cause repetitive firing (DeCoursey et al. 1981). In fact, appropriate adjustment of the sodium channel activating kinetics was necessary for a correct simulation of the experimentally observed lack of accommodation. Interestingly, corresponding abnormalities in the properties of the sodium channels were noted in myotonic muscle, both from the goat (Adrian and Marshall 1976) and humans (Lipicky 1977; Franke et al. 1991; Iaizzo et al. 1991). Variation of the degree of artificial block of the chloride conductance of normal rat muscle (Bryant 1982) showed that, for myotonia to occur, the fraction of the chloride conductance may not be larger than 30% of the total WT membrane conductance (the rest is mainly potassium conductance). Determination of this fraction in a large number of patients having either dominant or recessive myotonia congenita showed that this fraction ranged from 0% to 66% (Franke et al. 1991). Functional gene dosage compensation on the channel protein seems to determine the sarcolemmal density of chloride channels and the reduction of other ion conductances may be secondary effects of the myotonia (Chen et al. 1996, see also below: Myotonic Mice).

This fairly coherent "low-chloride conductance" hypothesis was tested with many animal models, for example, the myotonic goat, rats that had been fed on a diet containing blockers of the cholesterol synthesis (see Kwiecinski 1981), and muscles made myotonic in vitro by replacing chloride in the extracellular fluid by an impermeant anion, or by blocking the chloride channels with monocarboxylic aromatic acids, the most potent of them being 9-anthracene carbonic acid (Bryant and Morales-Aguilera 1971). Up to the era of molecular biology it was unknown, however, whether in the hereditary myotonias the muscle chloride channels were defective or altogether missing. This question was solved when *CLCN1*, the gene encoding sarcolemmal chloride channels, was discovered.

5.4 Myotonia in the Goat

About 30 years after the first description of myotonia in man (Thomsen 1876), White and Plaskett (1904) described a breed of "fainting" goats raised in Tennessee. The animals tended to have attacks of extreme muscle stiffness when attempting a quick forceful motion, so that they often fell to the ground for 5–20 s with extension of the limbs and neck. Clark et al. (1939) were the first to refer to the disease as "a form of congenital myotonia in goats." It was excised external intercostal muscle from this strain, that the American pharmacologist Shirley H. Bryant used for his famous electrophysiological studies of the membrane conductance (reviewed by Bryant 1973). These studies led him to the conclusion that the fundamental electrical abnormality of resting myotonic fibers is a reduced chloride conductance. Studies on the same muscle from the myotonic goat also first elucidated the role of the T-tubular system in generating repetitive action potentials (Adrian and Bryant 1974). Curiously, the myotonic goat did not play a role for the finding of the genetic defect causing the reduced chloride conductance. Only long after the gene encoding the muscle chloride channel was localized and cloned for mouse (Steinmeyer et al. 1991a) and man (Koch et al. 1992), was the goat mutation in the homologous gene determined (Beck et al. 1996). It causes an Ala-885-Pro substitution in the C terminal end of the chloride channel protein (Fig. 4) that results in a right shift of the activation curve.

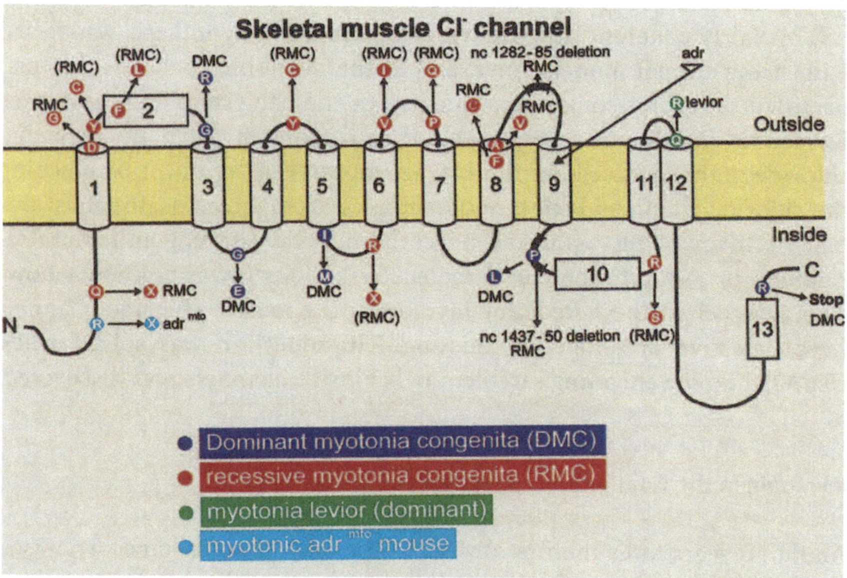

Fig. 4. Substitutions predicted in the voltage-gated chloride channel protein of skeletal muscle (modified after Middleton et al. 1994). *adr*, Positions of mutations in adr myotonic mouse; *DMC*, dominant myotonia congenita; *RMC*, recessive myotonia congenita

5.5 Myotonic Mice

In the late 1970s two spontaneous mouse mutations were detected (as reviewed by Rüdel 1990), one in the A2G strain in London (Watkins and Watts 1984), the other in the SWR/J strain in Bar Harbor/Maine (Heller et al. 1982). The behavioral abnormalities of the affected animals were very similar, and in both mutations the traits were transmitted as an autosomal recessive trait. The British scientists were struck by the observation that from days 10–12 onwards the affected animals had difficulty in righting themselves when placed supine and therefore called the mutation "ADR" for "arrested development of righting response." The Americans observed that shaking the cage provoked sustained extension of an animal's hind limbs, and since electrical myotonia was recorded in the EMG from the stiff muscles, this strain was called "MTO," for "myotonic." Thorough investigations of the time course of twitches and tetanic contractions and of the excitability properties of the muscle fiber membranes showed for both the MTO (Entrikin et al. 1987) and the ADR (Reininghaus et al. 1988) mice that the typical myotonic features were present even in the isolated muscles. As far as the phenotype is concerned, the two models of myotonia are virtually indistinguishable (Reininghaus et al. 1988; Költgen et al. 1989). Transplantation experiments with ADR and wild type grafts showed that the ADR phenotype is based on an intrinsic muscle property (Füchtbauer et al. 1988) and electrophysiology proved for MTO (Bryant et al. 1987) and ADR (Mehrke et al. 1988) that, as in the myotonic goat, the reason for the abnormal excitability is a reduced chloride conductance. Intercrosses of the type A2G *adr*/+alSWR/J *mto*/+ lead to the diseased phenotype indicating that *adr* and *mto* are allelic (Jockusch and Bertram 1986). The latter allele was then renamed adr^{mto}. Three further mutations in the adr allelomorphic series have been reported (Adkison et al. 1989; Neumann and Weber 1989), and the *mto* allele was mapped by conventional markers to chromosome 6 (Davisson et al. 1989). The assumption of conservation of the genomic structure in the vicinity of the *Hox 1.1* gene led Jockusch (1990) to predict that the Becker myotonia gene is located on the human chromosome 7 before the major chloride channel, Clc-1, of mammalian skeletal muscle was cloned.

When this was finally accomplished (Steinmeyer et al. 1991a; see below), it was soon shown that the spontaneous mutation of the ADR mouse consists in the insertion of a retroposon of the ETn family in the gene coding for Clc-1. This destroys the gene's coding potential for several membrane spanning domains (Steinmeyer et al. 1991b). From this and the lack of recombination between the *Clc-1* gene and the *adr* locus, it was concluded that a lack of functional chloride channels is the primary cause of mouse myotonia. Meanwhile, it was found that the adr^{mto} allele has a stop codon introduced close to the N terminus, and that the ethylnitrosourea-induced mutation adr^{K} causes the substitution Ile-553-Thr (Gronemeier et al. 1994).

Tocainide, an antiarrhythmic drug often used in human patients for the temporary abolition of myotonia (Rüdel et al. 1980) was administered to the animals by incorporating it in food pellets. It readily prevented the development of myotonia in young animals and suppressed an existing myotonia in older animals (Jokusch et al. 1988). At the same time as myotonia was suppressed the abnormalities in proportion of fiber types, myosin light chain phosphorylation and parvalbumin content were reversed. Thus, it is obvious that the biochemical abnormalities were just secondary effects of the hyperactivity. Neuromuscular transmission was found unimpaired both in ADR and MTO mice (Költgen et al. 1991).

Myotonic mouse models, mostly the ADR mouse, were also used to study the role of chloride conductance in myotubes and neonatal muscle (Wischmeyer et al. 1993) and the role of innervation on the expression of the *Clc-1* gene (Klocke et al. 1994). No indication of electrical myotonia was detected in cultured ADR myotubes and it was concluded that the low chloride conductance of myotubes is not controlled by the *Clc-1* gene as the presence of *Clc-1* mRNA is % (Bardouille et al. 1996). In contrast, in the neonatal mouse, normal excitability depends on a normal dosage of functional *Clc-1* alleles. *Clc-1* expression was found to depend on the innervation by the motor nerves, and the level of *Clc-1* mRNA was higher in fast (extensor digitorum longus) than in slow (soleus) muscle. In heterozygous adult *adr*/+ mice, the macroscopic chloride conductance of the sarcolemma is not significantly different from wild type in spite of the presence of only 50% functional *Clc-1* RNA. This indicates that a regulatory mechanism acts on the protein level limiting the density of Clc-1 channels (Chen et al. 1996).

Wischmeyer et al. (1995) were successful in recording chloride and potassium currents from single channels in lipid-supplemented sarcolemmal vesicles prepared from mouse skeletal muscle. Five different chloride channels were repeatedly identified and designated Clc-a to Clc-e, Clc-b being presumably identical with a known chloride channel (Chua and Betz 1991).

The function of all these channels is so far unknown, none of them seems identical with Clc-1. Interestingly, Clc-d and Clc-e, showed some properties with the macroscopic chloride conductance and were not detected in vesicles made from ADR mouse muscle. The authors speculate that the chloride channels in the vesicles are made up of subunits of various chloride channel proteins (for multimeric structure of Clc-1, see below). Comparison of heterologously expressed ClC-1 channels with native channels (Bretag 1987) has not revealed substantial functional differences. Thus, for the reconstituted channels a physiological function has yet to be found.

5.6
The ClC Multigene Family Encoding Voltage-Gated Chloride Channels

ClC-1, the chloride channel of the skeletal muscle fiber membrane, belongs to a family of voltage-gated chloride channels whose members are encoded by a gene family that has no relation with any other known ion channel gene family. Thus, ClC-1 is structurally entirely different from the epithelial chloride channels affected in cystic fibrosis.

Several electric fish of the species *Torpedo* contain at high density chloride channels in the cells of electroplax, the electric organ that develops from skeletal muscle tissue (White and Miller 1979). The cDNA encoding this channel was cloned by Jentsch et al. (1990). Expressed in *Xenopus* oocytes, it gave rise to large chloride currents. The channel conducting these currents was called ClC-0. It was the very first recognized gene product of a member of the newly detected *ClC* gene family (Steinmeyer et al. 1991b).

Meanwhile at least nine different *ClC* genes (reviewed by Pusch and Jentsch 1994 and Jentsch et al. 1995) have been discovered in mammals, several of which seem to be expressed ubiquitously. *CLCN1* encodes the major chloride channel in skeletal muscle, to be extensively discussed below. *CLC-2*, having an overall identity of about 50% with *CLCN1* codes for a swelling-activated chloride channel (Gründer et al. 1992; Thiemann et al. 1992). Related to the same branch of the gene family are the kidney-specific channel genes *ClC-K1* and *ClC-K2* (Uchida et al. 1993) and their human counterparts *hCLC-Ka* and *hClc-Kb* (Kieferle et al. 1994). They are about 80%–90% homologous to each other, but the channels are differentially distributed along the nephron. A more distant branch of the family contains *ClC-3* (Kawasaki et al. 1994, 1995) and *ClC-4* (van Slegtenhorst et al. 1994). *ClC-3* is rather ubiquitously expressed, its function is not yet known. With *CLC-4*, expression in man is highest in muscle and brain (van Slegtenhorst et al. 1994), whereas in rat it is most prominent in liver and brain (Jentsch et al. 1995). It may play a role in patients with recessive myotonia congenita (Mailänder et al. 1996).

5.7 CLCN1, the Human Gene Encoding ClC-1

Two independent lines of experiments led to the localization of *CLCN1*, the human gene encoding ClC-1. One line made use of the myotonic mouse mutant adr (for "arrested development of righting response," Heller et al. 1982; see above). Conventional mapping techniques and the use of restriction fragment length polymorphisms (RFLPs) resulted in localization of the *adr* locus on chromosome 6, close to the locus for the T cell receptor β (Tcrb; Adkison et al. 1989). This made it very probable that the Becker myotonia gene is localized in the homologous region on human chromosome 7q32-ter. Subsequently, tight linkage was shown between the TCRB locus and Thomsen myotonia (Abdalla et al. 1992).

The other, more direct way to localizing HUMCLC started by expression cloning of ClC-0, the chloride channel of the electric fish, *Torpedo marmorata* (Jentsch et al. 1990). Partial cDNAs of chloride channels of rat and mouse (Steinmeyer et al. 1991b) and human (Koch et al. 1992) skeletal muscle, ClC-1, were then cloned by homology screening. The probes obtained in this way were used to show linkage to the *adr* locus and to both Becker and Thomsen myotonia, respectively (Koch et al. 1992). The gene, located at 7q35, spans at least 40 kb and contains 23 exons whose boundaries have been located (Lorenz et al. 1994). It consists of 2964 base pairs and codes for a protein of 988 amino acids. The cDNA was functionally expressed in *Xenopus* oocytes (Lorenz et al. 1994). Primer sequences for the amplification of all exons are given in Lehmann-Horn et al. 1995.

5.8 Naturally Occurring Mutations in CLCN1

Fourteen missense mutations, three nonsense mutations, and two deletions in various exons of *CLCN1* have been discovered (Fig. 4, Table 4. Which contains editional mutation very resently reported). Six point mutations exert dominant or partly dominant effects; five of them are missense mutations such as Gly-230-Glu (George et al. 1993a), Pro-480-Leu (Steinmeyer

Table 2. Mutations of *CLCN1*, the gene encoding the chloride channel of human adult skeletal muscle. T = terminus; E = exon

Genotype	E	Substitution	Domain	Mode of inheritance	Genetic status of propositus	First report
C220T	2	Gln-74-Stop	N-T	Recessive	Homozygous	Mailänder et al. 1996
C313T	3	Arg-105-Cys	N-T	Recessive	Heterozygous	Meyer-Kleine et al. 1995
A407G	3	Asp-136-Gly	1	Recessive	Homozygous	Heine et al. 1994
A449G	4	Tyr-150-Cys	1/2	Recessive	Compound heterozygous	Mailänder et al. 1996
T494G	4	Val-165-Gly	2	Recessive	Heterozygous	Meyer-Kleine et al. 1995
C501G	4	Phe-167-Leu	2	Recessive	Heterozygous	George et al. 1994b
G598A	5	Gly-200-Arg	2/3	Dominant	Heterozygous	Mailänder et al. 1996
G689A	5	Gly-230-Glu	3/4	Dominant	Heterozygous	George et al. 1993a
A782G	7	Tyr-261-Cys	4/5	Recessive	Compound heterozygous	Mailänder et al. 1996
C870G	8	Ile-290-Met	5/6	Dominant	Heterozygous	Lehmann-Horn et al. 1995a
G871A	8	Glu-291-Lys	5/6	Recessive	Compound heterozygous	Meyer-Kleine et al. 1995
C898T	8	Arg-300-Stop	5/6	Recessive	Compound heterozygous	George et al. 1994b
G950A	8	Arg-317-Gln	6	Dominant	Heterozygous	Meyer-Kleine et al. 1995
G979A-1	8	Splice mutation	6/7	Recessive	Compound heterozygous	Lorenz et al. 1994
8T986C	9	Ile-329-Thr	6/7	Recessive	Heterozygous	Meyer-Kleine et al. 1995
G1013A	9	Arg-338-Gln	6/7	Recessive	Compound heterozygous	George et al. 1994b

Table 2. Continued.

Genotype	domain	Channel	Substitution	no.	Exon inheritance	Mode of of propositus
2 bp deletion	7	fs 387-Stop	10	Recessive	Heterozygous	Meyer-Kleine et al. 1995
T1238G	8	Phe-413-Cys	11	Recessive	Homozygous/ heterozygous	Koch et al. 1992
C1244T	8	Ala-415-Val	11	Recessive	Homozygous/ heterozygous	Mailänder et al. 1996
1262insC	8/9	fs 429-Stop	12	Recessive	Heterozygous	Meyer-Kleine et al. 1995
4 bp deletion 1282–85	8/9	fs 433-Stop	12	Recessive	Homozygous	Heine et al. 1994
C1439T	9/10	Pro-480-Leu	13	Dominant	Heterozygous	Steinmeyer et al. 1994
14 bp deletion 1437–50	9/10	fs 503-Stop	13	Recessive	Homozygous/ heterozygous	Meyer-Kleine et al. 1994b
G1444A	9/10	Gly-482-Arg	13	Recessive	Heterozygous et al. 1995	Meyer-Kleine
A1453G	9/10	Met-485-Val	13	Recessive	Compound heterozygous	Meyer-Kleine et al. 1995
G1471A	10	splice mutation	13	Recessive	Compound heterozygous	Meyer-Kleine et al. 1995
G1488T	10/11	Arg-496-Ser	14	Recessive	Compound heterozygous	Lorenz et al. 1994
A1655G	12	Gln-552-Arg	15	Dominant "levior"	Heterozygous	Lehmann-Horn et al. 1995a
C2680T	C-T.	Arg-894-Stop	23	Dominant recessive	Compound heterozygous Homozygous	George et al. 1994b Meyer-Kleine et al. 1995

et al. 1994), Ile-290-Met (Lehmann-Horn et al. 1995a), Gln-552-Arg (Lehmann-Horn et al. 1995a), and Gly-200-Arg (Mailänder et al. 1996); and the sixth is a nonsense mutation which causes truncation at the very end of the protein: Arg-894-Stop (George et al. 1994b). Pro-480-Leu is present in Dr. Thomsen's offspring; Gln-552-Arg was found in a family with *myotonia levior*, a term coined by deJong for a dominant myotonia variant characterized by mild symptoms, late onset of myotonia and absence of muscle hypertrophy. With the exception of Gly-230-Glu which was detected in three Canadian families, each mutation was only detected in one single family.

The other point mutations, i.e., Phe-413-Cys (Koch et al. 1992), Val-327-Ile, Arg-496-Ser (Lorenz et al. 1994), Phe-167-Leu, Arg-300-Stop, Arg-338-Gln (George et al. 1994b), Asp-136-Gly (Heine et al. 1994), Gln-74-Stop, Tyr-150-Cys, Tyr-261-Cys, and Ala-415-Val (Mailänder et al. 1996) and the 4 bp (Heine et al. 1994) and 14 bp deletions (Meyer-Kleine et al. 1994b) were detected in (approximately 60) Becker-type patients. The majority of them were heterozygous for a mutation and supposed to carry a second, not yet identified mutation. Only thirteen index patients, most of them offspring of consanguineous parents, were homozygous (Koch et al. 1992; Heine et al. 1994; Meyer-Kleine et al. 1994b; Lehmann-Horn et al. 1995a; Mailänder et al. 1996) or compound heterozygous (George et al. 1994b; Lorenz et al. 1994; Mailänder et al. 1996). Ten further mutations that have been published very recently (Meyer-Kleine et al. 1995) are shown and in Table 2.

5.9
Analysis of the CLCN1 Gene Product

ClC-1 (predicted molecular weight 110 kDa) is functional when expressed in *Xenopus* oocytes (Steinmeyer et al. 1991b) or human embryonic kidney cells (Pusch et al. 1994; Fahlke et al. 1995) without any other subunits. Its properties are very similar to those of the channel protein, expressed in native muscle (Fahlke and Rüdel 1995). The channel conducts over the whole physiological voltage range, showing inward rectification in the negative potential range. It is activated upon depolarization and deactivated with hyperpolarizing voltage steps to a nonzero steady-state level. It is selective for chloride against bromide and iodide. Its single-channel conductance, estimated from noise analysis is very low, near 1 pS (Pusch et al. 1994). The large macroscopic chloride conductance, therefore, must result from a high channel density in the membrane. As already known from macroscopic experiments (Bryant and Morales-Aguilera 1971; Palade and Barchi 1977) the channel can be blocked by external I^- and by low concentrations of 9-anthracene carboxylic acid (Steinmeyer et al. 1991b; Pusch et al. 1994).

When the Gly-230-Glu substitution of the Canadian family (George et al. 1993a) or the Pro-480-Leu substitution of the Thomsen family (Steinmeyer et al. 1994) were inserted in the oocytes instead of wild type ClC-1, the chloride currents were completely missing (Lorenz et al. 1994; Steinmeyer et al. 1994).

Any mutation may cause either a gain of function or a loss of function. Gene dosage effects of loss-of-function mutations may lead to a recessive or dominant phenotype, depending on whether 50% of the gene product (supplied by the normal allele) is or is not sufficient for normal function. The effect of a particular mutation on the inheritance pattern depends on the ability of mutant ClC-1 to interact with other monomers and change the function of the channel complex. Mutant ClC-1 unable to polymerize, for example, severely truncated proteins, allow normal ClC-1 monomers (expressed by the other allele) to form normal complexes, although reduced in number (50%). If there are no compensatory mechanisms effective, clinically normal heterozygous carriers of such mutations would have 50% muscle chloride conductance; effects of such mutations would be recessive. Interestingly, the mutation predicting Arg-894-Stop was found to be transmitted either as dominant or recessive trait "probably depending on the genetic background" (Meyer-Kleine et al. 1995).

The number of ClC-1 monomers that make up the channel is still a matter of dispute. In experiments coexpressing wild-type and the two disease-producing mutants Gly-230-Glu and Pro-480-Leu, Steinmeyer et al. (1994) found evidence suggesting the channel to be a tetramer. Middleton et al. (1994), purifying native *Torpedo* chloride channels (ClC-0), found sedimentation properties of a dimeric protein.

In contrast to severely truncated proteins, mutant ClC-1 may be able to interact with normal ClC-1 and thus destroy or change the function of the complex. If all monomers need to be mutants for an effect, mutation of one allele leaves the majority of the complexes functional (75% with dimers, 94% with tetramers) and exerts recessive effects. If one mutant monomer is sufficient for an effect, only a minority of complexes will be functional (25% with dimers, 6% with tetramers). Such mutations exert dominant effects unless the mutant complexes function partially (Lehmann-Horn and Rüdel 1995).

Experiments with myotonia-generating drugs (Kwiecinski 1981) have shown that blockade of 50% of the physiological chloride current is not sufficient to produce myotonic activity. This then explains the existence of recessive transmission in the case of mutations that completely destroy the gene's coding functions, such as the two myotonia-generating deletions reported above. Dominant inheritance would be explained by a mutant gene product that can bind to another protein and, in doing so, changes its function.

5.10 Structure–Function Relationship of ClC-1

Our knowledge of this relationship is still very limited compared to that of voltage-gated cation channels, and relevant experimental findings have only recently begun to emerge. Based on a hydrophobicity analysis of ClC-0, Jentsch et al. (1990) originally proposed 12 or 13 transmembrane spanning segments (D1–D13). The lack of a signal sequence, combined with mutational results of ClC-2, another family member, suggests a cytoplasmic location of both the N- and the C-terminals as well as segment D13 (Gründer et al. 1992). The linker region between D8 and D9 is most probably extracellularly located, because it contains a potential site for N-linked glycosylation, which has been demonstrated to be, indeed, glycosylated (Kieferle et al. 1994; Middleton et al. 1994). Accordingly, an odd number of transmembrane spanning segments between this linker region and both the N-terminus and D13 had to be assumed and a revision of the original model was necessary. Two new models have been proposed one of them placing segment D2 outside the membrane and D10 in the cytoplasm (Middleton et al. 1994), the other placing D4 outside and leaving the topology of region D9-D12 unanswered (Pusch and Jentsch 1994). In the latter work, it was also claimed that a lysine residue in segment D12 is located at the cytoplasmic end of the channel pore.

The currents through ClC-1 channels expressed in HEK 293 cells recorded at varied extracellular and intracellular chloride concentrations and pH values were used to develop a model for ClC-1 gating (Fahlke et al. 1996). A set of two identical residues carrying a single negative charge, are postulated to function as voltage sensor. Each of the three possible conformations of this voltage sensor determines another affinity of the intracellular mouth of the pore for a gating particle that mediates the transitions between the open and closed channel. As a consequence of the three different configurations of this set of voltage sensors, current activation as well as deactivation is composed of three different components, with fast, slow and time-independent time course. The rearrangements that the two voltage sensors undergo both upon hyperpolarizing and depolarizing voltage steps, are very rapid. This model is able to explain all macroscopic gating properties described so far.

The missense mutation (Asp-136-Gly) leading to recessive myotonia congenita (Heine et al. 1994) considerably affects voltage-dependent gating without altering permeation properties. This led to the suggestion that the aspartic acid in position 136 is the voltage sensor (Fahlke et al. 1995).

The channel can be phosphorylated by activation of protein kinase C and the site of phosphorylation was postulated to be located at the vestibule of the cytoplasmic face of the pore (Rosenbohm et al. 1996).

5.11
Do Other Chloride Channels Influence the Degree of Myotonia?

Heterozygous carriers of Becker mutations do not display any clinical symptoms of myotonia. However, heterozygous males often exhibit myotonic discharges in the electromyogram while heterozygous females usually do not (Mailänder et al. 1996). The cause of this apparent sex predominance of subclinical myotonic signs is unclear. Sex hormones or the product of another muscle chloride channel gene, *ClC-4* (Kawasaki et al. 1994, 1995), encoded on the X-chromosome, could contribute to this preference. Although probably not of importance in normal muscle fibers, i.e., in the presence of abundant ClC-1 channels, chloride channels encoded by *ClC-4* expressed on both X-chromosomes could have a compensatory effect when the number of functional ClC-1 channels is reduced. A copy of *ClC-4* on the Y-chromosome has been excluded (van Slegtenhorst et al. 1994) and its location on distal Xp may allow it to escape X-chromosomal inactivation (Disteche 1995).

The importance of muscular chloride channels other than ClC-1 was also suggested by results obtained with the missense mutation Asp-136-Gly (Heine et al. 1994) expressed in human embryonic kidney cells. This recessive mutation is associated with more severe myotonia than usually presented by Becker patients. The mutant differs from the wild type ClC-1 in the gating properties causing pronounced inward rectification that permits chloride influx but prevents chloride efflux. Thus, electrical activity would cause intracellular chloride depletion and a shift of the chloride equilibrium potential to very negative potentials. With this depletion, any additional chloride conductance that is operant in myotonia patients having mutations that do not cause this pronounced inward rectification becomes ineffective. Thus, the finding that myotonia is clinically so severe in Asp-136-Gly patients was taken as suggestive that in human skeletal muscle chloride channels other than ClC-1 (e.g., as described by Chua and Betz 1991) may be effective (Fahlke et al. 1995).

6 Conclusion

In addition to the clarification of the pathology of a whole group of hereditary diseases, the study of the consequences of these mutations at the levels of the whole system (patient), organ and cells (excised muscle specimens), and of the channel proteins has taught us that our current opinions on channel structure-function relations are far from being comprehensive. For instance it had been assumed that as in the potassium channel, also in the sodium and calcium channel proteins, the S4 unit is mainly responsible for channel activation. This notion has to be corrected as mutations in S4 of repeat IV were found to affect mainly channel inactivation. Moreover mutations affecting other channel domains, such as interlinkers or other intramembraneous subunits cause virtually the same alterations, not only when tested with the limited probe of the patch clamp but also on the level of the patient. The lack of mutations in other parts of the genes, in particular those coding for sodium or calcium channels, may indicate that proper function of the corresponding protein domains is essential for life. Thus the knowledge derived from the experiments of Nature, as these diseases may be looked upon by the cell biologist, provides a valuable addition to the results from site-directed mutagenesis. For a final understanding of the pathology of the diseases, for example, triggering effects of cold or potassium, it seems that the regression from the proteins back to the cellular or even systemic levels is unavoidable.

Acknowledgement. We wish to thank Drs. C. Fahlke, H. Jockusch, L. Kürz, H. Lerche, W. Melzer, and K. Ricker for valuable comments on the manuscript.

References

Abdalla JA, Casley WL, Cousin HK, Hudson AJ, Murphy EG, Cornélis FC, Hashimoto L, Ebers GC (1992) Linkage of Thomsen disease to the T-cell-receptor beta (TCRB) locus on chromosome 7q35. Am J Hum Genet 51:579–584

Adams BA, Beam KG (1989) A novel calcium current in dysgenic skeletal muscle. J Gen Physiol 94:429–444

Adkison L, Harris B, Lane PW, Davisson MT (1989) New alleles of "arrested development of righting response (adr)." Mouse News Lett 84:89–90

Adrian RH, Bryant SH (1974) On the repetitive discharge in myotonic muscle fibres. J Physiol Paris 240:505–515

Adrian RH, Marshall MW (1976) Action potentials reconstructed in normal and myotonic muscle fibres. J Physiol Paris 258:125–143

Ahmed CH, Ware DH, Lee SC, Patten CD, Ferrer-Montiel AV, Schinder AF, McPherson JD, Wagner-McPherson CB, Wasmuth JJ, Evans GA, Montal M (1992) Primary structure, chromosomal location, and functional expression of a voltage-gated sodium channel from human brain. Proc Natl Acad Sci USA 89:8220–8224

Angaut-Petit D, McArdle JJ, Mallart A, Bournaud R, Pinhon-Raymond M, Rieger F (1982) Electrophysiological and morphological studies of a motor nerve in "motor endplate disease" of the mouse. Proc R Soc Lond B Biol Sci 215:117–125

Armstrong CM, Bezanilla F (1977) Inactivation of the sodium channel: II. Gating current experiments. J Gen Physiol 70:567–590

Armstrong CM, Bezanilla F, Rojas E (1973) Destruction of sodium conductance inactivation in squid axons perfused with pronase. J Gen Physiol 62:375–391

Backx PH, Yue DT, Lawrence JH, Marban E, Tomaselli GF (1992) Molecular localization of an ion-binding site within the pore of mammalian sodium channels. Science 257:248–51

Baquero JL, Ayala RA, Wang J, Curless RG, Feero WG, Hoffman EP, Ebeid MR (1995) Hyperkalemic periodic paralysis with cardiac dysrhythmia: a novel sodium channel mutation? Ann Neurol 37:408–411

Barchi RL (1995) Molecular pathology of the skeletal muscle sodium channel. Annu Rev Physiol 57:355–385

Bardouille C, Vullhorst D, Jockusch H (1996) Chloride channel 1 mRNA as a marker of myotube maturation. J Cell Sci (in preparation)

Beam KG, Knudson CM (1988a) Calcium currents in embryonic and neonatal mammalian skeletal muscle. J Gen Physiol 91:781–798

Beam KG, Knudson CM (1988b) Effect of postnatal development on calcium currents and slow charge movement in mammalian skeletal muscle. J Gen Physiol 91:799–815

Beam KG, Knudson CM, Powell JA (1986) A lethal mutation in mice eliminates the slow calcium current in skeletal muscle cells. Nature 320:168–170

Beam KG, Brett AA, Niidome T, Numa S, Tanabe T (1992) Function of a truncated dihydropyridine receptor as both voltage sensor and calcium channel. Nature 360:169–171

Beck C, Fahlke C, George AL (1996) Altered voltage-dependent activation of skeletal muscle Cl-channel causes myotonia in the fainting goat. Biophys J 70:A69

Becker PE (1957) Zur Frage der Heterogenie der erblichen Myotonien. Nervenarzt 28:455–460

Becker PE (1970) Paramyotonia congenita (Eulenburg). Fortschritte der allgemeinen und klinischen Humangenetik. Thieme, Stuttgart

Becker PE (1977) Myotonia congenita and syndromes associated with myotonia. Thieme, Stuttgart

Bennett PB, Yazawa K, Makita N, George AL Jr (1995) Molecular mechanism for an inherited cardiac arrhythmia. Nature 376:683–685

Bezanilla F, Armstrong CM (1977) Inactivation of the sodium channel I. Sodium current experiments. J Gen Physiol 70:549–566

Blaxter AC, Lievesley P, Gruffydd-Jones T, Wotton P (1986) Periodic muscle weakness in burmese kittens. Vet Rec (May 31) pp 619–620

Block BA, Imagawa T, Campbell KP, Franzini-Armstrong C (1988) Structural evidence for direct interaction between the molecular components of the transverse tubule/sarcoplasmic reticulum junction in skeletal muscle. J Cell Biol 107:2587–2600

Bosse E, Regulla S, Biel M, Ruth P, Meyer HE, Flockerzi V, Hofmann F (1990) The cDNA and deduced amino acid sequence of the gamma subunit of the L-type calcium channel from rabbit skeletal muscle. FEBS Lett 267:153–156

Bradley W, Taylor R, Rice D, Hausmanowa-Petruzewicz I, Adelman L, Jenkison M, Jedrzejowska H, Drac H, Pendlebury W (1989) Progressive myopathy in hyperkalemic periodic paralysis. Arch Neurol 47:1013–1017

Bretag AH (1987) Muscle chloride channels. Physiol Rev 67:618–725

Browne DL, Gancher ST, Nutt JG, Brunt ERP, Smith EA, Kramer P, Litt M (1994) Episodic ataxia/myokymia syndrome is associated with point mutations in the human potassium channel gene, KCNA1. Nat Genet 8:136–140

Browne DL, Brunt ERP, Griggs RC, Nutt JG, Gancher ST, Smith EA, Litt M (1995) Identification of two new KCNA1 mutations in episodic ataxia/myokymia families. Hum Mol Genet 4:1671–1672

Brunt ERP, van Weerden TW (1990) Familial paroxysmal kinesigenic ataxia and continuous myokymia. Brain 113:1361–1382

Bryant SH (1969) Cable properties of external intercostal muscle fibres from myotonic and non-myotonic goats. J Physiol Paris 204:539–550

Bryant SH (1973) The electrophysiology of myotonia with a review of congenital myotonia of goats. In: Desmedt JE (ed) New developments in electromyography and clinical neurophysiology, vol 1. Karger, Basel, pp 407–412

Bryant SH (1976) Altered membrane properties in myotonia. In: Bolis L, Hoffman LP, Leaf A (eds) Membranes and diseases. Raven, New York, pp 197–206

Bryant SH (1982) Physical basis of myotonia. In: Schotland DL (ed) Disorders of the motor unit. Wiley, New York, pp 381–389

Bryant SH, Morales-Aguilera A (1971) Chloride conductance in normal and myotonic muscle fibres and the action of monocarboxylic aromatic acids. J Physiol (Paris) 219:361–383

Bryant SH, Mambrini M, Entrikin RK (1987) Chloride and potassium membrane conductances are decreased in skeletal muscle fibers from the (mto) myotonic mouse. Neuroscience 465:17

Burgess DL, Kohrman DC, Galt J, Plummer NW, Jones JM, Spear B, Meisler MH (1995) Mutation of a new sodium channel gene, Scn8a, in the mouse mutant "motor endplate disease". Nat Genet 10:461–465

Burton F, Dörstelmann U, Hunter OF (1988) Single-channel activity in sarcolemmal vesicles from human and other mammalian muscles. Muscle Nerve 11:1029–1038

Cannon SC, Corey DP (1993) Loss of Na^+ channel inactivation by anemone toxin (ATXII) mimics the myotonic state in hyperkalemic periodic paralysis. J Physiol (Lond) 466:501–520

Cannon SC, Strittmatter SM (1993) Functional expression of sodium channel mutations identified in families with periodic paralysis. Neuron 10:317-326

Cannon SC, Brown RH Jr, Corey DP (1991) A sodium channel defect in hyperkalemic periodic paralysis: potassium-induced failure of inactivation. Neuron 6:619–626

Cannon SC, Brown RH Jr, Corey DP (1993a) Theoretical reconstruction of myotonia and paralysis caused by incomplete inactivation of sodium channels. Biophys J 65:270–288

Cannon SC, McClatchey AI, Gusella JF (1993b) Modification of the Na^+ current conducted by the rat skeletal muscle α subunit by coexpression with a human brain β subunit. Pflügers Arch 423:155–157

Castellano A, Wei X, Birnbaumer L, Perez-Reyes E (1993) Cloning and expression of a neuronal calcium channel β subunit. J Biol Chem 268:12359–66

Catterall WA (1991) Functional subunit structure of voltage-gated calcium channels. Science 253:1499–1500

Catterall WA (1992) Cellular and molecular biology of voltage-gated sodium channels. Physiol Rev 72:S15-S48

Catterall WA, Striessnig J (1992) Receptor sites for Ca^{2+} channel antagonists. Trends Pharmacol Sci 13:256–262

Chahine M, George AL, Zhou M, Ji S, Sun W, Barchi RL, Horn R (1994) Sodium channel mutations in paramyotonia congenita uncouple inactivation from activation. Neuron 12:281-294

Chandy KG, Gutman GA (1995) Voltage-gated K^+ channel genes. In: North RA (ed) Handbook of receptors and channels: ligand- and voltage-gated ion channels. CRC Press, Boca Raton, pp 1–71

Chaudhari N (1992) A single nucleotide deletion in the skeletal muscle-specific calcium channel transcript of muscular dysgenesis (mdg) mice. J Biol Chem 267:25636–25639

Chen C, Cannon SC (1995) Modulation of Na^+ channel inactivation by the β_1 subunit: a deletion analysis. Pflügers Arch 431:186–195

Chen L-Q, Chahine M, Kallen RG, Barchi RL, Horn R (1992) Chimeric study of sodium channels from rat skeletal and cardiac muscle. FEBS Lett 309:253–57

Chen M, Niggeweg R, Jockusch H (1996) Chloride conductance in mutant mouse muscle is subject to functional gene dosage compensation. J Physiol (Lond) (submitted)

Chin H, Kozak CA, Kim H-L, Mock B, McBride OW (1991) A brain L-type calcium channel α1 subunit gene (CCHL1A2) maps to mouse chromosome 14 and human chromosome 3. Genomics 11:914–919

Chua M, Betz WJ (1991) Characterisation of ion channels on the surface membrane of adult rat skeletal muscle. Biophys J 59:1251–1260

Clark SL, Luton FH, Cutler JT (1939) A form of congenital myotonia in goats. J Nerv Ment Dis 90:297–309

Clausen T, Kohn PG (1977) The effect of insulin on the transport of sodium and potassium in rat soleus muscle. J Physiol Paris 265:19–42

Cognard C, Lazdunski M, Romey G (1986) Different types of Ca^{2+} channels in mammalian skeletal muscle cells in culture. Proc Natl Acad Sci USA 83:517–521

Cognard C, Constantin B, Rivet M, Imbert N, Besse C, Raymond G (1992) Excitation-contraction coupling in rat skeletal muscle cells: evolution during in vitro myogenesis. In: Frank GB (ed) Excitation-contraction coupling in skeletal, cardiac and smooth muscle. Plenum, New York, pp 73–89

Cognard C, Constantin B, Rivet-Bastide M, Raymond G (1993a) Intracellular calcium transients induced by different kinds of stimulus during myogenesis of rat skeletal muscle cells studied by laser cytofluorimetry with Indo-1. Cell Calcium 14:333–348

Cognard C, Lazdunski M, Romey G (1993b) Appearance and evolution of calcium currents and contraction during the early post-fusional stages of rat skeletal muscle cells developing in primary culture. Development 117:1153–1161

Cooper DN, Krawczak M (1990) The mutational spectrum of single base-pair substitutions causing human genetic disease; patterns and predictions. Hum Genet 85:55–74

Cota G, Stefani E (1986) A fast activated inward calcium current in twitch muscle fibres of the frog (*Rana montezume*). J Physiol (Lond) 370:151–163

Cox JH (1985) An episodic weakness in four horses associated with intermittent serum hyperkalemia and the similarity of the disease to hyperkalemic periodic paralysis in man. Proc Am Ass Equine Practice 31:383–390

Cummins TR, Zhou J, Sigworth FJ, Ukomadu C, Stephan M, Ptáček LJ, Agnew WS (1993) Functional consequences of a Na^+ channel mutation causing hyperkalemic periodic paralysis. Neuron 10:667–678

Curran ME, Splawski I, Timothy KW, Vincent GM, Green ED, Keating MT (1995) A molecular basis for cardiac arrhythmia: HERG mutations cause long QT syndrome. Cell 80:795–803

Davisson MT, Harris B, Lane PW (1989) Mouse News Lett 83:167

DeCoursey TE, Bryant SH, Owenburg KM (1981) Dependence of membrane potential on extracellular ionic concentrations in myotonic goats and rats. Am J Physiol 240:C56-C63
DeJong JGY (1966) Myotonia levior. In: Kuhn E (ed) Progressive Muskeldystrophie – Myotonie – Myasthenie. Springer Verlag, Berlin Heidelberg New York, pp 255–259
DeJongh KS, Warner C, Colvin AA, Catterall WA (1991) Characterization of the two size forms of the α1 subunit of skeletal muscle L-type calcium channels. Proc Natl Acad Sci USA 88:10778–82
Denborough MA, Lovell RRH (1960) Anaesthetic deaths in a family. Lancet 2:45
DeSilva SM, Kuncl RW, Griffin JW, Cornblath DR, Chavoustie S (1990) Paramyotonia congenita or hyperkalemic periodic paralysis? Clinical and electrophysiological features of each entity in one family. Muscle Nerve 13:21–26
Deufel T, Golla A, Iles D, Meindl A, Meitinger T, Schindelhauer D, DeVries A, Pongratz D, MacLennan D, Johnson KJ, Lehmann-Horn F (1992) Evidence for genetic heterogeneity of malignant hyperthermia susceptibility. Am J Hum Genet 50:1151–1161
Disteche CM (1995) Escape from X inactivation in human and mouse. Trends Genet 11:17–22
Drouet B, Garcia L, Simon-Chazottes D, Mattei MG, Guénet J-L, Schwartz A, Varadi G, Pinçon-Raymond M (1993) The gene encoding for the α1 subunit of the skeletal dihydropyridine receptor (Cchlla3=mdg) maps to mouse chromosome 1 and human 1q32. Mamm Genome 4:499–503
Duchen LW (1970) Hereditary motor end-plate disease in the mouse: light and electron microscopic studies. J Neurol Neurosurg Psychiatry 33:238–250
Duncan BK, Miller JH (1980) Mutagenic deamination of cytosin residues in DNA. Nature 287:560–561
Durell SR, Guy HR (1992) Atomic scale structure and functional models of voltage-gated potassium channels. Biophys J 62:238–247
Ebers GC, George AL, Barchi RL, Ting-Passador MS, Kallen RG, Lathrop GM, Beckmann JS, Hahn AF, Brown WF, Campbell RD, Hudson AJ (1991) Paramyotonia congenita and hyperkalemic periodic paralysis are linked to the adult muscle sodium channel gene. Ann Neurol 30:810–816
Elbaz A, Vale-Santos J, Jurkat-Rott K, Lapie P, Ophoff RA, Bady B, Links TP, Puissan C, Villa A, Monnier N, Padberg GW, Abe K, Feingold N, Guimaraes J, Wintzen AR, Van der Hoeven JH, Saudubray JM, Grundfeld JP, Lenoir G, Nivet H, Echenne B, Frants RR, Fardeu M, Lehmann-Horn F, Fontaine B (1995) Hypokalemic periodic paralysis (hypoPP) and the dihydropyridine receptor (CACNL1A3): genotype/phenotype correlations for two predominant mutations and evidence for the absence of a founder effect in 16 Caucasian families. Am J Hum Genet 56:374–380
Ellis SB, Williams ME, Ways NR, Brenner R, Sharp AH, Leung AT, Campbell KP, McKenna E, Koch WJ, Hui A, Schwartz A, Harpold MM (1988) Sequence and expression of mRNAs encoding the α1 and α2 subunits of a DHP-sensitive calcium channel. Science 241:1661–64

Engel AG, Lambert EH, Rosevaer JW, Newlon TW (1965) Clinical and electromyographic studies in a patient with primary hypokalemic periodic paralysis. Am J Med 38:626–640

Entrikin RK, Abresch RT, Larson DB, Levine NA, Sharman RB (1987) Neuromuscular disorders: myotonia and muscular dystrophy in mice. Soc Neurosci Abstr 10:785

Erdmann R, Lüttgau HC (1989) The effect of the phenylalkylamine D888 (Devapamil) on force and Ca^{2+} current in isolated frog skeletal muscle fibres. J Physiol (Lond) 413:521–541

Eulenburg A (1886) Über eine familiäre durch 6 Generationen verfolgbare Form congenitaler Paramyotonie. Neurol Zentralbl 5:265–272

Fahlke C, Rüdel R (1995) Chloride currents across the membrane of mammalian skeletal muscle fibres. J Physiol (Lond) 484:355–368

Fahlke C, Zachar E, Rüdel R (1992) Single-channel recordings of chloride currents in cultured human skeletal muscle. Pflügers Arch 421:108–116

Fahlke C, Rüdel R, Mitrovic N, Zhou M, George AL Jr (1995) An aspartic acid residue important for voltage-dependent gating of human muscle chloride channels. Neuron 15:1–20

Fahlke C, Rosenbohm A, Mitrovic N, George AL, Rüdel R (1996) Mechanism of voltage-dependent gating in skeletal muscle chloride channels. Biophys J 70:August issue

Feero WG, Wang J, Barany F, Zhou J, Todorovic SM, Conwit R, Galloway G, Hausmanowa-Petrusewicz I, Fidzianska A, Arahata K, Wessel HB, Wadelius C, Marks HG, Hartlage P, Hayakawa H, Hoffman EP (1993) Hyperkalemic periodic paralysis: rapid molecular diagnosis and relationship of genotype to phenotype in 12 families. Neurology 43:668–673

Flatman JA, Clausen T (1979) Combined effects of adrenaline and insulin on active electrogenic Na^+-K^+ transport in rat soleus muscle. Nature 281:580–581

Fontaine B, Khurana TS, Hoffman EP, Bruns GAP, Haines JL, Trofatter J, Hanson MP, Rich J, McFarlane H, McKenna-Yasek D, Romano D, Gusella JF, Brown RH Jr (1990) Hyperkalemic periodic paralysis and the adult muscle sodium channel alpha-subunit gene. Science 250:1000-1003

Fontaine B, Vale Santos JM, Jurkat-Rott K, Reboul J, Plassart E, Rime CS, Elbaz A, Heine R, Guimaraes J, Weissenbach J, Baumann N, Fardeau M, Lehmann-Horn F (1994) Mapping of hypokalemic periodic paralysis (HypoPP) to chromosome 1q31-q32 by a genome-wide search in three European families. Nat Genet 6:267-272

Francini-Armstrong C (1970) Studies on the triad. I. Structure of the junction in frog twitch fibers. J Cell Biol 47:488–499

Francini-Armstrong C, Jorgensen AO (1994) Structure and development of E-C coupling units in skeletal muscle. Annu Rev Physiol 56:509–534

Franke C, Iaizzo PA, Hatt H, Spittelmeister W, Ricker K, Lehmann-Horn F (1991) Altered Na channel activity and reduced Cl conductance cause hyperexcitability in recessive generalized myotonia (Becker). Muscle Nerve 14:762–770

Füchtbauer E-M, Reininghaus J, Jockusch H (1988) Developmental control of the excitability of muscle: transplantation experiments on a myotonic mutant. Proc Natl Acad Sci USA 85:3880–3884

Gamstorp I (1956) Adynamia episodica hereditaria. Acta Paediatr (Uppsala) Suppl 45:1–126

Gancher ST, Nutt JG (1986) Autosomal dominant episodic ataxia: a heterogeneous syndrome. Mov Disord 1:239–253

Gautron S, Dos Santos G, Pinto-Henrique D, Koulakoff A, Gros F, Bergwald-Netter Y (1992) The glial voltage-gated sodium channel: cell- and tissue-specific mRNA expression. Proc Natl Acad Sci USA 89:7272–7276

Gellens ME, George AL Jr, Chen L, Chahine M, Horn R, Barchi RL, Kallen RG (1992) Primary structure and functional expression of the human cardiac tetrodotoxin-insensitive voltage-dependent sodium channel. Proc Natl Acad Sci USA 89:554–558

George AL Jr, Knittle T, Tamkun MM (1992a) Molecular cloning of an atypical voltage-gated sodium channel expressed in human heart and uterus: evidence for a distinct gene family. Proc Natl Acad Sci USA 89:4893–4897

George AL Jr, Komisarof J, Kallen RG, Barchi RL (1992b) Primary structure of the adult human skeletal muscle voltage-dependent sodium channel. Ann Neurol 31:131–137

George AL Jr, Crackover MA, Abdalla JA, Hudson JA, Ebers GC (1993a) Molecular basis of Thomsen's disease (autosomal dominant myotonia congenita). Nat Genet 3:305-310

George AL Jr, Iyer GS, Kleinfeld R, Kallen RG, Barchi RL (1993b) Genomic organization of the human skeletal muscle sodium channel gene. Genomics 15:598–606

George AL Jr, Knops JF, Han J, Finley WH, Knittle TJ, Tamkun MM, Brown GB (1994a) Assignment of a human voltage-dependent sodium channel α-subunit gene (SCN6A) to 2q21q23. Genomics 19:395–397

George AL Jr, Sloan-Brown K, Fenichel GM, Mitchell GA, Spiegel R, Pascuzzi RM (1994b) Nonsense and missense mutations of the muscle chloride channel gene in patients with myotonia congenita. Hum Mol Genet 3:2071–2072

George AL Jr, Varkony TA, Drabkin HA, Han J, Knops JF, Finley WH, Brown GB, Ward DC, Haas M (1995) Assignment of the human heart tetrodotoxin-resistant voltage-gated Na^+ channel α-subunit gene (SCN5A) to band 3p21. Cytogenet Cell Genet 68:67–70

Gillard EF, Otsu K, Fujii J, Khanna VK, DeLeon S, Derdemezi J, Britt BA, Duff CL, Worton RG, MacLennan DH (1991) A substitution of cysteine for arginine 614 in the ryanodine receptor is potentially causative of human malignant hyperthermia. Genomics 11:751–755

Gonoi T, Hasegawa S (1988) Post-natal disappearance of transient calcium channels in mouse skeletal muscle: effects of denervation and culture. J Physiol (Lond) 401:617–637

Grafe P, Quasthoff S, Strupp M, Lehmann-Horn F (1990) Enhancement of K^+ conductance improves in vitro the contraction force of skeletal muscle in hypokalemic periodic paralysis. Muscle Nerve 13:451–457

Gregg RG, Couch F, Hogan K, Powers PA (1993) Assignment of the human gene for the α1-subunit of the skeletal muscle DHP-sensitive calcium channel (CACNL1A3) to chromosome 1q31-32. Genomics 15:107–112

Grissmer S, Dethlefs B, Wasmuth JJ, Goldin AL, Gutman GA, Cahalan MD, Chandy KG (1990) Expression and chromosomal localization of a lymphocyte K^+ channel gene. Proc Natl Acad Sci USA 87:9411–9416

Grob D, Johns RJ, Liljestrand A (1957) Potassium movement in patients with familial periodic paralysis. Am J Med 23:356–375

Gronemeier M, Condie A, Prosser J, Steinmeyer K, Jentsch TJ, Jockusch H (1994) Nonsense and missense mutations in the muscular chloride channel gene *Clc-1* of myotonic mice. J Biol Chem 269:5963–5967

Grosson CLS, Esteban J, McKenna-Yasek D, Gusella JF, Brown RH Jr (1996) Hypokalemic periodic paralysis mutations: confirmation of mutation and analysis of founder effect. Neuromuscul Disord 6:27–31

Gründer S, Thiemann A, Pusch M, Jentsch TJ (1992) Regions involved in the opening of ClC-2 chloride channel by voltage and cell volume. Nature 360:759–763

Gurnett CA, Kahl SD, Anderson RD, Campbell KP (1995) Absence of the skeletal muscle sarcolemma chloride channel ClC-1 in myotonic mice. J Biol Chem 270:9035–9038

Gutman GA, Chandy KG (1993) Nomenclature of mammalian voltage-dependent potassium channel genes. Neuroscience 5:101–106

Hartshorne RP, Messner DJ, Coppersmith JC, Catterall WA (1982) The saxitoxin receptor of the sodium channel from rat brain. Evidence for two nonidentical beta subunits. J Biol Chem 257:13888–13891

Heine R (1995) Korrelationen von Genotyp und Phänotyp bei Natrium- und Chloridkanalkrankheiten. Dissertation, University of Ulm

Heine R, Pika U, Lehmann-Horn F (1993) A novel SCN4A mutation causing myotonia aggravated by cold and potassium. Hum Mol Genet 2:1349–1353

Heine R, George AL, Pika U, Deymeer F, Rüdel R, Lehmann-Horn F (1994) Proof of a non-functional muscle chloride channel in recessive myotonia congenita (Becker) by detection of a 4 base pair deletion. Hum Mol Genet 3:1123-1128

Heine R, Herzog J, Deymeer F, Michaels J, Moog U, Lehmann-Horn F (1996) Genotype-phenotype relations in paramyotonia congenita. Am J Hum Genet (submitted)

Heinemann SH, Terlau H, Stühmer W, Imoto K, Numa S (1992) Calcium channel characteristics conferred on the sodium channel by single mutations. Nature 356:441–443

Heller AH, Eicher EM, Hallett M, Sidman RL (1982) Myotonia, a new inherited muscle disease in mice. J Neurosci 2:924–933

Hirning LD, Fox AP, McCleskey EW, Olivera BM, Thayer SA, Miller RJ, Tsien RW (1988) Dominant role of N-type Ca^{2+} channels in evoked release of norepinephrine from sympathetic neurons. Science 239:57–61

Hoffman EP, Lehmann-Horn F, Rüdel R (1995) Overexcited or inactive: ion channels in muscle disease. Cell 80:681–686

Hofmann F, Biel M, Flockerzi V (1994) Molecular basis for Ca^{2+} channel diversity. Annu Rev Neurosci 17:399–418

Hogan K, Powers PA, Gregg RG (1994) Cloning of the human skeletal muscle α1 subunit of the dihydropyridine-sensitive L-type calcium channel (CACNL1A3). Genomics 24:608–609

Hoshi T, Zagotta WN, Aldrich RW (1990) Biophysical and molecular mechanisms of *Shaker* potassium channel inactivation. Science 250:533–538

Hudson AJ, Ebers GC, Bulman DE (1995) The skeletal muscle sodium and chloride channel disease. Brain 118:547–563

Hullin R, Singer-Lahat D, Freichel M, Biel M, Dascal N, Hofmann F, Flockerzi V (1992) Calcium channel β subunit heterogeneity: functional expression of cloned cDNA from heart, aorta and brain. EMBO J 11:885–90

Iaizzo PA, Lehmann-Horn F (1995) Anesthetic complications in muscle disorders. Anesthesiology 82:1093–1096

Iaizzo PA, Franke C, Hatt H, Spittelmeister W, Ricker K, Rüdel R, Lehmann-Horn F (1991) Altered sodium channel behaviour causes myotonia in dominantly inherited myotonia congenita. Neuromuscul Disord 1:47–53

Iles DE, Segers B, Sengers RCA, Monsieurs K, Heytens L, Halsall PJ, Hopkins PM, Ellis FR, Hall-Curran JL, Stewart AD, Wieringa B (1993a) Genetic mapping of the β1- and γ-subunits of the human skeletal muscle L-type voltage-dependent calcium channel on chromosome 17q and exclusion as candidate genes for malignant hyperthermia susceptibility. Hum Mol Genet 2:863–868

Iles DE, Segers B, Weghuis DO, Suikerbuijk R, Wieringa B (1993b) Localization of the γ-subunit of the skeletal muscle L-type voltage-dependent calcium channel gene (CACNLG) to human chromosome band 17q24 by in situ hybridization and identification of a polymorphic repetitive DNA sequence at the gene locus. Cytogenet Cell Genet 64:227–230

Iles D, Lehmann-Horn F, Deufel T, Scherer SW, Tsui L-C, Olde Weghuis D, Suijkerbuijk RF, Heytens L, Mikala G, Schwartz A, Ellis FR, Stewart AD, Wieringa B (1994) Localization of the gene encoding the α2/δ-subunits of the L-type voltage-dependent calcium channel to chromosome 7q and segregation of flanking markers in malignant hyperthermia susceptible families. Hum Mol Genet 3:969–975

Isom LL, DeJongh KS, Patton DE, Reber RFX, Offord J, Charbonneau H, Walsh K, Goldin AL, Catterall WA (1992) Primary structure and functional expression of the β1 subunit of the rat brain sodium channel. Science 256:839–842

Isom LL, DeJongh KS, Catterall WA (1994) Auxiliary subunits of voltage-gated ion channels. Neuron 12:1183–1194

Isom LL, Ragsdale DS, DeJongh KS, Westenbroek RE, Reber BFX, Scheuer T, Catterall A (1995a) Structure and function of the β2 subunit of brain sodium channels, a transmembrane glycoprotein with a CAM motif. Cell 83:443–442

Isom LL, Scheuer T, Brownstein AB, Ragsdale DS, Murphy BJ, Catterall WA (1995b) Functional co-expression of the β1 and type IIA α subunits of sodium channels in a mammalian cell line. J Biol Chem 270:3306–3312

Jay SD, Ellis SB, McCue AF, Williams ME, Vedvick TS, Harpold MM, Campbell KP (1990) Primary structure of the gamma-subunit of the DHP-sensitive calcium channel from skeletal muscle. Science 248:490–492

Jay SD, Sharp AH, Kahl SD, Vedvick TS, Harpold MM, Campbell KP (1991) Structural characterization of the dihydropyridine-sensitive calcium channel α2 subunit and the associated δ peptides. J Biol Chem 266:3287–93

Jentsch TJ, Steinmeyer K, Schwarz G (1990) Primary structure of *Torpedo marmorata* chloride channel isolated by expression cloning in *Xenopus* oocytes. Nature 348:510–514

Jentsch TJ, Günther W, Pusch M, Schwappach B (1995) Properties of voltage-gated chloride channels of the ClC gene family. J Physiol (Lond) 482.P:19S-25S

Ji S, Sun W, George AL Jr, Horn R, Barchi RL (1994) Voltage-dependent regulation of modal gating in the rat SkM1 expressed in Xenopus oocytes. J Gen Physiol 104:625–643

Jiang C, Atkinson D, Towbin JA, Splawski I, Lehmann MH, Li H, Timothy K, Taggart RT, Schwartz PJ, Vincent GM, Moss AJ, Keating MT (1994) Two long QT syndrome loci map to chromosomes 3 and 7 with evidence for further heterogeneity. Nat Genet 8:141–147

Jockusch H (1990) Molecular aspects of myotonia: the ADR mouse as a model. J Neurol Sci (Suppl) 88:9

Jockusch H, Bertram K (1986) "Arrested development of righting response" (adr) and "myotonia" (mto) are allelic. Mouse News Lett 75:19–20

Jockusch H, Bertram K, Schenk S (1988) The genes for two neuromuscular diseases of the mouse, "arrested development of righting response," adr, and "myotonia," mto, are allelic. Genet Res 52:203–205

Ju YK, Saint DA, Gage PW (1994) Inactivation-resistant channels underlying the persistent sodium current in rat ventricular myocytes. Proc R Soc Lond B Biol Sci 256:163–168

Jurkat-Rott K, Lehmann-Horn F, Elbaz A, Heine R, Gregg RG, Hogan K, Powers P, Lapie P, Vale-Santos JE, Weissenbach J, Fontaine B (1994) A calcium channel mutation causing hypokalemic periodic paralysis. Hum Mol Genet 3:1415-1419

Kawasaki M, Uchida S, Monkawa T, Miyawaki A, Mikoshiba K, Marumo F, Sasaki S (1994) Cloning and expression of a protein kinase C-regulated chloride channel abundantly expressed in rat brain neuronal cells. Neuron 12:597–604

Kawasaki M, Suzuki M, Uchida S, Sasaki S, Marumo F (1995) Stable and functional expression of the ClC-3 chloride channel in somatic cell lines. Neuron 14:1285–1291

Keynes RD (1994) The kinetics of voltage-gated ion channels. Q Rev Biophys 27:339–434

Kieferle S, Bens M, Vandewalle A, Jentsch T (1994) Two highly homologous members of the ClC chloride channel family in both rat and human kidney. Proc Natl Acad Sci USA 91:6943–6947

Klocke R, Steinmeyer K, Jentsch TJ, Jockusch H (1994) Role of innervation, excitability, and myogenic factors in the expression of the muscular chloride channel ClC-1. J Biol Chem 269:27635–27639

Knudson CM, Chaudhari N, Sharp AH, Powell JA, Beam KG, Campbell KP (1989) Specific absence of the α1 subunit of the dihydropyridine receptor in mice with muscular dysgenesis. J Biol Chem 264:1345–1348

Koch MC, Ricker K, Otto M, Grimm T, Bender K, Zoll B, Harper PS, Lehmann-Horn F, Rüdel R, Hoffman E (1991) Linkage data suggesting allelic heterogeneity for paramyotonia congenita and hyperkalemic periodic paralysis on chromosome 17. Hum Genet 88:71–74

Koch MC, Steinmeyer K, Lorenz C, Ricker K, Wolf F, Otto M, Zoll B, Lehmann-Horn F, Grzeschik KH, Jentsch TJ (1992) The skeletal muscle chloride channel in dominant and recessive human myotonia. Science 257:797–800

Koch MC, Baumbach K, George AL, Ricker K (1995) Paramyotonia congenita without paralysis on exposure to cold: a novel mutation in the SCN4A gene (Val1293Ile). Neuroreport 6:2001–2004

Költgen D, Reininghaus J, Jockusch H (1989) More evidence against the involvement of neuromuscular transmission in mouse myotonia. Mouse News Lett 83:151

Költgen D, Brinkmeier H, Jockusch H (1991) Myotonia and neuromuscular transmission in the mouse. Muscle Nerve 14:775–780

Krafte DS, Snutch TP, Leonard JP, Davidson N, Lester HA (1988) Evidence for the involvement of more than one mRNA in controlling the inactivation process of rat and rabbit brain Na channels expressed in Xenopus oocytes. J Neurosci 8:2859–2868

Krafte DS, Goldin AL, Auld VJ, Dunn RJ, Davidson N, Lester HA (1990) Inactivation of cloned Na channels expressed in xenopus oocytes. J Gen Physiol 96:689–706

Kraner SD, Tanaka JC, Barchi RL (1985) Purification and functional reconstitution of the voltage-sensitive sodium channel from rabbit T-tubular membranes. J Biol Chem 260:6341–47

Kwiecinski H (1981) Myotonia induced by chemical agents. Crit Rev Toxicol 8:279–310

Kwiecinski H, Lehmann-Horn F, Rüdel R (1984) Membrane currents in human intercostal muscle at varied extracellular potassium. Muscle Nerve 7:465–469

Lacerda AE, Kim HS, Ruth P, Perez-Reyes E, Flockerzi V, Hofmann F, Birnbaumer L, Brown AM (1991) Normalization of current kinetics by interaction between the α1 und β subunits of the skeletal muscle dihydropyridine-sensitive Ca^{2+} channel. Nature 352:527–530

Lamb GD (1992) DHP receptors and excitation-contraction coupling. J Muscle Res Cell Motil 13:394–405

Lehmann-Horn F, Rüdel R (1995) Hereditary nondystrophic myotonias and periodic paralyses. Curr Opin Neurol 8:402–410

Lehmann-Horn F, Rüdel R, Ricker K, Lorkovic H, Dengler R, Hopf HC (1983) Two cases of adynamia episodica hereditaria: in vitro investigation of muscle cell membrane and contraction parameters. Muscle Nerve 6:113–121

Lehmann-Horn F, Küther G, Ricker K, Grafe P, Ballanyi K, Rüdel R (1987a) Adynamia episodica hereditaria with myotonia: a non-inactivating sodium current and the effect of extracellular pH. Muscle Nerve 10:363–374

Lehmann-Horn F, Rüdel R, Ricker K (1987b) Membrane defects in paramyotonia congenita (Eulenburg). Muscle Nerve 10:633–641

Lehmann-Horn F, Iaizzo PA, Hatt H, Franke C (1991) Altered gating and reduced conductance of single sodium channels in hyperkalemic periodic paralysis. Pflugers Arch 418:297–299

Lehmann-Horn F, Rüdel R, Ricker K (1993) Workshop report: non-dystrophic myotonias and periodic paralyses. Neuromuscul Disord 3:161–168

Lehmann-Horn F, Engel AG, Ricker K, Rüdel R (1994) The periodic paralyses and paramyotonia congenita. In: Engel AG, Franzini-Armstrong C (eds) Myology, 2nd edn. McGraw-Hill, New York, pp 1303–1334

Lehmann-Horn F, Mailänder V, Heine R, George AL (1995a) Myotonia levior is a chloride channel myotonia. Hum Mol Genet 4:1397–1402

Lehmann-Horn F, Sipos I, Jurkat-Rott K, Heine R, Brinkmeier H, Fontaine B, Kovacs L, Melzer W (1995b) Altered calcium currents in human hypokalemic periodic paralysis myotubes expressing mutant L-type calcium channels. In: Dawson DC, Frizzell RA (eds) Ion channels and genetic diseases. Rockefeller University Press, New York, pp 101–113

Lennox G, Purves A, Marsden D (1992) Myotonia fluctuans. Arch Neurol 49:1010–1011

Lerche H, Heine R, Pika U, George AL, Mitrovic N, Browatzki M, Weiß T, Rivet-Bastide M, Franke C, Lomonaco M, Ricker K, Lehmann-Horn F (1993) Human sodium channel myotonia: slowed channel inactivation due to substitutions for glycine within the III/IV linker. J Physiol (Lond) 470:13–22

Lerche H, Klugbauer N, Lehmann-Horn F, Hofmann F, Melzer W (1996a) Expression and functional characterization of the cardiac L-type calcium channel carrying a skeletal muscle DHP-receptor mutation causing hypokalaemic periodic. Pflügers Arch 431:461–463

Lerche H, Mitrovic N, Dubowitz V, Lehmann-Horn F (1996b) Pathophysiology of paramyotonia congenita: the R1448P sodium channel mutation in adult human skeletal muscle. Ann Neurol 39:599–608

Levis R, O'Hare K, Rubin GM (1984) Effects of transposable element insertions on RNA encoded by the *white* gene of Drosophila. Cell 38:471–481

Links TP, Smit AJ, Molenaar WM, Zwarts MJ, Oosterhuis HJGH (1994a) Familial hypokalemic periodic paralysis: clinical, diagnostic and therapeutic aspects. J Neurol Sci 122:33–43

Links TP, Vanderhoeven JH, Zwarts MJ (1994b) Surface EMG and muscle fibre conduction during attacks of hypokalaemic periodic paralysis. J Neurol Neurosurg Psychiatry 57:632–634

Lipicky RJ (1977) Studies in human myotonic dystrophy. In: Rowland LP (ed) Pathogenesis of human muscular dystrophy. Excerpta Medica, Amsterdam, pp 729–738

Lipicky RJ (1979) Myotonic syndromes other than myotonic dystrophy. In: Vinken PJ, Bruyn GW (eds) Handbook of clinical neurology, vol 40. Elsevier, Amsterdam, pp 533–571

Lipicky RJ, Bryant SH, Salmon JH (1971) Cable parameters, sodium, potassium, chloride, and water content and potassium efflux in isolated external intercostal muscle of normal volunteers and patients with myotonia congenita. J Clin Invest 50:2091–2103

Litt M, Kramer P, Browne D, Gancher S, Brunt ERP, Root D, Phromchotikul T, Dubay CJ, Nutt J (1994) A gene for episodic ataxia/myokymia maps to chromosome 12p13. Am J Hum Genet 55:702–709

Lorenz C, Meyer-Kleine C, Steinmeyer K, Koch MC, Jentsch TJ (1994) Genomic organization of the human muscle chloride channel ClC-1 and analysis of novel mutations leading to Becker-type myotonia. Hum Mol Genet 3:941-946

Loughney K, Kreber R, Ganetzky B (1989) Molecular analysis of the para locus, a sodium channel gene in drosophila. Cell 58:1143–1154

MacKinnon R, Aldrich RW, Lee AW (1993) Functional stoichiometry of Shaker potassium channel inactivation. Science 262:757–759

MacLennan DH, Phillips MS (1992) Malignant hyperthermia. Science 256:789–794

MacLennan DH, Duff C, Zorzato F, Fujii J, Phillips M, Korneluk R, Frodis W, Britt BA, Worton RG (1990) Ryanodine receptor gene is a candidate for predisposition to malignant hyperthermia. Nature 343:559-561

Mailänder V, Heine R, Deymeer F, Lehmann-Horn F (1996) Novel chloride channel mutations and their effects on heterozygous carriers. Am J Hum Genet 58:317–324

Makita N, Bennett PB Jr, George AL Jr (1994a) Voltage-gated Na^+ channel β1 subunit mRNA expressed in adult human skeletal muscle, heart, and brain is encoded by a single gene. J Biol Chem 269:7571–7578

Makita N, Sloan-Brown K, Weghuis DO, Ropers HH, George AL Jr (1994b) Genomic organization and chromosomal assignment of the human voltage-gated Na^+ channel β1 subunit gene (SCN1B). Genomics 23:628–634

Malo MS, Blanchard BJ, Andresen JM, Srivastava K, Chen XN, Li X, Wang E, Korenberg JR, Ingram VM (1994a) Localisation of a putative human brain sodium channel gene (SCN1A) to chromosome band 2q24. Cytogenet Cell Genet 67:178–186

Malo MS, Srivastava K, Andresen JM, Chen XN, Korenberg JR, Ingram VM (1994b) Targeted gene walking by low stringency polymerase chain reaction: assignment of a putative human brain sodium channel gene (SCN3A) to chromosome 2q24-31. Proc Natl Acad Sci USA 91:2975–2979

Malouf NN, McMahon DK, Heinsworth CN, Kay BK (1992) A two-motif isoform of the major calcium channel subunit in skeletal muscle. Neuron 8:899–906

Mason K (1988) A hereditary disease in burmese cats manifested as an episodic weakness with head nodding and neck ventroflexion. J Am Anim Hospital Ass 24:147–151

McCarthy TV, Healy JMS, Lehane M, Heffron JJA, Deufel T, Lehmann-Horn F, Farrall M, Johnson K (1990) Localization of the malignant hyperthermia susceptibility locus to human chromosome 19q11.2-13.2. Nature 343:562-563

McClatchey AI, Lin CS, Wang J, Hoffman EP, Rojas C, Gusella JF (1992a) The genomic structure of the human skeletal muscle sodium channel gene. Hum Mol Genet 1:521–527

McClatchey AI, McKenna-Yasek D, Cros D, Worthen HG, Kuncl RW, DeSilva SM, Cornblath DR, Gusella JF, Brown RH Jr (1992b) Novel mutations in families with unusual and variable disorders of the skeletal muscle sodium channel. Nat Genet 2:148–152

McClatchey AI, van den Bergh P, Pericak-Vance MA, Raskind W, Verellen C, McKenna-Yasek D, Rao K, Haines JL, Bird T, Brown RH Jr, Gusella JF (1992c) Temperature-sensitive mutations in the III-IV cytoplasmic loop region of the skeletal muscle sodium channel gene in paramyotonia congenita. Cell 68:769–774

McClatchey AI, Cannon SC, Slaugenhaupt SA, Gusella JF (1993) The cloning and expression of a sodium channel β–1-subunit cDNA from human brain. Hum Mol Genet 2:745–749

McKusick VA (1992) Mendelian inheritance in man. Hopkins University Press, Baltimore

Mehrke G, Brinkmeier H, Jockusch H (1988) The myotonic mouse mutant ADR: electrophysiology of the muscle fiber. Muscle Nerve 11:440–446

Melzer W, Herrmann-Frank A, Lüttgau HC (1995) The role of Ca^{2+} ions in excitation-contraction coupling of skeletal muscle fibres. Biochim Biophys Acta 1241:59–116

Meyer-Kleine C, Otto M, Zoll B, Koch MC (1994a) Molecular and genetic characterisation of German families with paramyotonia congenita and demonstration of founder effect in the Ravensberg families. Hum Genet 93:707–710

Meyer-Kleine C, Ricker K, Otto M, Koch MC (1994b) A recurrent 14 bp deletion in the CLCN1 gene associated with generalized myotonia (Becker). Hum Mol Genet 3:1015–1016

Meyer-Kleine C, Steinmeyer K, Ricker K, Jentsch TJ, Koch MC (1995) Spectrum of mutations in the major human skeletal muscle chloride channel gene (CLCNI) leading to myotonia. Am J Hum Genet 57:1325–1334

Middleton RE, Pheasant DJ, Miller C (1994) Purification, reconstitution, and subunit composition of a voltage-gated chloride channel from *Torpedo electroplax*. Biochemistry 33:13189–13198

Mikami A, Imoto K, Tanabe T, Niidome T, Mori Y, Takeshima H, Narumiya S, Numa S (1989) Primary structure and functional expression of the cardiac dihydropyridine-sensitive calcium channel. Nature 340:230–233

Mintz IM, Venema VJ, Swiderek KM, Lee TD, Bean BP, Adams ME (1992) P-type calcium channels blocked by the spider toxin ω-Aga-IVA. Nature 355:827–30

Mitchell G, Heffron JJA (1982) Porcine stress syndrome. Adv Food Res 28:167–230

Mitrovic N, George AL Jr, Heine R, Wagner S, Pika U, Hartlaub U, Zhou M, Lerche H, Fahlke C, Lehmann-Horn F (1994) Potassium-aggravated myotonia: the V1589 M mutation destabilizes the inactivated state of the human muscle sodium channel. J Physiol (Lond) 478:395-402

Mitrovic N, George AL Jr, Heine R, Lehmann-Horn F (1995) Potassium-aggravated myotonia: biophysical and clinical implications of the G1306A/V/E human muscle sodium channel mutations. J Physiol (Lond) 487:107–114

Mitrovic N, Lerche H, Heine R, Fleischhauer R, Pika-Hartlaub U, Hartlaub U, George Jr AL, Lehmann-Horn F. Role in fast inactivation of conserved amino acids in the IV/S4-S5 loop of the human muscle Na^+ channel. Neuroscience Letters, in press (1996)

Moorman JR, Kirsch GE, VanDongen AMJ, Joho RH, Brown AM (1990) Fast and slow gating of sodium channels encoded by a single mRNA. Neuron 4:243–252

Neumann P, Weber T (1989) Mouse News Lett 83:157

Nilius B (1988) Modal gating behavior of cardiac sodium channels in cell-free membrane patches. Biophys J 53:857–862

Noda M, Suzuki H, Numa S, Stühmer W (1989) A single point mutation confers tetrodotoxin and saxitoxin insensitivity on the sodium channel II. FEBS Lett 259:213–216

Nowycky MC, Fox AP, Tsien RW (1985) Three types of neuronal calcium channel with different calcium agonist sensitivity. Nature 316:440–443

Palade PT, Barchi RL (1977) On the inhibition of muscle membrane chloride conductance by aromatic carboxylic acids. J Gen Physiol 69:879–896

Patlak J, Ortiz M (1986) Two modes of gating during late Na^+ channel currents in frog sartorius muscle. J Gen Physiol 87:305–326

Patton DE, West JW, Catterall WA, Goldin AL (1992) Amino acid residues required for fast Na^+-channel inactivation: charge neutralizations and deletions in the III-IV linker. Proc Natl Acad Sci USA 89:10905–10909

Patton DE, West JW, Catterall WA, Goldin AL (1993) A peptide segment critical for sodium channel inactivation functions as an inactivation gate in a potassium channel. Neuron 11:967–974

Pelzer S, Sieber M, Hofmann F, Trautwein W (1989) Calcium channels reconstituted from the skeletal muscle dihydropyridine receptor protein complex and its $\alpha 1$ peptide subunit in lipid bilayers. Ann NY Acad Sci 560:138–154

Perez-Reyes E, Castellano A, Haeyoung SK, Bertrand P, Baggstrom E, Lacerda AE, Wei X, Birnbaumer L (1992) Cloning and expression of a cardiac/brain β subunit of the L-type calcium channel. J Biol Chem 267:1792–1797

Pickar JG, Spier SJ, Snyder JR, Carlsen RC (1991) Altered ionic permeability in skeletal muscle from horses with hyperkalemic periodic paralysis. Am J Physiol 260:C92-C933

Plassart E, Reboul J, Rime C-S, Recan D, Millasseau P, Eymard B, Pelletier J, Thomas C, Chapon F, Desnuelle C, Confavreux C, Bady B, Martin J-J, Lenoir G, Serratrice G, Fardeau M, Fontaine B (1994) Mutations in the muscle sodium channel gene (SCN4A) in 13 French families with hyperkalemic periodic paralysis and paramyotonia congenita: phenotype to genotype correlations and demonstration of the predominance of two mutations. Eur J Hum Genet 2:110–124

Pongs O (1992) Molecular biology of voltage-dependent potassium channels. Physiol Rev 72:S69-S88

Poskanzer DC, Kerr DNS (1961) A third type of periodic paralysis with normokalemia and favourable response to sodium chloride. Am J Med 31:328–342

Potts JF, Regan MR, Rochelle JM, Seldin MF, Agnew WS (1993) A glial-specific voltage-sensitive Na channel gene maps close to clustered genes for neuronal isoforms on mouse chromosome 2. Biochem Biophys Res Commun 197:100–107

Powers PA, Gregg RG, Lalley P, Liao M, Hogan K (1991) Assignment of the human gene for the α1 subunit of the cardiac DHP-sensitive Ca^{2+} channel to chromosome 12p12-pter. Genomics 10:835–839

Powers PA, Liu S, Hogan K, Gregg RG (1992) Skeletal muscle and brain isoforms of a β-subunit of human voltage-dependent calcium channels are encoded by a single gene. J Biol Chem 267:22967–22972

Powers PA, Liu S, Hogan K, Gregg RG (1993) Molecular characterization of the gene encoding the γ subunit of the human skeletal muscle 1,4-dihydropyridine-sensitive Ca^{2+} channel (CACNLG), cDNA sequence, gene structure, and chromosomal location. J Biol Chem 268:9275–9279

Powers PA, Scherer SW, Tsui L-C, Gregg RG, Hogan K (1994) Localization of the gene encoding the α2/δ subunit (CACNL2A) of the human skeletal muscle voltage-dependent Ca^{2+} channel to chromosome 7q21-q22 by somatic cell hybrid analysis. Genomics 19:192–193

Pragnell M, Sakamoto J, Jay SD, Campbell KP (1991) Cloning and tissue-specific expression of the brain calcium channel β-subunit. FEBS Lett 291:253–258

Ptáček LJ, George AL Jr, Griggs RC, Tawil R, Kallen RG, Barchi RL, Robertson M, Leppert MF (1991a) Identification of a mutation in the gene causing hyperkalemic periodic paralysis. Cell 7:1021–1027

Ptáček LJ, Trimmer JS, Agnew WS, Roberts JW, Petajan JH, Leppert M (1991b) Paramyotonia congenita and hyperkalemic periodic paralysis map to the same sodium channel gene locus. Am J Hum Genet 49:851–854

Ptáček LJ, George AL Jr, Barchi RL, Griggs RC, Riggs JE, Robertson M, Leppert MF (1992a) Mutations in an S4 segment of the adult skeletal muscle sodium channel cause paramyotonia congenita. Neuron 8:891–897

Ptáček LJ, Tawil R, Griggs RC, Storvick D, Leppert M (1992b) Linkage of atypical myotonia congenita to sodium channel locus. Neurology 42:431–433

Ptáček LJ, Gouw L, Kwiecinski H, McManis P, Mendell JR, Barohn RJ, George AL, Barchi RL, Robertson M, Leppert M (1993) Sodium channel mutations in paramyotonia congenita and hyperkalemic periodic paralysis. Ann Neurol 33:300–307

Ptáček L, Tawil R, Griggs RC, Engel A, Layzer RB, Kwiecinski H, McManis PG, Santiago L, Moore M, Fouad G, Bradley P, Leppert MF (1994a) Dihydropyridine receptor mutations cause hypokalemic periodic paralysis. Cell 77:863-868

Ptáček LJ, Tawil R, Griggs RC, Meola G, McManis P, Barohn RJ, Mendell JR, Harris C, Spitzer R, Santiago F, Leppert MF (1994b) Sodium channel mutations in acetazo-

lamide-responsive myotonia congenita, paramyotonia congenita and hyperkalemic periodic paralysis. Neurology 44:1500–1503
Pusch M, Jentsch TJ (1994) Molecular physiology of voltage-gated chloride channels. Physiol Rev 74:813–827
Pusch M, Steinmeyer K, Jentsch TJ (1994) Low single channel conductance of the major skeletal muscle chloride channel, ClC-1. Biophys J 66:149-152
Reininghaus J, Füchtbauer EM, Bertram K, Jockusch H (1988) The myotonic mouse mutant ADR: physiological and histochemical properties of muscle. Muscle Nerve 11:433–439
Ricker K, Camacho L, Grafe P, Lehmann-Horn F, Rüdel R (1989) Adynamia episodica hereditaria: what causes the weakness? Muscle Nerve 10:883–891
Ricker K, Lehmann-Horn F, Moxley RT (1990) Myotonia fluctuans. Arch Neurol 47:268–272
Ricker K, Moxley RT, Heine R, Lehmann-Horn F (1994) Myotonia fluctuans, a third type of muscle sodium channel disease. Arch Neurol 51:1095–1102
Rivet M, Cognard C, Imbert N, Rideau Y, Duport G, Raymond G (1992) A third type of calcium current in cultured human skeletal muscle cells. Neurosci Lett 138:97–102
Rogart RB, Cribbs LL, Muglia LK, Kephart DD, Kaiser MW (1989) Molecular cloning of a putative tetrodotoxin-resistant rat heart Na^+ channel isoform. Proc Natl Acad Sci USA 86:8170–8174
Rojas CV, Wang J, Schwartz L, Hoffman EP, Powell BR, Brown RH Jr (1991) A Met-to-Val mutation in the skeletal muscle sodium channel alpha-subunit in hyperkalemic periodic paralysis. Nature 354:387-389
Romey G, García L, Dimitriadou V, Pincon-Raymond M, Rieger F, Lazdunski M (1989) Ontogenesis and localization of Ca^{2+} channels in mammalian skeletal muscle in culture and role in excitation-contraction coupling. Proc Natl Acad Sci USA 86:2933–2937
Rosenbohm A, Rüdel R, George AL Jr, Fahlke C (1996) Regulation of the human muscle chloride channel, hClC-1, by protein kinase C. J Neurosci (submitted)
Rüdel R (1990) The myotonic mouse – a realistic model for the study of human recessive generalized myotonia. Trends Neurosci 13:1–3
Rüdel R, Dengler R, Ricker K, Haass A, Emser W (1980) Improved therapy of myotonia with the lidocaine derivative tocainide. J Neurol 222:275–278
Rüdel R, Lehmann-Horn F, Ricker K, Küther G (1984) Hypokalemic periodic paralysis: in vitro investigation of muscle fiber membrane parameters. Muscle Nerve 7:110–120
Rüdel R, Ricker K, Lehmann-Horn F (1988) Transient weakness and altered membrane characteristic in recessive generalized myotonia (Becker). Muscle Nerve 11:202–211
Rüdel R, Ricker K, Lehmann-Horn F (1993) Genotype-phenotype correlations in human skeletal muscle sodium channel diseases. Arch Neurol 50:1241-1248
Rüdel R, Lehmann-Horn F, Ricker K (1994) The non-progressive myotonias. In: Engel AG, Franzini-Armstrong C (eds) Myology, 2nd edn. McGraw-Hill, New York, pp 1291–1303

Rudolph JA, Spier SJ, Byrns G, Hoffmann EP (1992a) Linkage of hyperkalemic periodic paralysis in Quarter Horses to the horse adult skeletal muscle sodium channel gene. Anim Genet 23:241–250

Rudolph JA, Spier SJ, Byrns G, Rojas CV, Bernoco D, Hoffman EP (1992b) Periodic paralysis in quarter horses: a sodium channel mutation disseminated by selective breeding. Nat Genet 2:144–147

Rudy B (1988) Diversity and ubiquity of K channels. Neuroscience 25:729–749

Ruth P, Röhrkasten A, Biel M, Bosse E, Regulla S, Meyer HE, Flockerzi V, Hofmann F (1989) Primary structure of the β subunit of the DHP-sensitive calcium channel from skeletal muscle. Science 245:1115–1118

Saint DA, Ju YK, Gage PW (1992) A persistent sodium current in rat ventricular myocytes. J Physiol (Lond) 454:219–231

Salkoff L, Baker K, Butler A, Covarrubias M, Pak MD, Wei A (1992) An essential "set" of K^+ channels conserved in flies, mice and humans. Trends Neurosci 15:161–166

Sanguinetti MC, Jiang C, Curran ME, Keating MT (1995) A mechanistic link between an inherited and an acquired cardiac arrhythmia: HERG encodes the IKr potassium channel. Cell 81:299–307

Satin J, Kyle JW, Chen M, Bell P, Cribbs LL, Fozzard HA, Rogart RB (1992) A mutant of TTX-resistant cardiac sodium channels with TTX-sensitive properties. Science 256:1202–1205

Schreibmayer W, Wallner M, Lotan I (1994) Mechanism of modulation of single sodium channels from skeletal muscle by the β1-subunit from rat brain. Pflugers Arch 426:360–362

Schultz D, Mikala G, Yatani A, Engle DB, Iles DE, Segers B, Sinke RJ, Weghuis DO, Klöckner U, Wakamori M, Wang J-J, Melvin D, Varadi G, Schwartz A (1993) Cloning, chromosomal localization, and functional expression of the α1 subunit of the L-type voltage-dependent calcium channel from normal human heart. Proc Natl Acad Sci USA 90:6228–6232

Seino S, Chen L, Seino M, Blondel O, Takeda J, Johnson JH, Bell GI (1992a) Cloning of the α1 subunit of a voltage-dependent calcium channel expressed in pancreatic β cells. Proc Natl Acad Sci USA 89:584–588

Seino S, Yamada Y, Espinosa III R, Le Beau MM, Graeme IB (1992b) Assignment of the gene encoding the α1 subunit of the neuroendocrine/brain-type calcium channel (CACNL1A2) to human chromosome 3, band p14.3. Genomics 13:1375–1357

Singer D, Biel M, Lotan I, Flockerzi V, Hofmann F, Dascal N (1991) The roles of the subunits in the function of the calcium channel. Science 253:1553–1556

Sipos I, Melzer W (1995) Calcium currents in cultured human skeletal muscle cells. Pflugers Arch 426:R85

Sipos I, Jurkat-Rott K, Harasztosi CS, Fontaine B, Kovacs L, Melzer W, Lehmann-Horn F (1995) Skeletal muscle DHP receptor mutations alter calcium currents in human hypokalaemic periodic paralysis myotubes. J Physiol (Lond) 483:299–306

Spaans F, Theunissen P, Reekers A, Smit L, Veldman H (1990) Schwartz-Jampel syndrome: part I. Clinical, electromyographic, and histologic studies. Muscle Nerve 13:516–527

Steinmeyer K, Klocke R, Ortland C, Gronemeier M, Jockusch H, Gründer S, Jentsch TJ (1991a) Inactivation of muscle chloride channel by transposon insertion in myotonic mice. Nature 354:304–308

Steinmeyer K, Ortland C, Jentsch TJ (1991b) Primary structure and functional expression of a developmentally regulated skeletal muscle chloride channel. Nature 354:301–304

Steinmeyer K, Lorenz C, Pusch M, Koch MC, Jentsch TJ (1994) Multimeric structure of ClC-1 chloride channel revealed by mutations in dominant myotonia congenita (Thomsen). EMBO J 13:737-743

Stühmer W, Conti F, Suzuki H, Wang X, Noda M, Yahagi N, Kubo H, Numa S (1989) Structural parts involved in activation and inactivation of the sodium channel. Nature 339:597–603

Tahmoush A, Schaller KL, Zhang P, Hyslop T, Heiman-Patterson T, Caldwell JH (1994) Muscle sodium channel inactivation defect in paramyotonia congenita with the Thr1313Met mutation. Neuromuscul Disord 4:447–454

Tanabe T, Takeshima H, Mikami A, Flockerzi V, Takahashi H, Kangawa K, Kojima M, Matsuo H, Hirose T, Numa S (1987) Primary structure of the receptor for calcium channel blockers from skeletal muscle. Nature 328:313–318

Tanabe T, Beam KG, Powell JA, Numa S (1988) Restoration of excitation-contraction coupling and slow calcium current in dysgenic muscle by dihydropyridine receptor complementary DNA. Nature 336:134–139

Terlau H, Heinemann SH, Stühmer W, Pusch M, Conti F, Imoto K, Numa S (1991) Mapping the site of block by tetrodotoxin and saxitoxin of sodium channel II. FEBS Lett 293:93–96

Thiemann A, Gründer S, Pusch M, Jentsch TJ (1992) A chloride channel widely expressed in epithelial and non-epithelial cells. Nature 356:57–60

Thomasen E (1948) Myotonia. Universitätsforlaget, Aarhus

Thomsen J (1876) Tonische Krämpfe in willkürlich beweglichen Muskeln in Folge von ererbter psychischer Disposition. Arch Psychiatr Nervenkrankheiten 6:702–718

Trimmer JS, Cooperman SS, Tomiko SA, Zhou J, Crean SM, Boyle MB, Kallen RG, Sheng Z, Barchi RL, Sigworth FJ, Goodman RH, Agnew WS, Mandel G (1989) Primary structure and functional expression of mammalian skeletal muscle sodium channel. Neuron 3:33–49

Trudell RG, Kaiser KK, Griggs RC (1987) Acetazolamide responsive myotonia congenita. Neurology 37:488–491

Tyler FH, Stephens FE, Gunn FD, Perkoff GT (1951) Studies in disorders of muscle. VII. Clinical manifestations and inheritance of a type of periodic paralysis without hypopotassemia. J Clin Invest 30:492–502

Uchida S, Sasaki S, Furukawa T, Hiraoka M, Imai T, Hirata Y, Marumo F (1993) Molecular cloning of a chloride channel that is regulated by dehydration and expressed predominantly in kidney medulla. J Biol Chem 268:3821–3824

Ukomadu C, Zhou J, Sigworth FJ, Agnew WF (1992) μ1 Na^+ channels expressed transiently in human embryonic kidney cells: biochemical and biophysical properties. Neuron 8:663–676

Vahedi K, Joutel A, Van Bogaert P, Ducros A, Maciazeck J, Bach JF,Bousser MG, Tournier-Lasserve E (1995) A gene for hereditary paroxysmal cerebellar ataxia maps to chromosone 19p. Ann Neurol 37:289–293

van Slegtenhorst MA, Bassi MT, Borsani G, Wapenaar MC, Ferrero GB, de Concillis L, Rugarli EI, Grillo A, Franco B, Zoghbi HY, Ballabio A (1994) A gene from the Xp22.3 region shares homology with voltage-gated chloride channels. Hum Mol Genet 3:547–552

Varadi G, Lory P, Schultz D, Varadi M, Schwartz A (1991) Acceleration of activation and inactivation by the β subunit of the skeletal muscle calcium channel. Nature 352:159–162

Vassilev PM, Scheuer T, Catterall WA (1988) Identification of an intracellular peptide segment involved in sodium channel inactivation. Science 241:1658–1661

Vassilev P, Scheuer T, Catterall WA (1989) Inhibition of inactivation of single sodium channels by a site-directed antibody. Proc Natl Acad Sci USA 86:8147–8151

Vita GM, Olckers A, Jedlicka AE, George AL, Heiman-Patterson T, Rosenberg H, Fletcher JE, Levitt RC (1995) Masseter muscle rigidity associated with glycine1306-to-alanine mutation in adult muscle sodium channel α-subunit gene. Anesthesiology 82:1097–1103

Wallner M, Weigl L, Meera P, Lotan I (1993) Modulation of the skeletal muscle sodium channel α-subunit by the β1-subunit. FEBS Lett 336:535–539

Wang J, Rojas CV, Zhou J, Schwartz LS, Nicholas H, Hoffman EP (1992) Sequence and genomic structure of the human adult skeletal muscle sodium channel α subunit gene on 17q. Biochem Biophys Res Commun 182:794–801

Wang J, Zhou J, Todorovic SM, Feero WG, Barany F, Conwit R, Hausmanowa-Petrusewicz I, Fidzianska A, Arahata K, Wessel HB, Hartladge P, Ricker K, Lehmann-Horn F, Hayakawa H, Hoffman EP (1993) Molecular genetic and genetic correlations in sodium channelopathies: lack of founder effect and evidence for a second gene. Am J Hum Genet 52:1074–1084

Wang J, Dubowitz V, Lehmann-Horn F, Ricker K, Michaels J, Ptáček LJ, Hoffman EP (1995) In vivo structure/function studies: consecutive Arg1448 changes to Cys, His and Pro at the extracellular surface of IVS4. In: Proceedings of the society of general physiologists. Rockefeller University Press, New York, pp 77–88

Wang Q, Shen J, Splawski I, Atkinson D, Li Z, Robinson JL, Moss AJ, Towbin JA, Keating MT (1995) SCN5A mutations associated with an inherited cardiac arrhythmia, long QT syndrome. Cell 80:805–811

Watkins WJ, Watts DC (1984) Biological features of the new A2G-adr mouse with abnormal muscle function. Lab Anim 18:1–6

Wei X, Perez-Reyes E, Lacerda AE, Schuster G, Brown AM, Birnbaumer L (1991) Heterologous regulation of the cardiac Ca^{2+} channel α1 subunit by skeletal muscle β and γ subunits. J Biol Chem 266:21943–21947
West JW, Patton DE, Scheuer T, Wang Y, Goldin AL, Catterall WA (1992) A cluster of hydrophobic amino acid residues required for fast Na^+ channel inactivation. Proc Natl Acad Sci USA 89:10910–10914
White MM, Miller C (1979) A voltage-gated anion channel from the electric organ of *Torpedo californica*. J Biol Chem 254:10161–10166
White GR, Plaskett J (1904) "Nervous," "stiff-legged," or "fainting" goats. Am Vet Rev 28:556–560
Williams ME, Feldman DH, McCue AF, Brenner R, Velicelebi G, Ellis SB, Harpold MM (1992) Structure and functional expression of α1, α2 and β subunits of a novel human neuronal calcium channel subtype. Neuron 8:71–84
Wischmeyer E, Nolte E, Klocke R, Jockusch H, Brinkmeier H (1993) Development of electrical myotonia in the adr mouse: role of chloride conductance in myotubes and neonatal animals. Neuromuscul Disord 3:267–274
Wischmeyer E, Weber-Schürholz S, Jockusch H (1995) Sarcolemmal chloride and potassium channels from normal and myotonic mouse muscle studied in lipid supplemented vesicles. Biochem Biophys Res Commun 213:513–518
Yang N, Horn R (1995) Evidence for voltage-dependent S4 movement in sodium channels. Neuron 15:213–218
Yang N, Ji S, Zhou M, Ptáček LJ, Barchi RL, Horn R, George AL Jr (1994) Sodium channel mutations in paramyotonia congenita exhibit similar biophysical phenotypes in vitro. Proc Natl Acad Sci USA 91:12785–12789
Yellen G, Jurman ME, Abramson T, MacKinnon R (1991) Mutations affecting internal TEA blockade identify the probable pore-forming region of a K^+ channel. Science 251:939–942
Zhou J, Potts JF, Trimmer JS, Agnew WS, Sigworth FJ (1991) Multiple gating modes and the effect of modulating factors of the μ1 sodium channel. Neuron 7:775–785
Zhou J, Spier SJ, Beech J, Hoffman EP (1994) Pathophysiology of sodium channelopathies: correlation of normal/mutant mRNA ratios with clinical phenotype in dominantly-inherited periodic paralysis. Hum Mol Genet 3:1599-1603
Zierler KL, Andres R (1957) Movement of potassium into skeletal muscle during spontaneous attack in family periodic paralysis. J Clin Invest 36:730–737

Editor-in-charge: Professor D. Pette

Springer Verlag and the environment

We at the Springer-Verlag firmly believe that an international science publisher has a special obligation to the environment, and our corporate policies consistently reflect this conviction. We also expect our business partners – paper mills, printers, packaging manufacturers, etc. – to commit themselves to using environmentally friendly materials and production processes. The paper in this book is made from low- or no-chlorine pulp and is acid free, in conformance with international standards for paper permanency.

GPSR Compliance

The European Union's (EU) General Product Safety Regulation (GPSR) is a set of rules that requires consumer products to be safe and our obligations to ensure this.

If you have any concerns about our products, you can contact us on ProductSafety@springernature.com

In case Publisher is established outside the EU, the EU authorized representative is:

Springer Nature Customer Service Center GmbH
Europaplatz 3
69115 Heidelberg, Germany

Batch number: 10620141

Printed by Printforce, the Netherlands